ORGANIC
POLLUTANTS

ORGANIC POLLUTANTS

AN ECOTOXICOLOGICAL PERSPECTIVE

C. H. WALKER

First published 2001 by Taylor & Francis
11 New Fetter Lane, London EC4P 4EE

Simultaneously published in the USA and Canada
by Taylor & Francis Inc
29 West 35th Street, New York, NY 10001

Taylor & Francis is an imprint of the Taylor & Francis Group

© 2001 C. H. Walker

Typeset in Copperplate Gothic and Garamond by
Prepress Projects Ltd, Perth, Scotland
Printed and bound in Great Britain by
TJ International Ltd, Padstow, Cornwall

British Library Cataloguing in Publication Data
A catalogue record for this book is available from the British Library

Library of Congress Cataloging in Publication Data

Walker, C. H. (Colin Harold), 1936–
    Organic pollutants: an ecotoxicological perspective/C.H. Walker.
      p. cm.
    Includes bibliographical references and index.
    ISBN 0-7484-0961-0 (alk. paper) – ISBN 0-7484-0962-9 (pbk. : alk. paper)
    1. Organic compounds – toxicology. 2. Organic compounds – Environmental aspects.
    3. Environmental toxicology. I. Title.
    RA1235.W35 2001
    615.9′5–dc21

Contents

Preface

This book is intended to be a companion volume to *Principles of Ecotoxicology*, first published in 1996 and now in its second edition. Both texts have grown out of teaching material used for the MSc course Ecotoxicology of Natural Populations, which was taught at Reading University between 1991 and 1997. At the time of writing both of these books, a strong driving force was the lack of suitable teaching texts in the areas covered by the course. Although this shortcoming is beginning to be redressed in the wider field of ecotoxicology, with the recent appearance of some valuable new teaching texts, this is not evident in the more focused field of the ecotoxicology of organic pollutants viewed from a mechanistic biochemical point of view. Matters are further advanced in the field of medical toxicology, and there are now some very good teaching texts in biochemical toxicology.

Principles of Ecotoxicology deals in broad brush strokes with the whole field, giving due attention to the 'top-down' approach – considering adverse changes at the levels of population, community and ecosystem, and relating them to the effects of both organic and inorganic pollutants. The present text gives a much more detailed and focused account of major groups of organic pollutants and adopts a 'bottom-up' approach. The fate and effects of organic pollutants are seen from the point of view of the properties of the chemicals, and their biochemical interactions. Particular attention is given to comparative metabolism and mechanism of toxic action and these are related, where possible, to consequent ecological effects. Biomarker assays that provide measures of toxic action are given some prominence because they have the potential to link the adverse effects of particular types of pollutant at the cellular level to consequent effects at the levels of population and above. In this way the top-down approach is complementary to the bottom-up approach; biomarker assays can provide evidence of causality when adverse ecological effects in the field are associated with measured levels of pollutants. Under field conditions, the discovery of a relationship between the level of a pollutant and an adverse effect upon a population is no proof of causality. Many other factors (including other pollutants not determined in the analysis) can have ecological effects, and these factors may happen to correlate with the concentrations of pollutants determined in ecotoxicological studies. The text will also address the question, 'To what extent can ecological effects be predicted from the chemical properties and the biochemistry of pollutants?', which is relevant to the utility or otherwise of use of quantitative structure–activity relationships (QSARs) of chemicals in ecotoxicology.

The investigation of the effects of chemicals upon the numbers and genetic composition of populations has inevitably been a long-term matter, the fruits of which are now becoming more evident with the passage of time. The emergence of resistant strains in response to the selective pressure of pesticides and other pollutants has given insights into the evolutionary process. The evolution of detoxifying enzymes such as the monooxygenases which have cytochrome P450 at their active centre is believed to have occurred in herbivores and omnivores with the movement from water to land. The development of detoxifying mechanisms to protect animals against plant toxins is a feature of 'plant–animal warfare', and is mirrored in the resistance mechanisms developed by invertebrates against pesticides. In the present text, the ecological effects of organic pollutants are seen against the background of the evolutionary history of chemical warfare.

The text is divided into three parts. The first deals with the basic principles underlying the environmental behaviour and effects of organic pollutants; the second describes the properties and ecotoxicology of major pollutants in reasonable detail; the last discusses some issues that arise after consideration of the material in the second part of the text and looks at future prospects. The groups of compounds represented in the second part of the book are all regarded as pollutants rather than simply contaminants, because they have the potential to cause adverse biological effects at realistic environmental levels; in most cases these effects have been well documented under environmental conditions. The term 'adverse effects' includes harmful effects upon individual organisms as well as effects at the level of population and above.

The layout of Chapters 5–12, which constitute Part 2, follows the structure of the text *Principles of Ecotoxicology* as far as possible. Where there is sufficient evidence to do so, the presentations for individual groups of pollutants are arranged as follows:

Topic in this book	Part in *Principles of Ecotoxicology*
Chemical properties Metabolism Environmental fate }	1 Pollutants and their fate in ecosystems
Toxicity	2 Effects of pollutants on individual organisms
Ecological effects	3 Effects on populations and communities

C. H. Walker
Colyton
January 2001

Acknowledgements

Many people have contributed, in various ways, to this book, and it is not feasible in limited space to mention them all. Over a period of nearly 40 years colleagues at Monks Wood have given valuable advice on a variety of subjects. At Reading, colleagues and students have given much good advice, critical discussion and encouragement over many years. Working visits to the research group of Professor Franz Oesch in the Pharmacology Institute at the University of Mainz were stimulating and productive. Advanced courses such as the Ecotoxicology course at Ecomare, Texel, The Netherlands, run by the European Environmental Research Organisation (Professor and Mrs Koeman), and the Summer School on Multidisciplinary approaches in Environmental Toxicology at the University of Siena, Italy (Professor Renzoni), did much to advance knowledge of the subject – not least for those who were fortunate enough to be invited to contribute! To all of these grateful thanks are due.

David Peakall has been a continuing source of good advice and critical comment throughout the writing of this book – not least for compensating for some of the inadequacies of my computer system! Richard Sibly and Steve Hopkin continued to give advice and encouragement after completion of *Principles of Ecotoxicology*. I have benefited from the expert knowledge of the following: Gerry Brooks (organochlorine insecticides), Martin Johnson (organophosphorous insecticides), Ian Newton (ecology of raptors), David Livingstone and Peter Donkin (marine pollution), Frank Moriarty (bioaccumulation and kinetic models), Ken Hassall (biochemistry of herbicides), Mike Depledge (biomarkers), Bram Brouwer (PCB toxicology), Alistair Dawson (endocrine disruptors), Jean-Louis Riviére (avian toxicology), Laurent Lagadic (mesocosms), Alan McCaffery (resistance to insecticides) and Demetris Savva (DNA technology). My gratitude to all of them.

Last but not least, I am grateful to all the research students and postdoctoral research workers at Reading who have contributed in so many ways to the production of this text.

I am grateful for all those who granted permission to reproduce figures used in this book:

Figure 1.2 is reproduced from Lewis, D. *Cytochromes P450*, © (1996) Taylor & Francis Ltd.

Figure 2.8a is reproduced from Moriarty, F. M. and Walker, C. H. (1987) *Ecotoxicology and Environmental Safety*, Figure 2, p. 211, with permission of Academic Press Ltd.

Figure 2.8b is reproduced from Ronis, M. J. J. and Walker, C. H. (1989) *Reviews in Biochemical Toxicology* 10, Figure 16.1, p. 304, with permission of IOS Press, Amsterdam.

Figure 2.10 is reproduced from Walker, C. H. (1994). In Hodgson, E. and Levi, P. (eds) *Introduction to Biochemical Toxicology*, with permission from Appleton-Lange, Norwalk, CT.

Figure 2.16 is reproduced from Moriarty, F. M. (ed.) (1975) *Organochlorine Insecticides: Persistent Organic Pollutants*, Figure 13, p. 114, with permission of Academic Press.

Figure 5.4 is reproduced from Eldefrawi, M. E. and Eldefrawi, A. T. (1990) In Hodgson, E. and Levi, P. (eds) *Safer Insecticides*, with permission from Marcel Dekker.

Figure 5.7 is reproduced from Ratcliffe, D. (1993) *The Peregrine* (published by Poyser), Figure 18, with permission of Academic Press Ltd.

Figure 5.8 is reproduced from Newton, I. (1986) *The Sparrowhawk* (published by Poyser), Figures 82 and 84, with permission of Academic Press.

Figure 5.9 is reproduced from Walker, C. H. and Newton, I. (1999) with permission from Kluwer Academic Publishers.

Figure 6.2 is reproduced from Boon, J. P., *et al.* © (1992) with permission from Elsevier Science.

Figure 6.4 is reproduced from Boon, J. P., *et al.* © (1992) with permission from Elsevier Science.

Figure 9.4 is reproduced from Moore, M. N., *et al.* (1987) with permission from Springer-Verlag.

Figure 10.3 is reproduced from Sussman, J. L., *et al.* (1991) with permission from *Science*.

Figure 13.2 is reprinted from House, W. A., *et al.* © (1997) with permission from Elsevier Science.

PART 1

Basic principles

CHAPTER 1

Chemical warfare

1.1 Introduction

Chemical warfare has been taking place since very early in the history of life on earth, and the design of chemical weapons by humans is an extremely recent event on the evolutionary scale. The synthesis by plants of secondary compounds ('toxins') that are toxic to invertebrates and vertebrates that feed upon them, together with the development of detoxication mechanisms by the animals in response, has been termed a 'co-evolutionary arms race' (Ehrlich and Raven, 1964; Harborne, 1993). Animals, too, have developed chemical weapons, both for attack and for defence. Spiders, scorpions, wasps and snakes possess venoms that paralyse their prey. Bombardier beetles and some slow-moving herbivorous tropical fish produce chemicals that are toxic to other organisms that prey upon them (see Agosta, 1996). Microorganisms produce compounds that are toxic to other microorganisms that may compete with them (e.g. penicillin produced by the mould *Penicillium notatum*).

A very large number of natural chemical weapons have already been identified and characterised, and many more are being discovered with each passing year – in plants, marine organisms, etc. Like pesticides and other man-made chemicals, they are, in a biochemical sense, 'foreign compounds' (xenobiotics). They are 'normal' to the organism that synthesises them but 'foreign' to the organism against which they express toxicity. During the course of evolution, defence mechanisms have evolved to provide protection against the toxins of plants and other naturally occurring xenobiotics.

The use of pesticides and 'chemical warfare agents' by man should be seen against this background. Many defence mechanisms already exist in nature, mechanisms that have evolved to give protection against natural xenobiotics. These systems may work, to a greater or lesser extent, against man-made pesticides when they are first introduced into the environment. Many pesticides are not as novel as they may seem. Some, such as the pyrethroid insecticides, are modelled upon natural insecticides (in the present case pyrethrins) with which they share a common mode of action and similar routes of metabolic detoxication. Also, many detoxication mechanisms are relatively non-specific, operating against a wide range of compounds that have common structural features, e.g. benzene rings, methyl groups or ester bonds. Thus, they can metabolise both man-made and natural xenobiotics, even where overall structures are not closely related.

We are dealing here with an area of science where pure and applied approaches come together. The discovery of natural products with high biological activity (toxicity in the present case), the elucidation of their modes of action and of the defence systems that operate against them can all provide knowledge that aids the development of new pesticides, the development of mechanistic **biomarker assays** which can establish their side-effects on non-target organisms, and the elucidation of mechanisms of resistance that operate against them. Whether compounds are natural or man-made, the molecular basis of toxicity remains a fundamental issue; whether biocides are natural or unnatural, similar mechanisms of action and of metabolism apply. Much of what is now known about the structure and function of enzymes that metabolise xenobiotics has been elucidated during the course of 'applied' research with pesticides and drugs, and the knowledge gained from this is immediately relevant to the metabolism of naturally occurring compounds. The development of this branch of science illustrates how misleading the division between 'pure' and 'applied' science can be. Here, at a fundamental scientific level, they are one and the same; the difference is a question of motivation – whether or not the work is done with some view to a 'practical' outcome (e.g. development of a new pesticide).

The phenomenon of plant–animal warfare will now be discussed, before moving on to a brief review of toxins produced by animals.

1.2 *Plant–animal warfare*

1.2.1 TOXIC COMPOUNDS PRODUCED BY PLANTS

A formidable array of compounds of diverse structure which are toxic to invertebrates and/or vertebrates has been isolated from plants. Some examples are given in Figure 1.1. Many of the known toxic compounds produced by plants are described in Harborne and Baxter (1993). Information about the mode of action of a few of them is given in Table 1.1, noting cases where man-made pesticides act in a similar way.

Let us consider, briefly, the compounds featured in Table 1.1. Pyrethrins are lipophilic esters that occur in *Chrysanthemum* spp. Extracts of flower heads of *Chrysanthemum* spp. contain six different pyrethrins and have been used for insect control (Chapter 12). Pyrethrins act upon sodium channels in a similar manner to p,p'-DDT (p,p'-

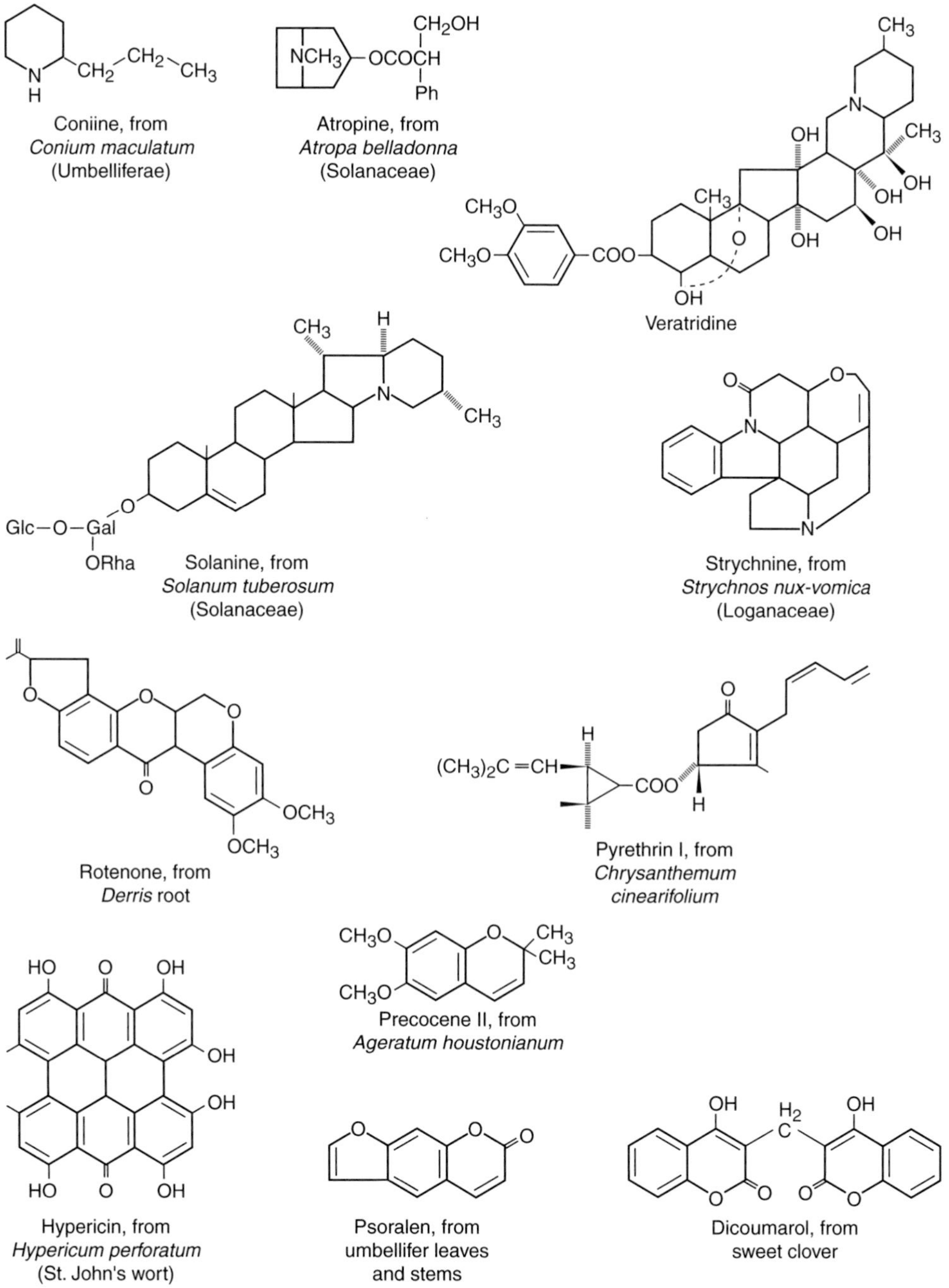

FIGURE 1.1 *Some toxins produced by plants.*

dichlorodiphenyltrichloroethane). The highly successful synthetic pyrethroid insecticides were modelled on natural pyrethrins.

TABLE 1.1 *Some toxins produced by plants*

Compound(s)	Mode of action	Pesticides acting against same target	Comments
Pyrethrins	Na⁺ channel of axonal membrane	Pyrethroids *p,p′*-DDT	Pyrethroids modelled on pyrethrins
Veratridine	Na⁺ channel of axonal membrane	Pyrethroids *p,p′*-DDT	Binding site appears to differ from the one occupied by pesticides
Dicoumarol	Vitamin K antagonist	Warfarin Superwarfarin	All act as anticoagulants (Chapter 11)
Strychnine	Acts on CNS		Used to control some vertebrate pests
Rotenone	Inhibits mitochondrial electron transport		Used as an insecticide (Derris powder)
Physostigmine	Cholinesterase inhibitor	Carbamate insecticides	A model for carbamate insecticides
Nicotine	Acts on nicotinic receptor for acetyl choline	Neonicotinoids, e.g. imidacloprid	An insecticide in its own right and a model for neonicotinoids
Precocenes	Inhibit synthesis of juvenile hormone in some insects		A model for the development of novel insecticides

Data from Harborne (1993), Eldefrawi and Eldefrawi (1990), Brooks *et al.* (1979), Salgado (1999).

Veratridine is a complex lipophilic alkaloid that also binds to sodium channels, causing them to stay open and thereby disrupting the transmission of nerve action potential. *p,p′* -DDT and pyrethroid have similar effects to veratridine but evidently do not bind at the same location on the sodium channel (Eldefrawi and Eldefrawi, 1990).

Dicoumarol is found in sweet clover, and can cause haemorrhaging in cattle because of its anticoagulant action. It acts as a vitamin K antagonist, and has served as a model for the development of warfarin and related anticoagulant rodenticides (Chapter 11).

Strychnine is a complex lipophilic alkaloid from the plant *Strychnos nux-vomica*, which acts as a neurotoxin. It has been used to control vertebrate pests, including moles. The acute oral LD_{50} (median lethal dose) to the rat is 2 mg/kg.

Rotenone is a complex flavonoid found in the plant *Derris ellyptica*. It acts by inhibiting electron transport in the mitochondrion. Derris powder is an insecticidal preparation made from the plant that is highly toxic to fish.

Nicotine is a component of *Nicotiana tabacum*, the tobacco plant. It is toxic to many insects because of its action upon the nicotinic receptor of acetyl choline. It has served as a model for a new range of insecticides, the neonicotinoids, which also act upon the nicotinic receptor (Salgado, 1999).

Precocenes are found in *Ageratum houstonianum* and can cause premature moulting in milkweed bugs (*Oncopeltus fasciatus*) and locusts (*Locusta migratoria*). Precocene 2 is activated by monooxygenase attack to form a reactive metabolite (evidently an epoxide) that inhibits synthesis of juvenile hormone by the corpora allata of milkweed bugs and locusts, leading to atrophy of the organ itself (Brooks *et al.*, 1979).

Widening the range, some further toxic compounds are shown in Figure 1.1: coniine is a toxic compound in hemlock (*Conium maculaatum*), and solanine is the toxic component of green potatoes. Atropine is the principal toxin of deadly nightshade (*Atropa belladonna*). It acts as an antagonist of acetyl choline at muscarinic receptors, and it is used in small quantities as an antidote for poisoning by organophosphorous compounds (see Box 10.1). Hyperiicin is a toxic compound found in St. John's wort (*Hypericum* spp.); psoralen is a toxin found in the stems and leaves of umbellifers.

These are just a few examples among many. They are intended to illustrate the remarkable range of chemical structures among the toxic compounds produced by plants, which provide evidence of the intensity of plant–animal warfare during the course of evolution. In some cases they provide examples of how natural compounds have served as models for the development of pesticides.

1.2.2 ANIMAL DEFENCE MECHANISMS AGAINST TOXINS PRODUCED BY PLANTS

The toxicity of chemicals to living organisms is determined by the operation of both toxicokinetic and toxicodynamic processes (Chapter 2). The evolution of defence mechanisms involves changes in toxicokinetics and/or toxicodynamics that will reduce toxicity. Thus, at the toxicokinetic level, increased storage or metabolic detoxication will lead to reduced toxicity; at the toxicodynamic level, changes in the site of action that reduce affinity with a toxin will bring reduced toxicity.

Some insects can protect themselves against the toxins present in their food plants by storing them. One example is the monarch butterfly (*Daunus plexippus*), the caterpillars of which store potentially toxic cardiac glycosides obtained from a food plant, the milkweed (see Harborne, 1993). Subsequently, the stored glycosides have a deterrent effect upon blue jays that feed upon them!

Both direct and indirect evidence point to the importance of enzymic detoxication in protecting animals, vertebrates or invertebrates, against toxic chemicals produced by plants. In the first place, it is known that detoxication mechanisms operate against natural as well as man-made xenobiotics. Nicotine, for example, undergoes metabolic detoxication in the housefly. Grey kangaroos (*Macropus* spp.) defluorinate fluoracetate, a natural plant product that occurs in some thirty-four species of *Gastrolobium* and *Oxylobium* that grow in Western Australia. Rat kangaroos (*Bettongia* spp.) appear to have developed resistance to fluoracetate in Western Australia, where they are exposed to relatively high levels of the compound, but not in the east, where they are not (see Harborne, 1993).

There is increasing evidence that microsomal monooxygenases with cytochrome P450 as their active centre have a dominant role in the detoxication of the great majority of lipophilic xenobiotics, be they naturally occurring or man-made (Lewis, 1996; Chapter 2 of this book). Gene families 1, 2, 3 and 4 are all involved in xenobiotic metabolism. They have wide-ranging, yet overlapping, substrate specificities and, collectively, can detoxify nearly all lipophilic xenobiotics below a certain size. Some of them are inducible, so they can be 'up-regulated' when there is exposure to unduly high levels of xenobiotics.

The wide distribution of cytochrome P450 enzymes throughout all aerobic organisms clearly indicates a prokaryotic origin with increasing diversification of forms during the course of evolution of vertebrates. Attempts have been made to relate the appearance of different forms of members of P450 families 1–4 to evolutionary events, represented as an evolutionary tree originating from a primordial P450 gene (Nelson and Strobel, 1987; see Lewis, 1996). Particular interest centres on the radiation of the cytochrome P450 family 2 (CYP2; see Figure 1.2), which is believed to have commenced about 400 million years ago, thus coinciding with the movement of animals to land (Nebert and Gonzalez, 1987). Whereas most aquatic organisms can lose lipophilic compounds obtained in their food by diffusion across permeable membranes (especially respiratory membranes) into ambient water, this simple detoxication mechanism is not available to terrestrial animals. They have evolved detoxication systems (predominantly monooxygenases) that can convert lipophilic compounds to water-soluble products that are readily excreted into urine and faeces (Chapter 2). It therefore seems reasonable to suggest that the radiation of CYP2 represents an adaptation of herbivorous/omnivorous animals to life on land, where survival is dependent upon the ability to detoxify lipophilic toxins produced by plants.

This argument gains strength from a comparison of monooxygenase activities in different groups of vertebrates (see Chapter 2 and Walker, 1978, 1980; Ronis and Walker, 1989). Considering activities of microsomal monooxygenase towards a range of lipophilic xenobiotics, fish have much lower activities than herbivorous or omnivorous terrestrial mammals. Among birds, fish-eating birds and other specialist predators tend to have much lower activities than omnivorous/herbivorous birds or mammals (also see 'Monooxygenases' and Walker, 1998a). Most of the xenobiotics used in the assays in question are substrates for P450s belonging to CYP2. It therefore appears that certain P450s (principally CYP2 isoforms) have evolved in omnivorous/ herbivorous vertebrates with adaptation to land and do not occur in fish or in specialised predatory birds. The fish-eating birds represented in this study cannot use diffusion into the water as a mechanism of detoxication to any important extent because they do not have permeable respiratory membranes (in contrast to the gills of fish). Also, most of them spend long periods out of water on nesting sites and roosts. It appears that they have not developed certain P450-based detoxication systems because there

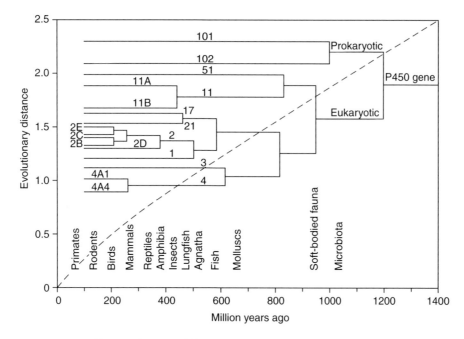

FIGURE 1.2 *An abbreviated version of the P450 phylogenetic tree compared with an evolutionary time scale. From Lewis (1996). The dashed line represents a plot of evolutionary distance (Nelson and Strobel, 1987).*

have been very few lipophilic xenobiotics in their food in comparison with the plants eaten by herbivores/omnivores.

There is growing evidence that different P450 forms of families 1 and 2 do have some degree of specialisation, notwithstanding the rather wide range of compounds that most of them can metabolise. Members of CYP1 specialise in planar compounds, a characteristic that is due to the structure of the binding site at the active centre (Chapter 2; Lewis, 1996). The CYP1 family can metabolise flavones and safroles. Coumarins are metabolised by CYP2A, pyrazines by CYP2E and quinoline alkaloids by CYP2D. The evolution of P450 forms within the general scenario of plant–animal warfare is a rich field for investigation.

The extent to which the evolution of defence systems against natural xenobiotics has involved alterations in toxicodynamics is an open question. Studies on the development of resistance by insects to insecticides (see section 2.4) have frequently established the existence of resistant strains possessing insensitive 'aberrant' forms of the target, frequently differing from the normal sensitive forms by only a single amino acid substitution. Included here are forms of acetylcholinesterase (AChE), the axonal sodium channel and the gamma-aminobutyric acid (GABA) receptor, which are insensitive to organophosphorous insecticides (OPs), pyrethroids and cyclodienes respectively. This indicates the existence of considerable genetic diversity in insect populations and the possibility of the emergence of resistant strains carrying genes coding for insensitive forms of target proteins under the selective pressure of toxic chemicals. As at least two of these targets are common to both man-made insecticides and naturally occurring ones, it seems probable that resistance of this type evolved in nature long before the appearance of commercial insecticides.

1.3 *Toxins produced by animals and microorganisms*

1.3.1 TOXINS PRODUCED BY ANIMALS

Animals use chemical weapons for both defence and attack. Considering defensive tactics first, bombardier beetles (*Brachinus* spp.) can fire a hot solution of irritant quinones at their attackers (see Agosta, 1996). The quinones are generated in abdominal glands by mixing phenols and hydrogen peroxide with catalases and peroxidases. Heat is generated by the reaction, and the cocktail is fired at the assailant with an audible pop. Silent, but more deadly, is the action of tetrodotoxin (Figure 1.3), found in the puffer fish (*Fugu vermicularis*). Tetrodotoxin is an organic cation which can bind to and consequently block sodium channels (Eldefrawi and Eldefrawi, 1990). Interestingly, tetrodotoxin is synthesised by microorganisms that exist on reefs and is evidently taken up and stored by puffer fish. Humans as well as other predators of puffer fish have died from tetrodotoxin poisoning. Saxitoxin, a toxin found in red tide, acts in the same way as tetrodotoxin (Figure 1.3). The use of chemical defence is not uncommon in small slow-moving herbivorous fish that live in enclosed spaces such as reefs. For them, avoidance of predation by rapid movement is not a viable strategy, and chemical weapons can be important for survival. On the other hand, chemical defence is not usually found in the fast-swimming predatory fish of open oceans, many of which are consumed by humans. Chemical defence is also important in immobile invertebrates such as sea anemones.

Turning now to chemical attack, many predators immobilise their prey by injecting toxins, often neurotoxins, into them. Examples include venomous snakes, spiders and scorpions. Some spider toxins (Figure 1.3) are neurotoxic through antagonistic action upon glutamate receptors. The venom of some scorpions contains polypetide neurotoxins that bind to the sodium channel.

A striking feature of the toxic compounds considered so far is that many of them are neurotoxic to vertebrates and/or invertebrates. The nervous system of animals appears to be a particularly vulnerable target in chemical warfare. Not altogether surprisingly, all the major types of insecticides that have been commercially successful are also neurotoxins!

1.3.2 MICROBIAL TOXINS

This is a large subject and can only be dealt with in the barest outline in the present text. Many antibacterial and antifungal compounds have been discovered in

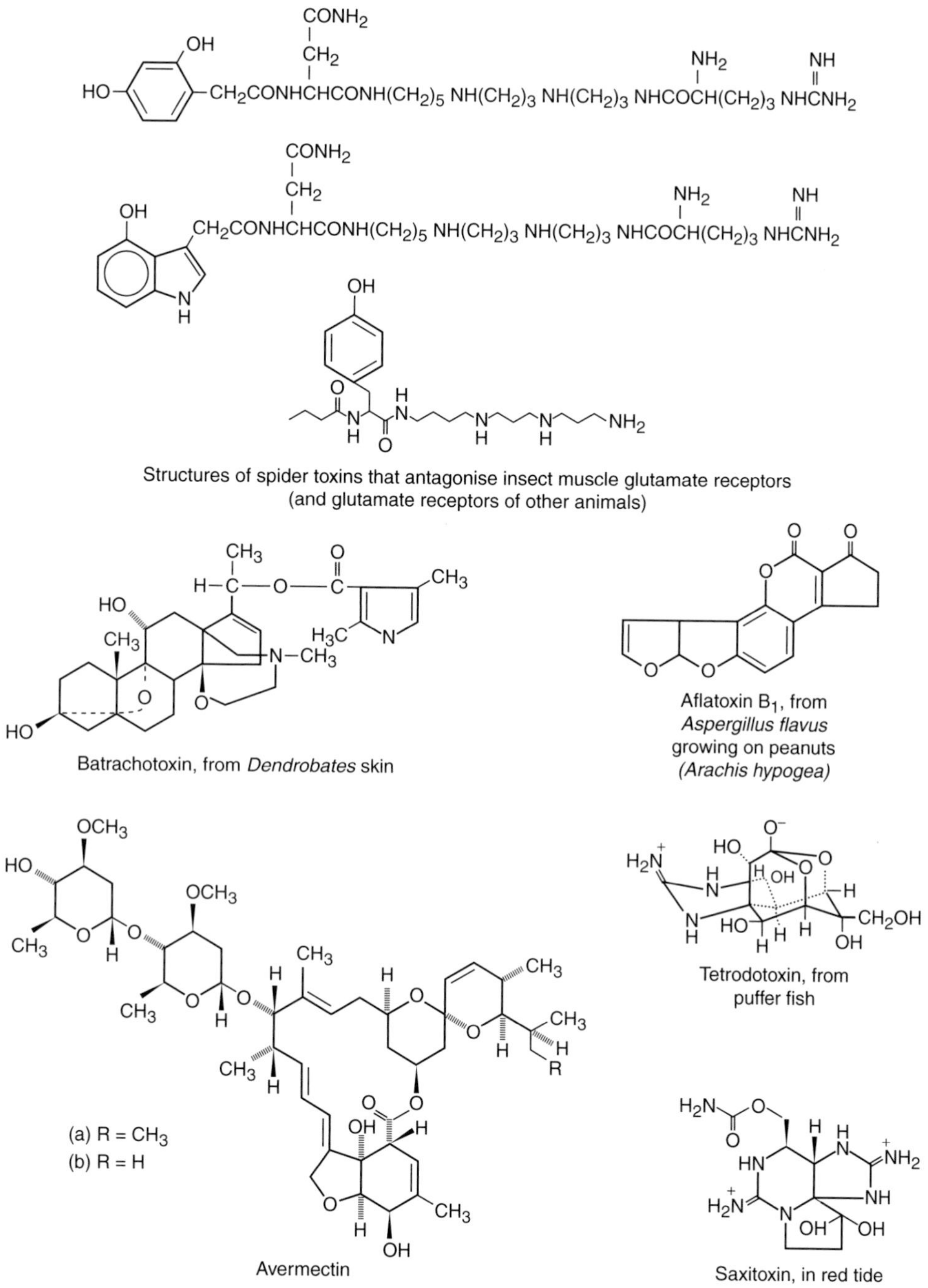

Structures of spider toxins that antagonise insect muscle glutamate receptors
(and glutamate receptors of other animals)

Batrachotoxin, from *Dendrobates* skin

Aflatoxin B$_1$, from
Aspergillus flavus
growing on peanuts
(Arachis hypogea)

(a) R = CH$_3$
(b) R = H

Avermectin

Tetrodotoxin, from
puffer fish

Saxitoxin, in red tide

FIGURE 1.3 *Some toxins from animals and microorganisms.*

microorganisms, and some of them have been successfully developed as antibiotics
for use in human and veterinary medicine. They lie outside the scope of this book. A
considerable number of other microbial compounds act as insecticides, acaricides or

herbicides, although few of them have been developed commercially (Copping and Menn, 2000). The following compounds are examples of microbial compounds that are toxic to vertebrates and/or invertebrates.

Aflatoxin (Figure 1.3) is a toxic compound that acts as a hepatocarcinogen. It is synthesised by a strain of the common mould, *Aspergillus flavus*, which grows on badly stored ground nuts. Its toxicity was recognised when turkeys that had been fed contaminated ground nuts died. Aflatoxin is converted into a highly reactive epoxide by P450-based monooxygenase attack. The epoxide form adducts with guanine residues of DNA.

Avermectins (Figure 1.3) are complex molecules, synthesised by the bacterium *Streptomyces avermitilis*, which have strong insecticidal, acaricidal and antihelminthic properties. Eight forms have been found to occur naturally, and a slightly modified structure, ivermectin, has been synthesised and marketed commercially. They are toxic because they stimulate the release of gamma-aminobutyric acid (GABA) from nerve endings, and so cause overstimulation of GABA (Copping and Menn, 2000). Avermectins have been used for insecticide and mite control and also for deworming cattle. In the last case they remain in faeces and effectively control the insects that inhabit dung pats! Ecologists have argued that large-scale use could have serious effects upon insect populations in grasslands where there are cattle.

Preparations of the bacterium *Bacillus thuringiensis* are applied as sprays to control insect pests on agricultural crops. The bacterium produces endotoxins that are highly toxic to insects.

1.4 Man-made chemical weapons

The synthesis of toxic organic compounds by man, and their release into the natural environment, began to assume significant proportions in the twentieth century, especially after the Second World War. Before that there was relatively little chemical industry, and the largest impact of man was probably the release of hydrocarbons, especially polycyclic aromatic hydrocarbons (PAHs), with the combustion of coal and other fuels. The pollutants to be described in this book are predominantly the products of human activity although a few, such as PAHs and methyl mercury, are also naturally occurring.

The harmful effects of these pollutants upon individual organisms, upon the numbers and genetic composition of populations and upon the structure and function of communities and ecosystems represent the basic material of this book. The main purpose of this introductory chapter is to put the matter into perspective. The effects to be described should always be seen against the background of the long and continuing history of chemical warfare on earth.

1.5 *Summary*

In this introductory chapter, a broad overview is given of the history of chemical warfare on earth, and the compounds, species and mechanisms involved. The impact of man-made compounds on the environment, which is the subject of this book, is an extremely recent event in evolutionary terms. It is important to take a holistic view, and to see the effects of man-made pollutants on the environment against the background of chemical warfare in nature.

1.6 *Further reading*

Agosta, W. (1996) *Bombadier Beetles and Fever Trees*. A very readable account of chemical warfare, without much detail on chemical or biochemical aspects.

Copping, L. G. and Menn, J. J. (2000) An up-to-date account of naturally occurring compounds that act as pesticides.

Harborne, J. B. (1993) *Introduction to Ecological Biochemistry*. A very popular text which contains valuable chapters on the toxic compounds of plants and the 'co-evolutionary arms race'.

Harborne, J. B. and Baxter, H. (1993) *Phytochemical Dictionary*. A dictionary that gives details of many toxins produced by plants.

Lewis, D. F. V. (1996) *Cytochromes P450*. This book has a very useful chapter on the evolution of forms of cytochrome P450.

CHAPTER **2**

Factors determining the toxicity of organic pollutants to animals and plants

2.1 *Introduction*

This chapter will consider the processes that determine the toxicity of organic pollutants to living organisms. The term 'toxicity' will encompass harmful effects in general and will not be restricted to lethality. With the rapid advances in mechanistic toxicology in recent years, it is increasingly possible to understand the underlying sequence of changes that leads to the appearance of symptoms of intoxication and how differences in the operation of these processes between species, strains, sexes and age groups can account for selective toxicity. Thus, in a text of this kind, it is convenient to deal with these principles at an early stage, because they underlie many of the issues to be discussed later. It is important to understand why chemicals are toxic and why they are selective, not only as a matter of scientific interest but also for more practical reasons. An understanding of mechanism can contribute to the development of new biomarker assays, the design of more environmentally friendly pesticides and the control of resistant pests.

Although many of the standard ecotoxicity tests use lethality as the end point, it is now widely recognised that sublethal effects may be at least as important as lethal ones in ecotoxicology. Pollutants that affect reproductive success can cause populations to decline. The persistent DDT metabolite *p,p'* -DDE (*p,p'* -dichlorodiphenyl-dichloroethylene) caused the decline of certain predatory birds in North America through eggshell thinning and consequent reduction in breeding success (see Chapter 5). The antifouling agent tributyltin (TBT) caused population decline in the dog whelk (*Nucella lapillus*) through making the females infertile (see Chapter 8). Neurotoxic compounds can have behavioural effects in the field (see Chapters 5 and 10), and these may reduce the breeding or feeding success of animals. A number of the examples that follow are of sublethal effects of pollutants. The occurrence of sublethal effects in natural populations is intimately connected with the question of persistence. Chemicals with long biological half-lives present a particular risk. The maintenance of substantial levels in individuals, and along food chains, over long periods of time maximises the risk of sublethal effects. Risks are fewer with less persistent compounds, which are rapidly eliminated by living organisms. As will be discussed later, biomarker assays are already making an important contribution to the recognition and quantification of sublethal effects in ecotoxicology (see section 15.4).

In ecotoxicology the primary concern is about effects seen at the level of population or above, and these can be the consequence of the indirect as well as the direct action of pollutants. Herbicides, for example, can indirectly cause the decline of animal populations by reducing or eliminating the plants upon which they feed. A well-documented example of this on agricultural land is the decline of insect populations and the grey partridges which feed upon them as a result of the removal of key weed species by herbicides (see Chapter 13). Thus, the toxicity of pollutants to plants can be critical in determining the fate of animal populations! When interpreting ecotoxicity data during the course of environmental risk assessment, it is very important to have an ecological perspective.

Toxicity is the outcome of interaction between a chemical and a living organism. The toxicity of any chemical depends upon its own properties, and the operation of certain physiological and biochemical processes within animals or plants that are exposed to it. These processes are the subject of the present chapter. They can operate very differently in different species, which is the main reason for the selective toxicity of chemicals between species. For the same reasons, chemicals show selective toxicity (henceforward simply 'selectivity') between groups of organisms (e.g. animals versus plants and invertebrates versus vertebrates), and also between sexes, strains and age groups of the same species.

Selectivity is a very important aspect of ecotoxicology. In the first place, there is immediate concern about the direct toxicity of any environmental chemical to the most sensitive species that will be exposed to it. Usually the most sensitive species is not known, because only a small number of species can ever be used for toxicity testing in the laboratory in comparison with the very large number at risk in the field. As with human toxicology, risk assessment depends upon the interpretation of toxicity data obtained with surrogate species. The problem comes in extrapolating between species. In ecotoxicology such extrapolations are often made very difficult because the surrogate species is only distantly related to the species of environmental concern. Predicting toxicity to predatory birds from toxicity data obtained with feral pigeons (*Columba livia*) or Japanese quail (*Coturnix coturnix japonica*) is not a straightforward matter. The great diversity of wild animals and plants cannot be overemphasised. For this reason large safety factors are often used when estimating 'environmental toxicity' from the very sparse ecotoxicity data. Understanding the mechanistic basis of selectivity can improve confidence in making interspecies comparisons in risk assessment. Knowing more about the operation of processes that determine toxicity in different species can give some insight into the question 'How comparable are different species?' when interpreting toxicity data. The presence of the same sites of action, or of similar levels of key detoxifying enzymes, may strengthen confidence when extrapolating from one species to another in the interpretation of toxicity data. Conversely, large differences in these factors between species discourage the use of one species as a surrogate for another.

Finally, selectivity is a vital consideration in relation to the safety and efficacy of pesticides. In designing new pesticides manufacturers seek to maximise toxicity to the target organism, which may be an insect pest, a vertebrate pest, a weed or a plant pathogen, while minimising toxicity towards humans or beneficial organisms. Beneficial organisms include farm animals, domestic animals, beneficial insects, fish and most species of wildlife (vertebrate pests such as rats not included). Understanding mechanisms of toxicity can lead manufacturers towards the design of safer pesticides. Physiological and biochemical differences between pest species and beneficial organisms can be exploited in the design of new and safer pesticides. Examples of this will be given in the following text. On the question of efficacy, the development of resistance is an inevitable consequence of the heavy and continuous use of pesticides. Understanding the factors responsible for resistance (e.g. enhanced detoxication or insensitivity of the site of action in a resistant strain) can point to ways of overcoming it. For example, alternative pesticides not susceptible to a resistance mechanism may be used. Also, new pesticides can be developed that overcome resistance mechanisms. In general, a better understanding of the mechanisms responsible for selectivity can facilitate the safer and more effective use of pesticides.

2.2 *Factors which determine toxicity and persistence*

The fate of a xenobiotic in a living organism, seen from a toxicological point of view, is summarised in Figure 2.1. This highly simplified diagram draws attention to the main processes that determine toxicity. Three main types of location are shown within the diagram.

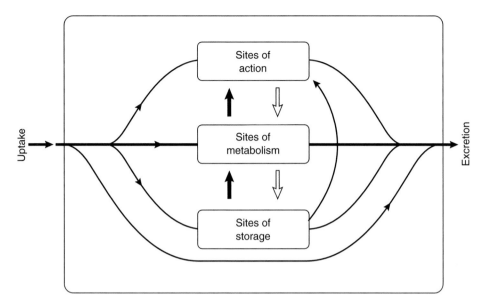

FIGURE 2.1 *Toxicokinetic model.*

1 Sites of action. When a chemical interacts with one or more of these, there will be a toxic effect on the organism if the concentration exceeds a certain threshold. The chemical has an effect upon the organism.

2 Sites of metabolism. When a chemical reaches one of these, it is metabolised. Usually this means detoxication, but sometimes (most importantly) the consequence is activation. The organism acts upon the chemical.

3 Sites of storage. When located in one of these, the chemical has no toxic effect, is not metabolised and is not available for excretion. However, after release from store it may travel to sites of action and sites of metabolism.

In reality, things are more complex than this. For some chemicals there may be more than one type of site in any of the three categories. Also, any particular type of site may exist in a number of different locations. Thus, some chemicals have more than one site of action. The organophosphorous insecticide mipafox, for example, can produce toxic effects by interacting with either AChE or neuropathy target esterase. Also, many organophosphorous insecticides can interact with AChE located in different tissues (e.g. brain and peripheral nervous system). Regarding sites of metabolism, many xenobiotics are metabolised by two or more enzyme systems. Pyrethroid insecticides, for instance, are metabolised by both monooxygenases and esterases. Also, lipophilic compounds can be both stored in fat depots and bound to 'inert' proteins (that is proteins which do not metabolise the xenobiotic or represent a site of action).

Despite these complicating factors, the model shown in Figure 2.1 identifies the main events that determine toxicity in general and selective toxicity in particular.

More sophisticated versions of it can be used to explain or predict toxicity and selectivity. It is important to see the wood despite the trees! For many lipophilic compounds, rapid conversion into more polar metabolites and conjugates leads to efficient excretion, and thus efficient detoxication. This is emphasised by the use of a broad arrow running through the middle of the diagram. Inhibition of this process can cause a very large increase in toxicity (see later discussion of synergism). For convenience, the processes identified in Figure 2.1 can be separated into two distinct categories – toxicokinetics and toxicodynamics. Toxicokinetics covers uptake, distribution, metabolism and excretion. These processes determine how much of the toxic form of a chemical (parent compound and/or active metabolite) will reach the site of action. Toxicodynamics is concerned with the interaction with the site(s) of action, leading to the expression of toxic effects. The interplay of the processes of toxicokinetics and toxicodynamics determine toxicity. The more of the toxic form of the chemical that reaches the site of action, and the greater the sensitivity of the site of action to the chemical, the more toxic it will be. In the following text, toxicokinetics and toxicodynamics will be dealt with separately.

2.3 *Toxicokinetics*

From a toxicological point of view, the critical issue is how much of the toxic form of the chemical reaches the site of action. This will be determined by the interplay of the processes of uptake, distribution, metabolism, storage and excretion. These processes will now be discussed in a little more detail.

2.3.1 UPTAKE AND DISTRIBUTION

The major routes of uptake of xenobiotics by animals and plants are discussed in section 4.2. With animals, there is an important distinction between terrestrial species, on the one hand, and aquatic invertebrates and fish on the other. The latter readily absorb many xenobiotics directly from ambient water or sediment across permeable respiratory surfaces (e.g. gills) Some amphibia (e.g. frogs) readily absorb such compounds across permeable skin. By contrast, many aquatic vertebrates, such as whales and seabirds, absorb little by this route. In lung-breathing organisms, direct absorption from water across exposed respiratory membranes is not an important route of uptake.

Once compounds have entered organisms, they are transported around in blood and lymph (vertebrates), haemolymph (invertebrates) and in the phloem or xylem of plants, eventually moving into organs and tissues. During transport, polar compounds will be dissolved in water, or associated with charged groups on proteins such as albumin, whereas non-polar lipophilic compounds may be associated with lipoprotein complexes or fat droplets. Eventually, the ingested pollutants will move into cells and tissues, to be distributed between the various subcellular compartments (endoplasmic reticulum, mitochondria, nucleus, etc.). In vertebrates, movement from circulating blood into tissues may be due to simple diffusion across membranes or to transport with macromolecules, which are absorbed unchanged into cells. This latter process occurs when, for example, lipoprotein fragments are absorbed intact into liver cells (hepatocytes). The processes of distribution are less well understood in invertebrates and plants than they are in vertebrates.

An important factor in determining the course of uptake, transport, and distribution of xenobiotics is their *polarity*. Compounds of low polarity tend to be lipophilic and of low water solubility. Compounds of high polarity tend to be hydrophilic and of low fat solubility. The balance between the lipophilicity and hydrophilicity of any compound is indicated by its octanol–water partition coefficient (K_{ow}), a value determined when equilibrium is reached between the two adjoining phases:

$$K_{OW} = \frac{\text{concentration of compound in octanol}}{\text{concentration of compound in water}}$$

Compounds with high K_{ow} values are of low polarity and are described as being lipophilic and hydrophobic. Compounds with low K_{ow} values are of high polarity and are hydrophilic. Although the partition coefficient between octanol and water is the one most frequently encountered, partition coefficients between other non-polar liquids (e.g. hexane, olive oil) and water also give a measure of the balance between lipophilicity and hydrophilicity. K_{ow} values for highly lipophilic compounds are very large, and they are commonly expressed as log values to the base 10 (log K_{ow}).

K_{ow} values determine how compounds will distribute themselves across polar–non-polar interfaces. Thus, in the case of biological membranes, lipophilic compounds of high K_{ow} below a certain molecular weight move from ambient water into the hydrophobic regions of the membrane, where they associate with lipids and hydrophobic proteins. Such compounds will show little tendency to diffuse out of membranes, i.e. they readily move into membranes but do not tend to cross into the compartment on the opposite side. Above a certain molecular mass (approximately 800 kDa), lipophilic molecules are not able to diffuse into biological membranes. That said, the great majority of pollutants described in the present text have molecular weights below 450 and are able to diffuse into membranes. By contrast, polar compounds with low K_{ow} values tend to stay in water and not move into membranes. The same arguments apply to other polar–non-polar interfaces within living organisms, e.g. lipoproteins in blood or fat droplets in adipose tissue. The compounds that diffuse most readily across membranous barriers are those with a balance between lipophilicity and hydrophilicity, having K_{ow} values of the order 0.1–1.

Some examples of log K_{ow} values of organic pollutants are given in Table 2.1. The compounds listed in the left-hand column are more polar than those in the right-hand column. They show less tendency to move into fat depots and bioaccumulate than compounds of higher K_{ow} do. That said, the herbicide Atrazine, which has the

TABLE 2.1 *Log K$_{ow}$ values of organic pollutants*

Low K$_{ow}$		High K$_{ow}$	
Hydrogen cyanide	−0.25	Malathion	2.89
Vinyl chloride	0.60	Lindane	3.78
Methyl bromide	1.19	Parathion	3.81
Phenol	1.45	2-Chlorobiphenyl	4.53
Chloroform	1.97	4,4-Dichlorobiphenyl	5.33
Trichlorofluoromethane	2.16	Dieldrin	5.48
Carbaryl	2.36	p,p′-DDT	6.36
Dichlorofluoromethane	2.53	Benzo(a)pyrene	6.50
Atrazine	2.56	TCDD (dioxin)	6.64

highest K_{ow} in the first group, has quite low water solubility (approximately 5 ppm) and is relatively persistent in soil. Turning to the second group, these tend to move into fat depots and bioaccumulate. Those that are resistant to metabolic detoxication have particularly long biological half-lives (e.g. dieldrin, p,p'-DDT and TCDD). Some of them, for example dieldrin and p,p'-DDT, have extremely long half-lives in soils (see section 4.2).

Before leaving the subject of polarity and K_{ow} in relation to uptake and distribution, mention should be made of weak acids and bases. The complicating factor here is that they exist in solution in different forms, the balance between which is dependent upon pH. The different forms have different polarities, and thus different K_{ow} values. In other words, the K_{ow} values measured are pH dependent. Take for example the plant growth regulator herbicide 2,4-D (2,4-dichlorophenoxyacetic acid). This is often formulated as the sodium or potassium salt, which has high water solubility. When dissolved in water, however, the following equilibrium is established:

$$R - COOH \rightleftharpoons ROO^- + H^+$$

Where R =

If the pH is reduced by adding an acid, the equilibrium moves from right to left, generating more of the undissociated acid. This has a higher K_{ow} than the anion from which it is formed. Consequently, it can move readily by diffusion, into and through hydrophobic barriers, which the anion cannot. If the herbicide is applied to plant leaf surfaces, absorption across the lipophilic cuticle into the plant occurs more rapidly at lower pH (e.g. in the presence of NH_4^+). The same argument applies to the uptake of weak acids such as aspirin (acetylsalicylic acid) across the wall of the vertebrate stomach. At the very low pH of the stomach contents, much of the aspirin exists in the form of the lipophilic undissociated acid, which readily diffuses across the membranes of the stomach wall, and into the bloodstream. A similar argument applies to weak bases, except that these tend to pass into the undissociated state at high rather than low pH. Substituted amides, for example, show the following equilibrium:

$$R - CO\ NH_3^+ \rightleftharpoons RNH_2 + H^+ \qquad R = alkyl\ or\ aryl\ group$$

As the pH increases, the concentration of OH^- also goes up. Hydrogen ions (H^+) are removed to form water, the equilibrium shifts from left to right and more relatively non-polar RNH_2 is generated.

Returning to the more general question of the movement of organic molecules through biological membranes during uptake and distribution – a major consideration, then, is movement through the underlying structure of the phospholipid bilayer. It should also be mentioned, however, that there are pores through membranes that are hydrophilic in character, through which ions and small polar organic molecules (e.g. methanol, acetone) may pass by diffusion. The diameter and characteristics of these pores varies between different types of membranes. Many of them have a critical role in regulating the movement of endogenous ions and molecules across membranes. Movement may be by diffusion, primary or secondary active transport, or facilitated diffusion. A more detailed consideration of pores would be inappropriate in the present context. Readers are referred to basic texts on biochemical toxicology (e.g. Timbrell, 1999) for a more extensive treatment. The main points to be emphasised here are that certain small, relatively polar, organic molecules can diffuse through hydrophilic pores, and that the nature of these pores varies between membranes of different tissues and different cellular locations.

Let us consider again movement across phospholipid bilayers; where only passive diffusion is involved, compounds below a certain molecular mass (approximately 800 kDa) with very high K_{ow} values tend to move into membranes, but show little tendency to move out again. In other words, they do not move across membranes, to any important extent, by passive diffusion alone. On the other hand, they may be co-transported across membranes by endogenous hydrophobic molecules with which they are associated, e.g. lipids or lipoproteins. There are transport mechanisms, e.g. phagocytosis (solids) and pinocytosis (liquids), that can move macromolecules across membranes. The particle or droplet is engulfed by the cell membrane and then extruded to the opposite side, carrying associated xenobiotics with it. The lipids associated with membranes are turned over, so lipophilic compounds taken into membranes and associated with them may be co-transported with the lipids to other cellular locations. Compounds of low K_{ow} do not tend to diffuse into lipid bilayers at all, and consequently they do not cross membranous barriers unless they are sufficiently small and polar to diffuse through pores (see p. 21). The blood–brain barrier of vertebrates is an example of a non-polar barrier between an organ and surrounding plasma that prevents the transit of ionised compounds in the absence of any specific uptake mechanism. The relatively low permeability of the capillaries of the central nervous system to ionised compounds is the consequence of two things:

1 the coverage of the basement membranes of the capillary endothelium by the processes of glial cells (astrocytes); and
2 the tight junctions that exist between capillaries, leaving few pores.

Lipophilic compounds (e.g. organochlorine insecticides, organophosphorous insecticides, organomercury compounds and organolead compounds) readily move into the brain to produce toxic effects, whereas many ionised compounds are excluded by this barrier.

General considerations

After uptake, lipophilic pollutants tend to move into hydrophobic domains within animals or plants (membranes, lipoproteins, depot fat, etc.) unless they are biotransformed into more polar and water-soluble compounds having very high K_{ow} values. Metabolism of lipophilic compounds proceeds in two stages:

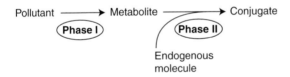

In phase 1, the pollutant is converted into a more water-soluble metabolite(s) by oxidation, hydrolysis, hydration or reduction. Usually phase 1 metabolism introduces one or more hydroxyl groups. In phase 2, a water-soluble endogenous species (usually an anion) is attached to the metabolite – very often through a hydroxyl group introduced during phase 1. Although this scheme describes the course of most biotransformations of lipophilic xenobiotics, there can be departures from it. Sometimes the pollutant is directly conjugated, for example by interaction of the endogenous molecule with the hydroxyl groups of phenols or alcohols. Phase 1 can involve more than one step, and sometimes yields an active metabolite that binds to cellular macromolecules without undergoing conjugation (as in the activation of benzo(*a*)pyrene and other carcinogens). A diagrammatic representation of metabolic changes linking them to detoxication and toxicity is shown in Figure 2.2. The description so far is based upon data for animals. Plants possess similar enzyme systems to animals, albeit at lower activities, but they have been little studied. The ensuing account is based on what is known of the enzymes of animals, especially mammals.

Many of the phase 1 enzymes are located in hydrophobic membrane environments. In vertebrates they are particularly associated with the endoplasmic reticulum of the

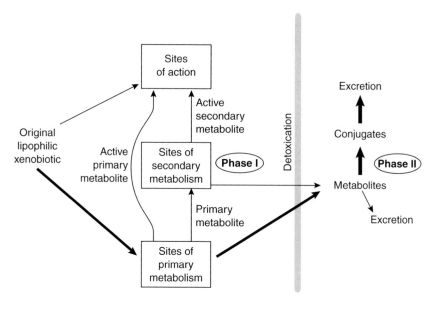

FIGURE 2.2 *Metabolism and toxicity.*

liver, in keeping with their role in detoxication. Lipophilic xenobiotics are moved to the liver after absorption from the gut, notably in the hepatic portal system of mammals. Once absorbed into hepatocytes, they will diffuse, or be transported, to the hydrophobic endoplasmic reticulum. Within the endoplasmic reticulum, enzymes convert them into more polar metabolites, which tend to diffuse out of the membrane and into the cytosol. Either in the membrane or more extensively in the cytosol, conjugases convert them into water-soluble conjugates that are ready for excretion. Phase 1 enzymes are located mainly in the endoplasmic reticulum, and phase 2 enzymes mainly in the cytosol.

The enzymes involved in the biotransformation of pollutants and other xenobiotics will now be described in more detail, starting with phase 1 enzymes, and then moving on to phase 2 enzymes. (For an account of the main types of enzymes involved in xenobiotic metabolism see Jakoby, 1980.)

Monooxygenases

Monooxygenases exist in a great variety of forms, with contrasting yet overlapping substrate specificities. Substrates include a very wide range of lipophilic compounds, both xenobiotics and endogenous molecules. They are located in membranes, most importantly in the endoplasmic reticulum of different animal tissues. In vertebrates, the liver is a particularly rich source, whereas in insects microsomes prepared from the midgut or the fat body contain substantial amounts of these enzymes. When lipophilic pollutants move into the endoplasmic reticulum, they are converted into more polar metabolites by monooxygenase attack, metabolites that partition out of the membrane into cytosol. Very often metabolism leads to the introduction of one or more hydroxyl groups, and these are available for conjugation with, for example, glucuronide or sulphate. Monooxygenases constitute the most important group of enzymes carrying out phase 1 biotransformation, and very few lipophilic xenobiotics are resistant to metabolic attack by them, the main exceptions being highly halogenated compounds, such as dioxin, p,p'-DDE and higher chlorinated polychlorinated biphenyls (PCBs).

Monooxygenases owe their catalytic properties to the haemprotein cytochrome P450 (Figure 2.3). Within the membrane of the endoplasmic reticulum (microsomal membrane), cytochrome P450 macromolecules are associated with another protein, NADPH-cytochrome P450 reductase. The latter enzyme is converted into its reduced form by the action of NADPH (reduced form of nicotinamide adenine dinucleotide phosphate). Electrons are passed from the reduced reductase to cytochrome P450,

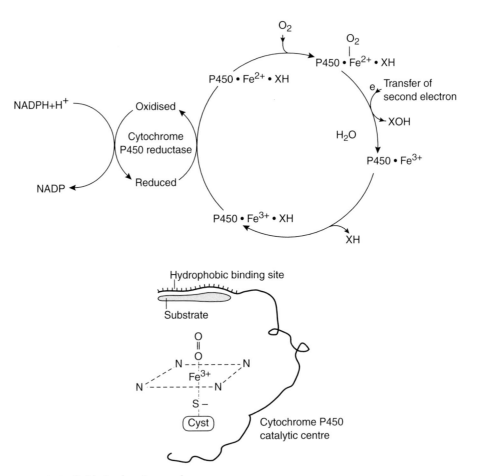

FIGURE 2.3 *Oxidation by microsomal monooxygenases.*

converting it to the Fe^{2+} state. Xenobiotic substrates attach themselves to the hydrophobic binding site of P450, where the iron of the haemprotein is in the Fe^{3+} state. After a single electron has been passed from the reductase to P450, the haemprotein moves into the Fe^{2+} state, and molecular oxygen can now bind to the enzyme–substrate complex. It binds to the free sixth ligand position of the iron, where it is now in close proximity to the bound lipophilic substrate (Figure 2.3). A further electron is then passed to P450, and this leads to the activation of the bound oxygen. This second electron may come from the same source as the first, or it may originate from another microsomal haemprotein, b5, which is reduced by NADH rather than NADPH. After this molecular oxygen is split, one atom being incorporated into the xenobiotic metabolite, the other into water. The exact mechanism involved in these changes is still controversial. However, a widely accepted version of the main events is shown in Figure 2.4. The uptake of the second electron leads to the formation of the highly reactive superoxide anion, O_2^-, after which the splitting of molecular oxygen and 'mixed function oxidation' immediately follow. The P450 returns to the Fe^{3+} state, and the whole cycle can begin again.

'Active' oxygen generated at the catalytic centre of cytochrome P450 can attack the great majority of organic molecules that become attached to the neighbouring substrate binding site (Figure 2.3). When substrates are bound, the position on the molecule that is attacked ('regioselectivity') will depend on the spatial relationship between the bound molecule and the activated oxygen. Active oxygen forms are most likely to attack the accessible positions on the xenobiotic that are nearest to them. Differences in substrate specificity between the many different P450 forms are due, very largely if not entirely, to differences in the structure and position of the binding site within the haemprotein. The mechanism of oxidation appears to be the same in the different forms of the enzyme, so could hardly provide the basis for substrate

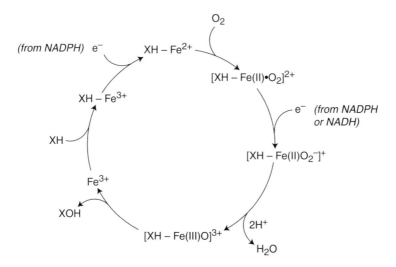

FIGURE 2.4 *Proposed mechanism for monooxygenation by cytochrome P450.*

1. Aromatic hydroxylation

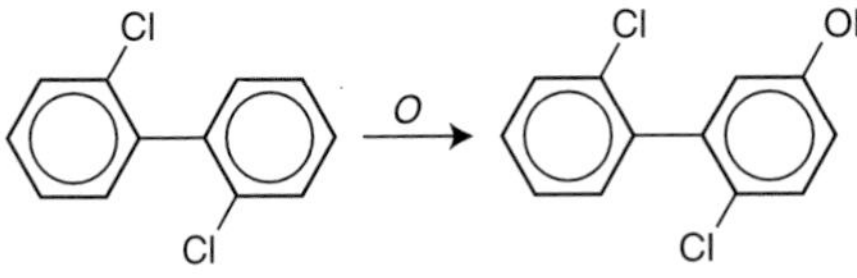

Dichlorophenyl

2. Aliphatic hydroxylation

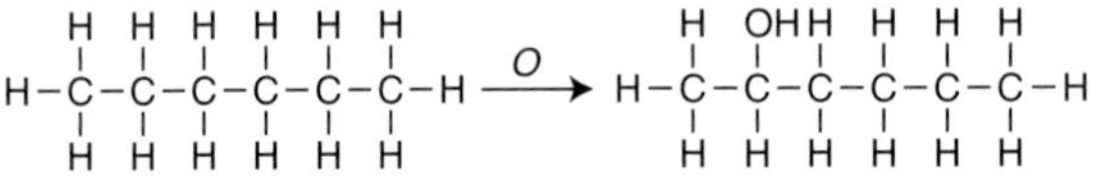

n-Hexane

3. Epoxidation

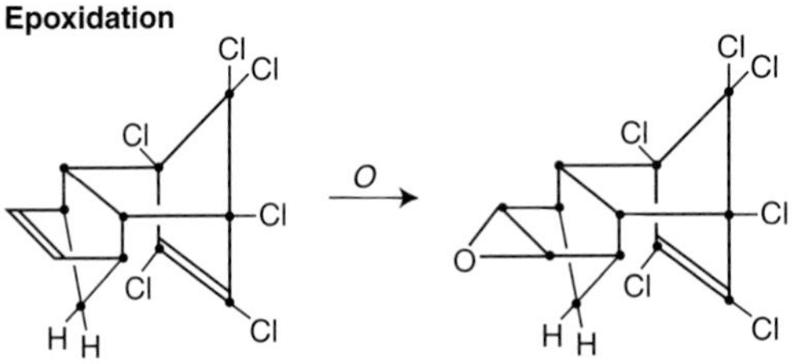

Aldrin   Dieldrin

4. O-Dealkylation

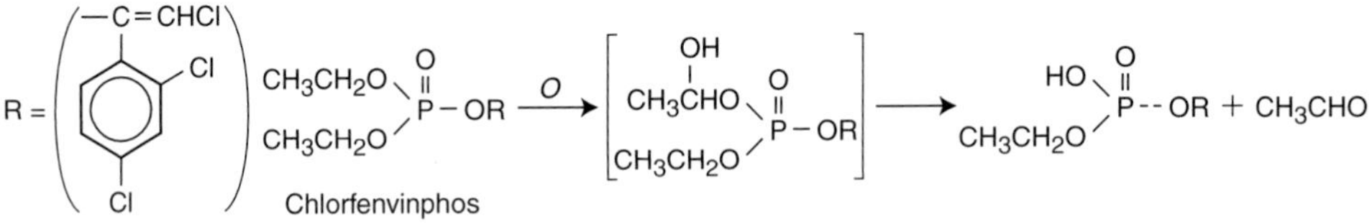

Chlorfenvinphos

5. N-Dealkylation

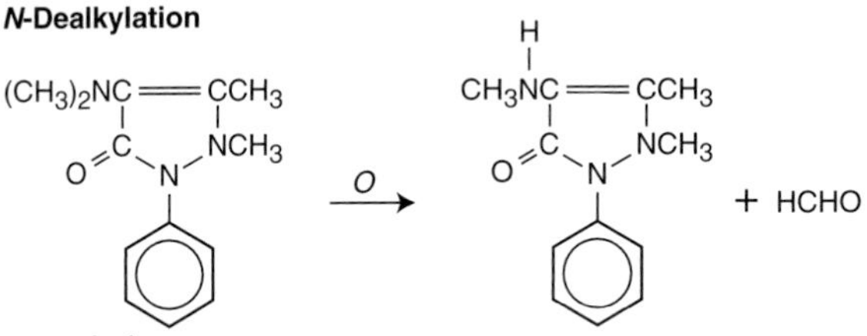

Aminopyrene   + HCHO

6. Oxidative desulphuration

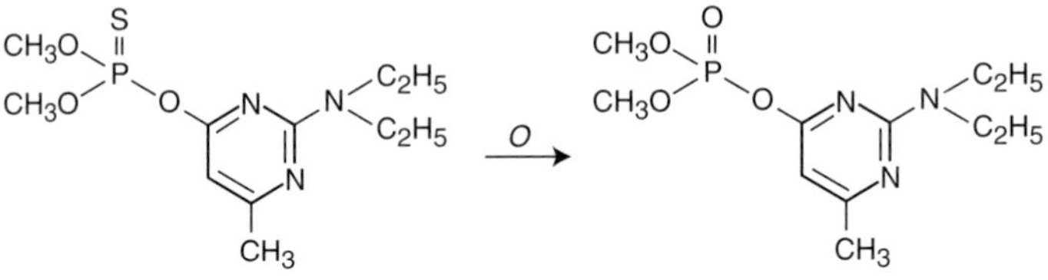

Diazinon   Diazoxon

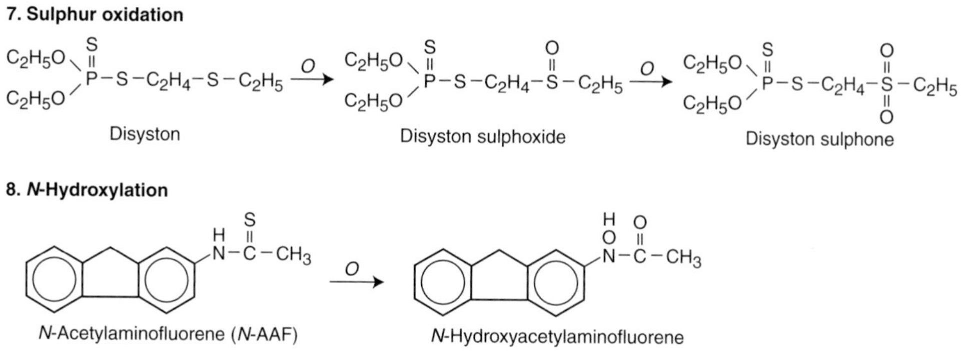

7. Sulphur oxidation

Disyston → Disyston sulphoxide → Disyston sulphone

8. *N*-Hydroxylation

N-Acetylaminofluorene (*N*-AAF) → *N*-Hydroxyacetylaminofluorene

FIGURE 2.5 *Biotransformations by cytochrome P450.*

specificity (see Trager, 1989). This explains regiospecific metabolism, where different forms of P450 attack the same substrate, but in different molecular positions. Regioselectivity is sometimes very critical in the activation of polycyclic aromatic hydrocarbons (PAHs) which act as carcinogens or mutagens (see Chapter 9). Cytochrome P4501A1, for example, tends to hydroxylate benzo(*a*)pyrene in the so-called bay region, yielding bay region epoxides that are highly mutagenic (Chapter 9). Other P450 forms attack different regions of the molecule, yielding less hazardous metabolites. The production of active forms of oxygen is, in itself, potentially hazardous, and it is very important that such reactive species do not escape from the catalytic zone of P450 to other parts of the membrane, where they could cause oxidative damage. There is evidence that, under certain circumstances, superoxide anions may escape in this way. This may occur where highly refractory substrates (e.g. higher chlorinated PCBs) are bound to P450 but resist metabolic attack (see section 14.3).

The wide range of oxidations catalysed by cytochrome P450 is illustrated by the examples given in Figure 2.5. Aromatic rings are hydroxylated, as in the case of 2,6'-dichlorobiphenyl. The initial product is usually an epoxide, but this rearranges to give a phenol. Alkyl groups can also be hydroxylated, as in the conversion of hexane to hexan-2-ol. If an alkyl group is linked to nitrogen or oxygen, hydroxylation may yield an unstable product. An aldehyde is released, leaving behind a proton attached to N or to O (N-dealkylation or O-dealkylation respectively). Thus, with the organophosphorous insecticide chlorfenvinphos, one of the ethoxy groups is hydroxylated, and the unstable metabolite so formed cleaves to release acetaldehyde and desethyl chlorfenviphos. In the case of the drug aminopyrene, a methyl group attached to N is hydroxylated, and the primary metabolite splits up to release formaldehyde and an amine. Sometimes the oxidation of C=C double bonds can generate stable epoxides, as in the conversion of aldrin to dieldrin, or heptachlor to heptachlor epoxide. Cytochrome P450s can also catalyse oxidative desulphuration. The example given is the organophosphorous insecticide diazinon, which is transformed into the active oxon, diazoxon. P=S is converted into P=O. With thioethers such as the organophosphorous insecticide disyston, P450 can catalyse the addition of oxygen to the sulphur bridge, generating sulphoxides and sulphones. P450s can also catalyse the N-hydroxylation of amines such as *N*-acetylaminofluorene (*N*-AAF).

This series of examples is by no means exhaustive, and others will be encountered in the later text. Although it is true that the great majority of oxidations catalysed by cytochrome P450 represent detoxication, in a small yet very important number of cases oxidation leads to activation. Activations are given prominence in the examples shown here because of their toxicological importance. Thus, among the examples given above, the oxidative desulphuration of diazinon and many other organophosphorous insecticides (OPs) causes activation; oxons are much more potent anticholinesterases than are thions! Some aromatic oxidations (e.g. of benzo(*a*)pyrene) yield highly reactive epoxides that are mutagenic. N-hydroxylation of some amines, e.g. *N*-AAF, can also yield mutagenic metabolites. Finally, the epoxidation of aldrin or heptachlor yields highly toxic metabolites, while sulphoxides and sulphones of OPs are sometimes more toxic than their parent compounds. Oxidation tends to increase polarity. Where this simply aids excretion, the result is detoxication. On the other hand, metabolic products are sometimes much more reactive than the parent compounds, and this can lead to interaction with cellular macromolecules, such as enzymes or DNA, with consequent toxicity.

Cytochrome P450 exists in a bewildering variety of forms, which have been assigned to 74 different gene families (Nelson *et al.*, 1998). The number of known isoforms described in the literature already exceeds 750 and continues to grow. In a recent review (Nelson, 1998) 37 families are described for metazoa alone. Although many of these appear to be primarily concerned with the metabolism of endogenous compounds, four families are strongly implicated in the metabolism of xenobiotics in animals. These are gene families *CYP1*, *CYP2*, *CYP3* and *CYP4* (see Table 2.2), which will shortly be described. A wider view of the different P450 forms and families was given earlier in Chapter 1 when considering evolutionary aspects of detoxifying enzymes. Differences in the form and function of P450s between the phyla will be discussed later in relation to the question of selectivity (section 2.5). Let us now consider P450 families that have an important role in xenobiotic metabolism. CYP1A1 and CYP1A2 are P450 forms that metabolise, and are inhibited by, planar molecules [e.g. planar polycyclic aromatic hydrocarbons (PAHs) and coplanar polychlorinated biphenyls (PCBs)]. This can be explained in terms of the deduced structure of the active site of CYP1A enzymes (Figure 2.6) (Lewis, 1996; Lewis and Lake, 1996). This takes the form of a rectangular slot, composed of several aromatic side chains, including the coplanar rings of phenylalanine 181 and tyrosine 437; these restrict the size of the cavity such that only planar structures of a certain rectangular dimension will be able to take up the binding position. Small differences in structure between the active sites of CYP1A1 and CYP1A2 may explain their differences in substrate preference, e.g. phenylalanine 259 (CYP1A1) vs. anserine 259 (1A2). CYP1A1 metabolises heterocyclic molecules, whereas CYP1A1 is more concerned with PAHs. By contrast, the active sites of families CYP2 and 3 have more open structures and are capable of

TABLE 2.2 *Some inhibitors of cytochrome P450*

Compound	Inhibitory action
Carbon monoxide	Inhibits all forms of P450 Competes with oxygen for haem binding site
Methylene dioxyphenyls	Carbene forms generated, which bind to haem Selective inhibitors
Imidazoles, triazoles and pyridines	Contain ring N, which binds to haem Selective inhibitors
Phosphorothionates	Oxidative desulphuration releases active sulphur, which binds to, and deactivates, P450 Selective inhibitors
1-Ethynyl pyrene	Specific inhibitor of 1A1
Furafylline	Specific inhibitor of 1A2
Diethyldithiocarbamate	Specific inhibitor of 2A6
Sulphenazole	Specific inhibitor of 2C9
Quinine	Specific inhibitor of 2D1
Disulfiram	Specific inhibitor of 2E1

binding a wide variety of different compounds, some planar but many of more globular shape. CYP2 is a particularly diverse family, whose rapid evolution coincides with the movement of animals from water to land (for discussion see Chapter 1). Very many lipophilic xenobiotics are metabolised by enzymes belonging to this family. Of particular interest from an ecotoxicological point of view, CYP2B is involved in the metabolism of organochlorine insecticides such as aldrin and endrin and some OPs, including parathion; CYP2C is involved with warfarin metabolism, and CYP2E with solvents of low molecular weight, including acetone and ethanol. CYP3 is noteworthy for the great diversity of substrates that it can metabolise, both endogenous and exogenous. Structural models indicate a highly unrestricted active site, in keeping with this characteristic (Lewis, 1996). This is in marked contrast to the highly restricted active sites proposed for the CYP1A family. Although CYP4 is especially involved in the endogenous metabolism of fatty acids, it does have a key role in the metabolism of a few xenobiotics, including phthalate esters.

The classification of P450s, which is based on amino acid sequencing, bears some relationship to metabolic function. That said, some xenobiotic molecules, especially when they are large and complex, are metabolised by several different P450 forms. Different forms of P450 tend to show regioselectivity, for example in the metabolism of PAHs such as benzo(*a*)pyrene and of steroids such as testosterone.

Oxidations catalysed by cytochrome P450 can be inhibited by many compounds.

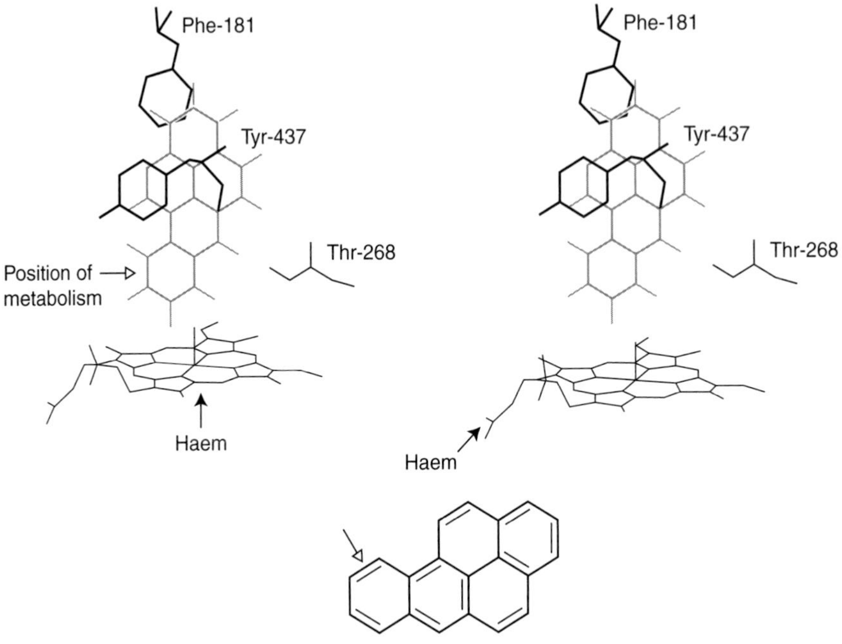

FIGURE 2.6 *The procarcinogen benzo(a)pyrene oriented in the CYP1A1 active site (stereo view) via π– π stacking between aromatic rings on the substrate and those of the complementary amino acid side chains, such that 7,8-epoxidation can occur. The substrate is shown with pale lines in the upper structures. The position of metabolism is indicated by an arrow in the lower structure. After Lewis (1996).*

Some of the more important examples are given in Table 2.2. Carbon monoxide inhibits all known forms of P450 by competing with oxygen for its binding position on haem. Indeed, this interaction was the original basis for the term 'cytochrome P450'. Interaction of CO with P450 in the Fe^{2+} state yields a complex that has an absorption maximum of ~ 450 nm. Many organic molecules act as inhibitors, but they are, in general, selective for particular forms of the haemprotein. Selectivity depends on structural features of the molecules: how well they fit into the active sites of particular forms, and the position in the molecule of functional groups that can interact with haem or with the substrate binding sites. To describe, briefly, some of the more important types of inhibitor – methylene dioxyphenyl compounds such as piperonyl butoxide act as suicide substrates. The removal of two protons leads to the formation of carbenes, which bind strongly to haem, thereby preventing the binding of oxygen (Figure 2.7). Compounds of this type have been used to synergise the effects of insecticides, such as pyrethroids and carbamates, which are subject to oxidative detoxication. A considerable number of compounds containing heterocyclic nitrogen are potent inhibitors (Figure 2.7). Included here are certain compounds containing the heterocyclic groupings imidazole, triazole and pyridine. Some compounds of this type have been successfully developed as antifungal agents as a result of their strong inhibition of CYP51, which has a critical role in ergosterol biosynthesis. Their inhibitory

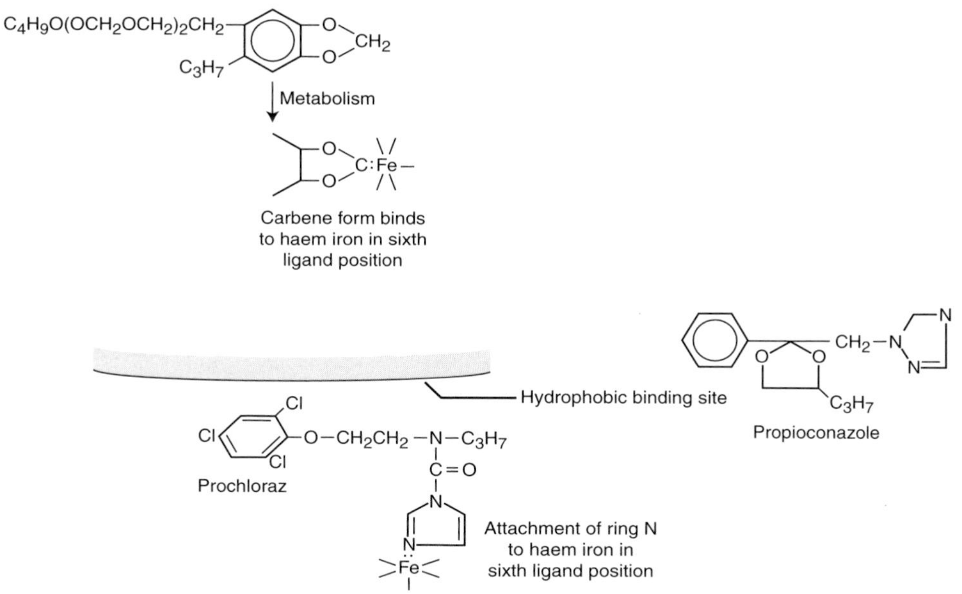

FIGURE 2.7 *Cytochrome P450 inhibitors.*

potency depends on the ability of the ring N to ligate to the iron of haem, thus preventing the activation of oxygen. One type of inhibition that is important in ecotoxicology is the deactivation of haem caused by the oxidative desulphuration of phosphorothionates (see 'Monooxygenases'). Sulphur atoms detached from phosphorothionates are bound in some form to cytochrome P450, destroying its catalytic activity. The exact mechanism for this is, at present, unknown. Apart from these broad classes of inhibitors, certain individual compounds are very selective for particular P450 forms, and thus are valuable for the purposes of identification and characterisation. Some examples are given in Table 2.2.

There are marked differences in hepatic microsomal monooxygenase (HMO) activities between different species and groups of vertebrates. Figure 2.8 summarises results from many studies reported in the general literature (Walker, 1980; Ronis and Walker, 1989). Mean activities for each species across a range of lipophilic xenobiotics are expressed relative to those of the male rat, making a correction for relative liver weight. Males and females of each species are each represented by a single point wherever possible. For some species there is just a single point because no distinction had been made between the sexes. The log relative activity is plotted against the log body weight.

The mammals, which are nearly all omnivorous, show a negative correlation between log relative HMO activity and log body weight. Thus, small mammals have much higher HMO per unit body weight than do large mammals. This is explicable in terms of the detoxifying function of P450 (much of the metabolism of these substrates is carried out by isoforms of CYP2). Small mammals have much larger surface area to

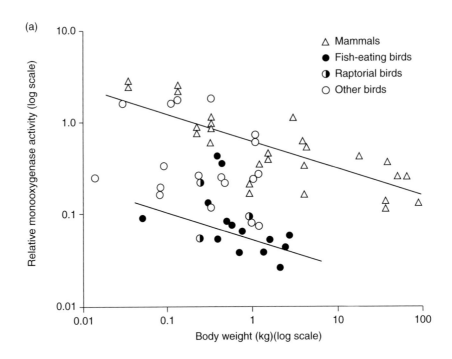

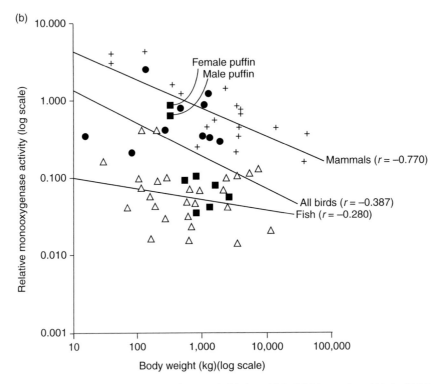

FIGURE 2.8 *Monooxygenase activities of mammals, birds and fish. (a) Mammals and birds. (b) Mammals, birds and fish. Activities are of hepatic microsomal monooxygenases to a range of substrates expressed in relation to body weight. Each point represents one species (males and females are sometimes entered separately). From Walker* et al. *(2000).*

body volume ratios than large mammals; thus they take in food and associated xenobiotics more rapidly in order to acquire sufficient metabolic energy to maintain their body temperatures. The birds studied differed widely in their type of food, ranging from omnivores and herbivores to specialised predators. Omnivorous and herbivorous birds had rather lower HMO activities than mammals of similar body size, with galliform birds showing similar activities to mammals. Fish-eating birds and raptors, however, showed lower HMO activities than other birds, and much lower activities than omnivorous mammals. This is explicable on the grounds that they have had little requirement for detoxication by P450 (e.g. isoforms of CYP2) during the course of evolution, in contrast to herbivores and omnivores, which have had to detoxify plant toxins. Fish-eating birds, like omnivorous mammals, show a negative correlation between log HMO activity and log body weight. The slopes are very similar in the two cases. The bird-eating sparrowhawk shows a very low value for HMO activity, similar to that of fish of similar body weight. Such a low detoxifying capability may well have contributed to the marked bioaccumulation of *p,p'*-DDE, dieldrin and heptachlor epoxide by this species (see Chapter 5).

Fish show generally low HMO activities that are not strongly related to body weight. The low activities may reflect a limited requirement of fish for metabolic detoxication; they are able to efficiently excrete many compounds by diffusion across the gills. The weak relationship of HMO activity with body weight is probably because they are poikilotherms and should not, therefore, have an energy requirement for the maintenance of body temperature that is srongly related to body size. In other words, the rate of intake of xenobiotics with food is unlikely to be strongly related to body size.

Esterases and hydrolases

Many xenobiotics, both man-made and naturally occurring, are lipophilic esters. They can be degraded to water-soluble acids and bases by hydrolytic attack. Two important examples of esteratic hydrolysis in ecotoxicology now follow:

$$R - \overset{\overset{\displaystyle O}{\|}}{C} - OX + H_2O \dashrightarrow R - \overset{\overset{\displaystyle O}{\|}}{C} - OH + XOH$$

Carboxyl ester   Carboxylic   Alcohol
                            acid

$$\overset{RO}{\underset{RO}{\diagdown}}\!\!\!\overset{\overset{\displaystyle O}{\|}}{P} - OX + H_2O \dashrightarrow \overset{RO}{\underset{RO}{\diagdown}}\!\!\!\overset{\overset{\displaystyle O}{\|}}{P} - OH + XOH$$

Organophosphate

Enzymes catalysing the hydrolysis of esters are termed esterases. Esterases belong to a larger group of enzymes termed hydrolases, which can cleave a variety of chemical bonds by hydrolytic attack. In the classification of hydrolases by the International Union of Biochemistry (IUB), the following categories are recognised:

3.1   acting on ester bonds (esterases)
3.2   acting on glyoacyl compounds
3.3   acting on ether bonds
3.4   acting on peptide bonds (peptidases)
3.5   acting on C–N bonds other than peptide bonds
3.6   acting on acid anhydrides (acid anhydrolases)
3.7   acting on C–C bonds
3.8   acting on halide bonds
3.9   acting on P–N bonds
3.10 acting on S–N bonds
3.11 acting on C–P bonds

Although it is convenient to define hydrolases according to their enzymic function, there is one serious underlying problem. Some hydrolases are capable of performing two or more of the above kinds of hydrolytic attack, so they do not fall simply into just one category. There are esterases, for example, that can also hydrolyse peptides, amides and halide bonds. The shortcomings of the early IUB classification, which was originally based on the measurement of activities in crude tissue preparation, have become apparent with the purification and characterisation of hydrolases. As yet, however, only limited progress has been made, and a comprehensive classification is still some distance away. In what follows, a simple and pragmatic classification will be described for esterases that hydrolyse xenobiotic esters (Figure 2.9). It should be emphasised that this is a classification seen from a toxicological point of view. Esterases are important both for their detoxifying function and as sites of action for toxic molecules. Thus, in Figure 2.9, esterases that degrade organophosphates serve a detoxifying function, whereas those inhibited by organophosphates often represent sites of action. The paradox of the latter is that esteratic hydrolysis leads to toxicity! Organophosphates behave as suicide substrates; during the course of hydrolysis the enzymes become irreversibly inhibited – or nearly so. The inhibitory action of organophosphates upon esterases will be discussed in section 2.4.

Looking at the classification shown in Figure 2.9, esterases that effectively detoxify organophosphorous compounds by continuing hydrolysis are termed 'A' esterases, following the early definition of Aldridge (1953). They fall into two broad categories – those that hydrolyse POC bonds (the oxon forms of many OPs are represented here), and those that hydrolyse P–F or P–CN bonds (a number of chemical warfare agents are represented here). Within the first category of A esterase, two main types have been recognised. First, arylesterase (EC 3.1.1.2) can hydrolyse phenylacetate as well as organophosphate esters. It occurs in a number of mammalian tissues, including liver and blood, and has been purified and characterised. It is found associated with the high-density lipoprotein of blood and in the endoplasmic reticulum of the liver. Other esterases that hydrolyse organophosphates but not phenylacetate have been partially purified and are termed aryldialkylphosphatases (EC 3.1.8.1) in recent versions of the IUB classification. These are also found in high-density lipoprotein of mammalian

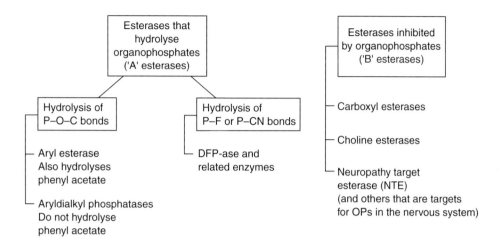

FIGURE 2.9 *Esterases that are important in ecotoxicology.*

blood and in the hepatic endoplasmic reticulum of vertebrates. Within the second category of A esterases are the diisopropylfluorophosphatases (EC 3.1.8.2), which catalyse the hydrolysis of chemical warfare agents ('nerve gases') such as DFP (diisopropylfluorophosphate), soman and tabun.

There are marked species differences in A esterase activity. Birds have very low, often undetectable, levels of activity in plasma towards paraoxon, diazoxon, pirimiphos-methyl oxon and chlorpyriphos oxon (Brealey *et al.*, 1980; Mackness *et al.*, 1987; Walker *et al.*, 1991) (Figure 2.10). Mammals have much higher plasma A esterase activities for all of these substrates. The toxicological implications of this are discussed in Chapter 10. Some species of insects have no measurable A esterase activity, even in strains that have resistance to organophosphorous pesticides (Mackness *et al.*, 1982; Walker, 1994a,b). These include the peach potato aphid (*Myzus persicae*) (Devonshire, 1991) and the rust-red flour beetle (*Tribolium castaneum*). Indeed, it has been questioned whether insects have A esterase at all; some studies have failed to make the distinction between this enzyme and high levels of 'B' esterase (Walker, 1994b).

Dealing now with the B esterases, the carboxylesterases (EC 3.1.1.1) represent a large group of enzymes that can hydrolyse both exogenous and endogenous esters. More than 12 different forms have been identified in rodents, and four of these have been purified from rat liver microsomes (Table 2.3, Mentlein *et al.*, 1987). The four forms shown have been characterised on the basis of their substrate specificities and their genetic classification. They have molecular weights of approximately 60 kDa when in the monomeric state. They are separable by isoelectric focusing, and the pI value for each is shown in the first column. The number assigned to each in the genetic classification is in the second column. As can be seen, they all show distinct ranges of substrate specificity with a certain degree of overlap. All four can hydrolyse both exogenous and endogenous esters. ES4 and ES15 have activities previously associated with earlier entries in the IUB classification, entries that were made on the

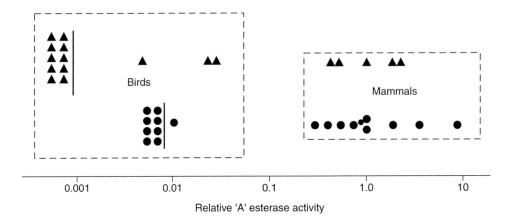

FIGURE 2.10 *Plasma A esterase activities of birds and mammals. Activities were originally measured as nanomoles product per millilitre of serum per minute, but have been converted to relative activities (male rat = 1) and plotted on a log scale. Each point represents a mean value for a single species. Substrates: ●, paraoxon; ▲, pirimiphos methyl oxon. Vertical lines indicate limits of detection, and all points plotted to the left of them are for species in which no activity was detected. (Activities in the male rat were 61 ± 4 and 2020 ± 130 for paraoxon and pirimiphos methyl oxon respectively.) From Walker (1994a) in Hodgson and Levi (1994).*

TABLE 2.3 *Types of carboxylesterase isolated from rat liver microsomes*

pI value	Genetic classification	Substrate	Comments
5.6	ES3	Simple aromatic esters, acetanilide, lysophospholipids, monoglycerides, long-chain acyl carnitines	Sometimes called lisophospholipase to distinguish it from other esterases featured here
6.2/6.4	ES4	Aspirin, malathion, pyrethroids, palmitoyl CoA, monoacylglycerol, cholesterol esters	May correspond to EC 3.1.2.2 and EC 3.1.1.23
6.0	ES8/ES10	Short-chain aliphatic esters, medium-chain acylglycerols, clofibrate, procaine	ES8 may be a monomer, ES10 a dimer
5.0/5.2	ES15	Mono- and diacylglycerols, acetyl carnitine, phorbol diesters	Correspond to acetyl carnitine hydrolase EC 3.1.1.28

basis of limited evidence. It may well be that some of these earlier entries can now be removed from the classification, the activities being due solely to members of EC 3.1.1.1. It is noteworthy that ES4 catalyses the hydrolysis of pyrethroid insecticides and malathion (Walker, 1994b). In mice, the carboxylesterases are tissue specific with a range of 10 different forms identified in the liver and kidney, but only a few in other tissues. Only three forms have been found in mouse serum. As with other enzymes that metabolise xenobiotics, the liver is a particularly rich source.

Cholinesterases are another group of B esterases. The two main types are acetylcholinesterase (AChE) (EC 3.1.1.7) and 'unspecific' or butyrylcholinesterase (EC 3.1.1.8). AChE is found in the postsynaptic membrane of cholinergic synapses of both the central and the peripheral nervous systems. It is the site of action of organophosphorous and carbamate insecticides, and will be described in more detail in section 2.4. Butyrylcholinesterase (BuChE) occurs in many vertebrate tissues, including blood and smooth muscle. Unlike AChE, it does not appear to represent a site of action for organophosphorous or carbamate insecticides. However, the inhibition of BuChE in blood has been used as a biomarker assay for exposure to OPs (see Thompson and Walker, 1994). Neuropathy target esterase (NTE) is another B esterase located in the nervous system. Inhibition of NTE can cause delayed neuropathy (see section 2.4). Finally, other hydrolases of the nervous system that are sensitive to inhibition by OPs have recently been identified (section 2.4).

The distinction between A and B esterases is based on the difference in their interaction with organophosphates. Cholinesterases have been more closely studied than other B esterases and are taken as models for the whole group. Cholinesterases contain serine at the active centre, and organophosphates phosphorylate this as the first stage in hydrolysis (Figure 2.11). This is a rapid reaction, which involves the splitting of the ester bond and the acylation of the serine hydroxyl. The leaving group, XO–, combines with a proton from the serine hydroxyl group to form an alcohol, XOH. The next stage in the process, the release of the phosphoryl moiety, the restoration of the serine hydroxyl and the reactivation of the enzyme, is usually very slow. Just how slow depends on the structure of the 'R' groups. The organophosphate has acted as a suicide substrate, inhibiting the enzyme during the course of hydrolytic attack. A further complication may be the 'ageing' of the bound phosphoryl moiety. The R group is lost, leaving behind a charged PO⁻ group. If this happens, the inhibition becomes irreversible, and the enzyme will not spontaneously reactivate. This process of ageing is believed to be critical in the development of delayed neuropathy, after

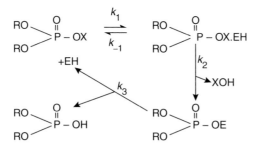

FIGURE 2.11 *Interaction between organophosphates and B esterases. R, alkyl group; E, enzyme.*

neuropathy target esterase (NTE) has been phosphorylated by an organophosphate (see section 2.4). It is believed that most, if not all, of the B esterases are sensitive to inhibition by organophosphates because they too have reactive serine at their active sites. It is important to emphasise that the interaction shown in Figure 2.11 occurs with organophosphates, i.e. OPs that contain an oxon group. Phosphorothionates, which contain a thion group instead, do not readily interact in this way. Many OPs are phosphorothionates, but these need to be converted to phosphate (oxon) forms by oxidative desulphuration before inhibition of AChE can proceed to any significant extent (see 'Monooxygenases').

The reason for the contrasting behaviour of A esterases is not yet clearly established. It has been suggested that the critical difference from B esterases is the presence of cysteine rather than serine at the active site. It is known that arylesterase, which hydrolyses organophosphates such as parathion, does contain cysteine, and that A esterase activity can be inhibited by agents that attack sulphhydryl groups (e.g. certain mercurial compounds). It may be that the acylation of cysteine rather than serine in the model shown in Figure 2.11 would be followed by rapid reactivation of the enzyme. In other words, $(RO)_2P(O)SE$ would be less stable than $(RO)_2P(O)OE$, readily breaking down to release the reactivated enzyme.

In addition to the hydrolases identified above, there are others that have been less well studied and are accordingly difficult to classify. Examples will be encountered later (see Chapters 5–12), when considering the ecotoxicology of various organic pollutants. In considering esterases, it is important to emphasise that we are only concerned with enzymes that split bonds by a hydrolytic mechanism. In early work on the biotransformation of xenobiotics, there was sometimes confusion between true hydrolases and other enzymes that can split ester bonds and yield the same products by different mechanisms. Thus, both monooxygenases and glutathione-S-transferases can break POC bonds of OPs and yield the same metabolites as esterases. The removal of alkyl groups from OPs can be accomplished by O-dealkylation, or by their transfer to the S group of glutathione. For further details, see 'Conjugases'. In early studies biotransformations were observed *in vivo* or in crude *in vitro* preparations such as homogenates, i.e. under circumstances in which it was not possible to establish the mechanism(s) by which biotransformations were being catalysed. What appeared to be hydrolysis was sometimes oxidation or group transfer. This complication needs to be borne in mind when looking at some papers in the older literature.

Epoxide hydrolase (EC 4.2.1.63)

Epoxide hydrolases hydrate epoxides to yield *trans*-dihydrodiols without any requirement for cofactors. Examples are given in Figure 2.12. Epoxide hydrolases are hydrophobic proteins of molecular mass $\sim$ 50 kDa; they are found, principally, in the endoplasmic reticulum of a variety of cell types. Vertebrate liver is a particularly rich source; appreciable levels are also found in the kidney, testis and ovary. A soluble epoxide hydrolase is found in some insects, in which it has the role of hydrating

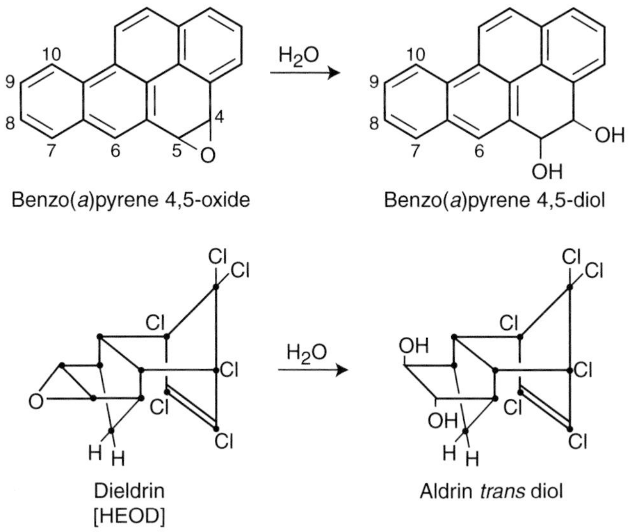

Benzo(*a*)pyrene 4,5-oxide   Benzo(*a*)pyrene 4,5-diol

Dieldrin
[HEOD]   Aldrin *trans* diol

FIGURE 2.12 *Epoxide hydration.*

epoxides of juvenile hormones. The microsomal epoxide hydrolases of vertebrate liver can degrade a wide range of epoxides, including those of PAHs, PCBs, cyclodienes (including dieldrin and analogues thereof) as well as certain endogenous steroids. Epoxide hydrolase can detoxify potentially mutagenic epoxides formed by the action of cytochrome P450 on, for example, PAHs. Benzo(*a*)pyrene 4,5-oxide is an example. Its rapid hydration within the endoplasmic reticulum, before it can migrate elsewhere, is important for the protection of the cell. In general, the conversion of epoxides into more polar *trans*-dihydrodiols serves a detoxifying function, although there are a few exceptions to this rule.

Reductases

A range of reductions of xenobiotics are known to occur, both in the endoplasmic reticulum and in the cytosol of a number of cell types. However, the enzymes (or other reductive agencies) responsible are seldom known in particular cases. Some reductions only occur at very low oxygen levels. Thus, they do not occur under normal cellular conditions, when there is a plentiful supply of oxygen.

Two important examples of reductive metabolism of xenobiotics are the reductive dehalogenation of organohalogen compounds and the reduction of nitroaromatic compounds. Examples of each are shown in Figure 2.13. Both types of reaction can take place in hepatic microsomal preparations at low oxygen tensions. Cytochrome P450 can catalyse both types of reduction. If a substrate is bound to P450 in the absence of oxygen, electrons can be passed from the iron atom of haem to the substrate. In the case of organohalogen compounds such as *p,p'*-DDT, carbon tetrachloride and

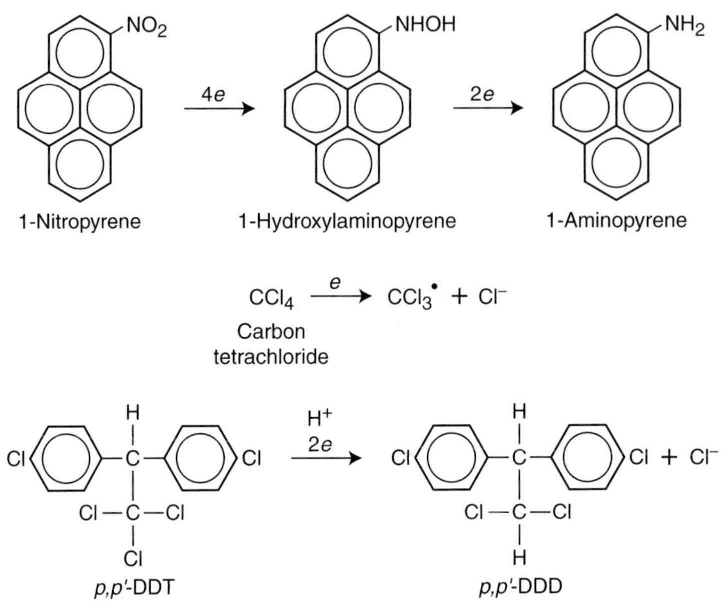

FIGURE 2.13 *Reductase metabolism.*

halothane, this leads to the loss of Cl⁻ and its replacement by hydrogen. If oxygen had been present, the electron would have passes to this (see Figure 2.4) and not directly to the substrate. With nitroaromatic compounds, reductions occur via an intermediate hydroxylamine stage, to yield an amine (Figure 2.13). Often, the second stage is rapid, and the intermediate form is not detectable.

Although the role of P450 in this type of reductive metabolism has been well established in *in vitro* studies, there are uncertainties about the course of events *in vivo*. In the first place, as mentioned above, oxygen levels may be high enough to prevent this type of reaction occurring (see discussion about the metabolism of p,p'-DDT in Chapter 5). Also, porphyrins other than P450 can catalyse reduction. So too can flavoprotein reductases such as NADPH-cytochrome P450 reductase. Even flavin adenine dinucleotide (FAD) can catalyse some reductions. Microorganisms in anaerobic soils and sediments can be very effective in degrading organohalogen compounds such as organochlorine insecticides, PCBs and dioxins. The metabolic degradation of polyhalogenated compounds is often difficult and slow under aerobic conditions. Effective aerobic detoxication enzymes have yet to evolve for many compounds of this type. On the other hand, reductive dehalogenation is often an effective mechanism for biodegradation, and it has been exploited in the development of genetically manipulated microorganisms for bioremediation.

The transfer of electrons in xenobiotic reactions is tied up with the problem of the generation of active radicals, including those of oxygen. CCl₄, for example, is reduced to the highly reactive CCl₃· radical. Some organonitrocompounds, such as the herbicide

Box 2.1 *Studying* in vitro *metabolism.*

Metabolism *in vivo* is studied by dosing animals with xenobiotics – often in radiolabelled form – and then extracting metabolites from urine, faeces, bile, blood and sometimes other tissues. In this way, a holistic picture of metabolism can be obtained, but there is little opportunity for the characterisation of individual enzymes and their kinetic properties. Characterisation of enzymic processes can be accomplished by *in vitro* studies in relatively simple systems under the close control of the experimenter. Examples of *in vitro* systems, in order of decreasing complexity, are tissue slices, tissue homogenates, subcellular fractions and purified enzymes. Following this sequence, the experimenter moves progressively away from the most complex systems, which are closest to *in vivo* conditions, to the simplest and furthest removed from the complex operation of the whole organism; this also represents progression from the least to the greatest experimental control. Thus, with the purified enzyme, there can be regulation not only of temperature, pH, ionic composition of medium, and levels of cofactors and inhibitors, but also of cooperating enzymes and the environment as well (e.g. enzymes that have been extracted from membranes can be returned to reconstituted membranes with regulated phospholipid composition).

In vertebrates the liver is a very rich source of enzymes that metabolise lipophilic xenobiotics, and subcellular fractions are prepared to study metabolism. Sometimes other tissues, e.g. brain, kidney, testis and ovary, are also treated in this way. A typical subcellular fractionation of liver might be as follows:

1 homogenisation of liver in buffer solution to give crude homogenate;
2 low-speed centrifugation of crude homogenate at 11 000 *g* to give 11 000 *g* supernatant + 11 000 *g* precipitate;
3 high-speed centrifugation of 11 000 *g* supernatant at 105 000 *g* to give 105 000 *g* supernatant + microsomal fraction.

The microsomal fraction consists mainly of vesicles (microsomes) derived from the endoplasmic reticulum (smooth and rough). It contains cytochrome P450 and NADPH-cytochrome P450 reductase (collectively the microsomal monooxygenase system), carboxylesterases, A esterases, epoxide hydrolases, glucuronyl transferases and other enzymes that metabolise xenobiotics. The 105 000 *g* supernatant contains 'soluble' enzymes such as glutathione-S-transferases, sulphotransferases and certain esterases. The 11 000 *g* supernatant contains all of the types of enzyme listed above.

Microsomes are widely used to study the metabolism of xenobiotics. Enzymes can be characterised on the basis of their requirement for cofactors (e.g. NADPH, UDPGA) and their response to inhibitors. Kinetic studies can be carried out, and kinetic constants determined. They are very useful in studies of comparative metabolism, where many species not available for *in vivo* experiment can be compared with widely investigated laboratory species such as rats, mice, feral pigeon, Japanese quail and rainbow trout.

paraquat, can undergo redox cycling. One-electron reduction of the paraquat yields an unstable radical. This radical passes an electron on to molecular oxygen, thereby generating the reactive superoxide ion and regenerating paraquat (see Chapter 14).

Conjugases

Conjugases catalyse phase 2 biotransformations: the coupling of xenobiotic metabolites (and sometimes original xenobiotics) with polar endogenous molecules, which are usually in the form of anions. Whereas phase 1 biotransformations of lipophilic compounds occur predominantly in the endoplasmic reticulum ('microsomal membrane'), phase 2 biotransformations often occur in the cytosol. Many different endogenous molecules are used for conjugation, and there can be large differences between groups and between species in the preferred metabolic pathway. The critical thing is that polar conjugates are produced that can be rapidly excreted. The following account will be mainly concerned with three groups of enzymes that are responsible for most of the conjugations in vertebrates – the glucuronyl transferases, the sulphotransferases and the glutathione-S-transferases. It should be emphasised that less is known about the conjugases of invertebrates and plants. Although conjugations are seen to be detoxifying, and in general protective towards the organism, in some instances they may be broken down to release potentially toxic compounds. For example, some glutathione conjugates can break down in the kidney, with toxic effects.

Uridine diphosphate (UDP)-glucuronyl transferases (henceforward simply glucuronyl transferases) exist in a number of different forms with contrasting, yet overlapping, substrate specificities. Both exogenous and endogenous substrates are metabolised. They are associated with the endoplasmic reticulum of many vertebrate tissues, notably liver, and have molecular weights of 50–58 kDa. They catalyse the transfer of glucuronate, from UDP-glucuronate (UDPGA), to a variety of substrates that possess functional groups with labile protons, principally –OH, but also –SH and –NH (see Figure 2.14). As noted earlier, these functional groups are often introduced during phase 1 metabolism. Sometimes they are present in original xenobiotics, in which case conjugation can proceed without any preliminary phase 1 metabolism. The products are β-D-glucuronides. For the reaction to proceed, glucuronate must de delivered by UDPGA. UDPGA is generated in the cytosol, not in the endoplasmic reticulum. It is synthesised in a two-step process:

1 Glucose 6-phosphate interacts with UTP (uridine triphosphate) to form UDPD-glucose and pyrophosphate.
2 UDPD glucose is oxidised to UDPGA by the action of UDPG dehydrogenase, with NAD as cofactor.

UDPGA then migrates to the membrane-bound glucuronyltransferase, where conjugation of xenobiotics can proceed (Figure 2.14). Glucuronyl transferases can be activated by *N*-acetylhexosamine.

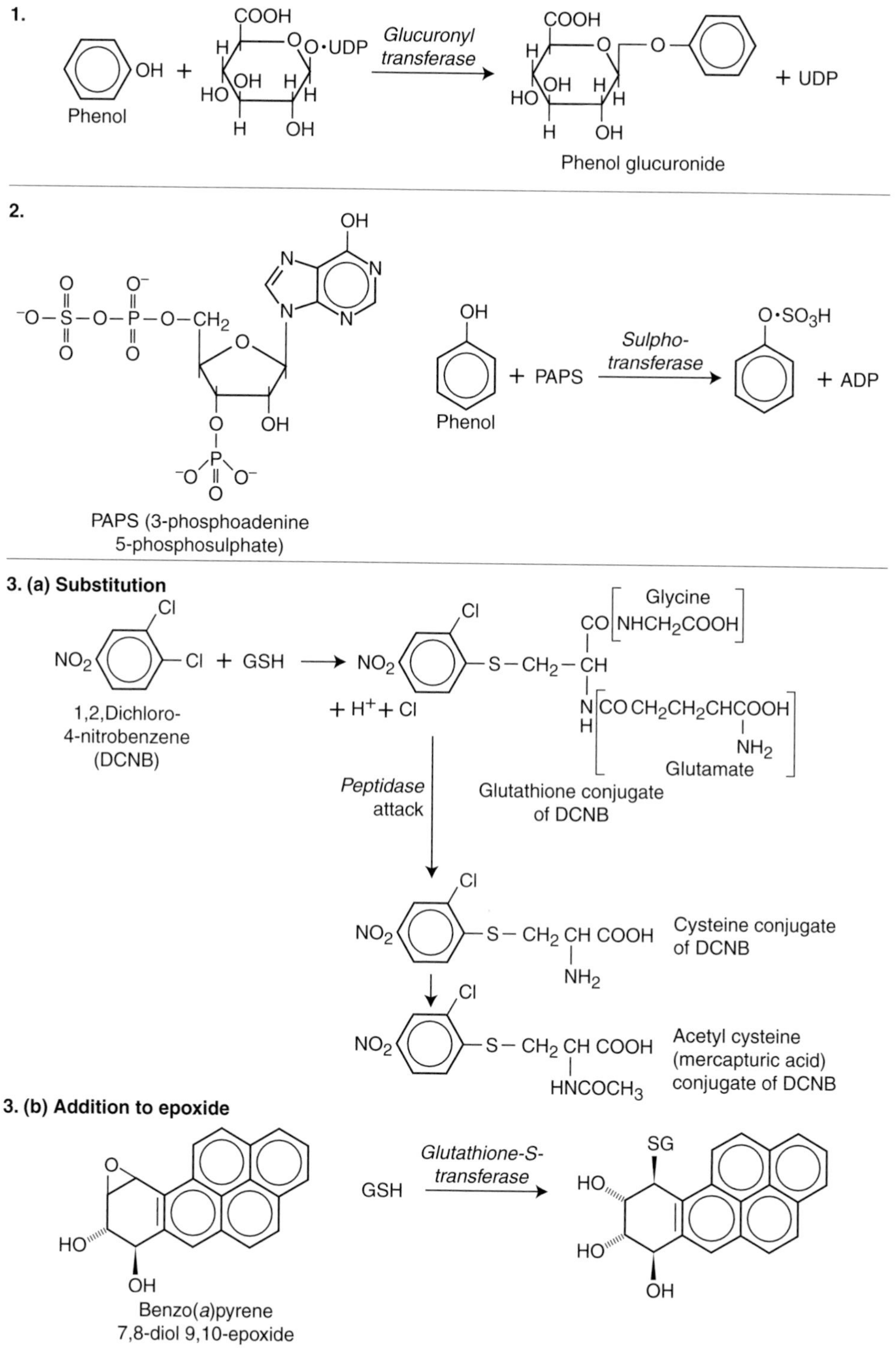

1.

Phenol + (UDP-glucuronic acid) → (Glucuronyl transferase) → Phenol glucuronide + UDP

2.

PAPS (3-phosphoadenine 5-phosphosulphate)

Phenol + PAPS → (Sulpho-transferase) → + ADP

3. (a) Substitution

1,2,Dichloro-4-nitrobenzene (DCNB) + GSH → Glutathione conjugate of DCNB + H⁺ + Cl

Glycine

Glutamate

Peptidase attack

Cysteine conjugate of DCNB

Acetyl cysteine (mercapturic acid) conjugate of DCNB

3. (b) Addition to epoxide

Benzo(a)pyrene 7,8-diol 9,10-epoxide + GSH → (Glutathione-S-transferase) →

FIGURE 2.14 *Phase II biotransformation–conjugation. 1. Glucuronide formation. 2. Sulphate formation. 3. Glutathione conjugation.*

Although glucuronyltransferases are well represented in mammals, birds, reptiles and amphibians, they are at low levels in fish. In insects, by contrast, conjugation tends to be with glucose rather than glucuronate.

Sulphotransferases represent an important group of conjugases in cytosol. They are present in many tissues, the liver once again being a rich source. As with the glucuronyltransferases, the conjugating group is transferred from a complex molecule to the substrate (Figure 2.14). Here, the sulphate is donated by 3-phosphoadenine 5-phosphosulphate (PAPS). PAPS is generated in the cytosol by a two-stage process.

1 ATP (adenosine triphosphate) interacts with a sulphate ion (SO_4^-), to form adenosine 5-phosphosulphate (APS) in a reaction mediated by the enzyme ATP sulphate adenylyl transferase.
2 ATP interacts with APS to form PAPS, catalysed by ATP adenylyl sulphate 3-phosphotransferase.

Sulphotransferases catalyse the transfer of sulphate from PAPS to a wide range of xenobiotics that possess hydroxyl groups. Steroid alcohols are among the endogenous substrates. The sulphotransferases exist in different forms.

Glutathione-S-transferases represent another group of enzymes located primarily in the cytosol, although one form occurs in microsomes. Glutathione conjugations depend on the capacity of glutathione to act as a nucleophile (GS^-) against xenobiotics which are electrophiles. Although many glutathione conjugations can proceed spontaneously, without the participation of the enzyme, the reaction tends to be slow. The glutathione-S-transferases are able to bind reduced glutathione (GSH) in close proximity to substrates held at a hydrophobic binding site and, by this means, increase the rate of conjugation.

Many xenobiotics and xenobiotic metabolites undergo glutathione conjugation. Indeed, it is one of the most important phase 1 transformations in both vertebrates and invertebrates. Two contrasting types of reaction are known (Figure 2.14). Nucleophilic attack can lead to the displacement of halide from organohalogen compounds, to be replaced by a sulphur linkage to glutathione. In the example shown, 1,2-dichloro-4-nitrobenzene, the initial glutathione conjugate, undergoes further biotransformation, involving the removal of the glycine and glutamate moieties by peptidase attack to yield a cysteine conjugate, which is then acetylated to form an acetyl cysteine conjugate (also referred to as a mercapturic acid conjugate). Mercapturic acid conjugates are often the main excreted forms after glutathione conjugation in mammals. However, cysteine conjugates are also excreted and, in insects, unchanged glutathione conjugates have been reported as the principal excreted forms. A contrasting type of glutathione conjugation involves attack upon epoxides. The epoxide ring is opened, a GS link is attached to one carbon and OH is attached to the other. The net effect is addition across the epoxide bond without substitution. This latter type of conjugation can be critical in removing newly generated mutagenic epoxides before they can cause cellular damage, e.g. by binding to DNA.

Glutathione-S-transferases are known to exist in a number of isoforms. These are homo- or heterodimers built from subunits of 22–28 kDa. In the rat, three classes of isoforms are known that are built upon subunits numbered 1–7.

Class	Constitution
α	1:1, 1:2 and 2:2
μ	3:3, 3:4, 4:4 and 6:6
π	7:7

There is less than 30% sequence homology between the different classes. Although glutathione-S-transferases are predominantly cytosolic enzymes, one microsomal form is known to exist as a trimer of molecular weight 51 kDa.

Some glutathione conjugates are unstable and are better regarded as intermediates in the process of biotranformation than as stable conjugates for excretion. Thus, with certain organohalogen compounds, conjugation with glutathione involves dehydrohalogenation, and the conjugate then breaks down to release a dehalogenated metabolite. One example is the conjugation of dichloromethane. The conjugate is hydrolysed and formaldehyde is released. (Figure 2.15). The dehydrochlorinations of

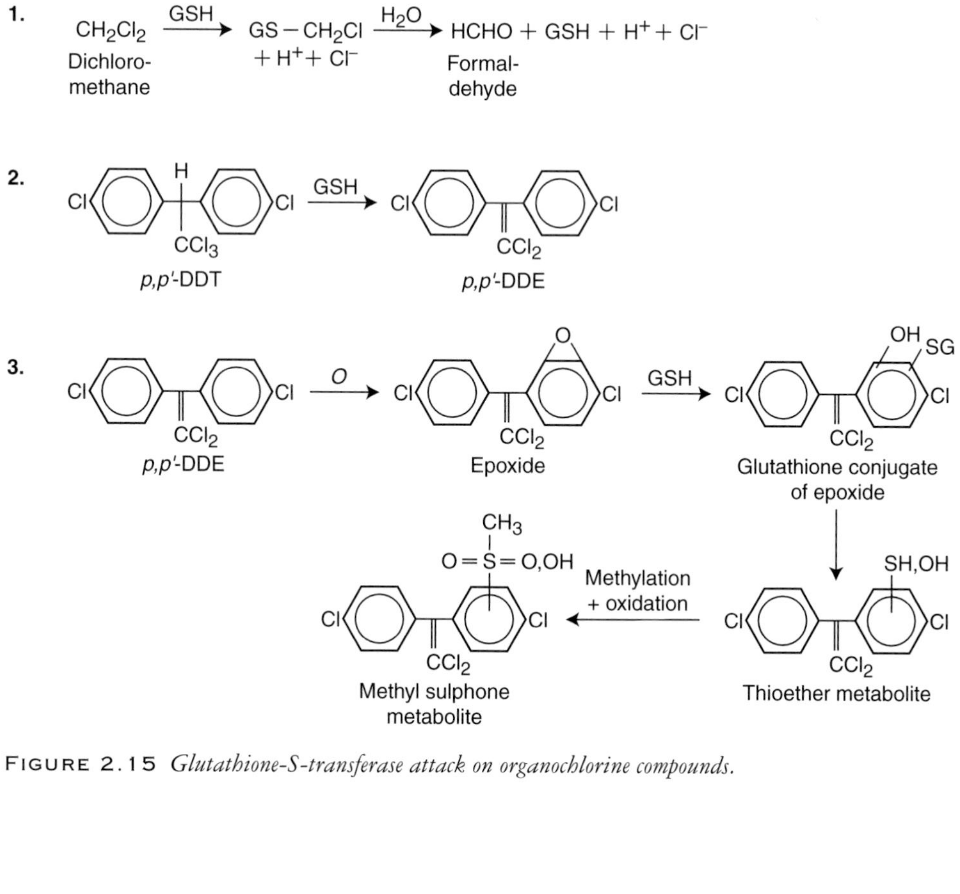

FIGURE 2.15 *Glutathione-S-transferase attack on organochlorine compounds.*

the organochlorine insecticides *p,p'*-DDT and γ-hexachlorocyclohexane (HCH) are mediated by a glutathione-S-transferase and are thought to proceed via a glutathione conjugate as intermediate (Figure 2.15). Finally, *p,p'*-DDE is epoxidised before glutathione conjugation. Some PCB congeners are metabolised in the same way (Bakke *et al.*, 1982). The glutathione conjugate is then hydrolysed, releasing a mercapto derivative of *p,p'*-DDE. The mercapto compound is methylated, and the resulting thiomethyl metabolite is oxidised to a sulphone. (Figure 2.15). This metabolite has been found at relatively high levels in marine mammals, notably in lung tissue.

Apart from their catalytic function, at least one form of glutathione-S-transferase has the function of simply binding xenobiotics and transporting them, without metabolism. In effect, this is an example of storage (see section 2.3.3). The form in question is termed 'ligandin', and binding is associated with one particular subunit. Binding is not associated with catalytic activity.

In addition to the above, a number of other conjugations have been identified. Peptide conjugations are very common throughout the animal kingdom, the preferred conjugating peptide varying greatly between species. It appears to be a question of ready availability. Glycine is very commonly involved, whereas ornithine is preferred in some species of birds. Some insects use mainly arginine. As with the foregoing examples, the conjugates are charged, water soluble and readily excretable. Acetylation is another conjugation reaction, but it differs from the foregoing examples in that the products tend to be less polar than the substrates.

Conjugates of xenobiotics can be hydrolytically degraded by enzymic attack. Very often the products are the original metabolites formed during phase 1 biotransformation. Thus, β-glucuronidases catalyse the hydrolysis of β-glucuronides, and sulphatases catalyse the hydrolysis of sulphate conjugates. If these biotransformations occur in the gut of vertebrates, reabsorption of the released metabolites (or original xenobiotics) may follow. This may promote enterohepatic circulation of metabolites that are excreted in bile (see section 2.3.4) and so delay final elimination from the body. Endogenous molecules, e.g. certain hydroxy metabolites of steroids, also undergo enterohepatic circulation.

Enzyme induction

The levels of enzymes existing in different tissues depend upon gene expression. Some enzymes that metabolise xenobiotics are just present at what are regarded as 'constitutive' levels and are not known to increase in quantity when the organism is exposed to chemicals. On the other hand, certain enzymes that have a role in xenobiotic metabolism increase in concentration, and consequently in activity, with exposure to chemicals. Such induction can be seen as advantageous to the organism. The quantity of enzyme synthesised is related to the requirement for detoxication, and energy is not wasted in the maintenance of unnecessarily high levels of enzyme. High levels of enzyme are only maintained so long as there is exposure to significant levels of the chemical; once the chemical disappears, the enzyme levels can fall again. Some P450s are readily inducible, as might be expected because of their critical role in detoxication, and have been studied in some detail. The following account will be principally concerned with them, making brief reference to some other enzymes that have been less well investigated.

Details of some inducible P450 forms that play a key role in the metabolism of xenobiotics are shown in Table 2.4. P450s belonging to family 1A are induced by various lipophilic planar compounds (Monod, 1997), including PAHs, coplanar PCBs, tetrachlorodibenzodioxin (TCDD) and other dioxins, and β-naphthoflavone. As noted earlier (p. 28), such planar compounds are also substrates for P4501A. In many cases the compounds induce the enzymes that will catalyse their own metabolism. Exceptions are refractory compounds like 2,3,7,8-TCDD, which is a powerful inducer for P4501A but a poor substrate. P450s belonging to family 2 are particularly important in the metabolism of a very wide range of non-planar lipophilic compounds, and a number of them are inducible. The CYP forms 2B1, 2B2 and 2C1–2C4 inclusive are all inducible by phenobarbital. DDT and dieldrin are inducers of CYP2B isozymes. CYP2E1 is inducible by ethanol, acetone, benzene and other small organic molecules. CYP3A isozymes are inducible by endogenous and synthetic steroids, phenobarbital and the antifungal agents clotrimazole and ketoconazole. Finally, CYP4A forms are inducible by clofibrate, di-2-ethylhexylphthalate and mono-2-ethylhexylphthalate. Induction of CYP4A isoforms is associated with peroxisome proliferation.

The induction of P450s belonging to family CYP2 by phenobarbital and other inducing agents is accompanied by proliferation of the endoplasmic reticulum and hypertrophy of hepatocytes. The total P450 content of microsomal membranes can increase by as much as twofold, and there can be very large increases in the specific content of particular P450 isoforms and associated enzyme activities. By contrast, induction of CYP1A forms is not associated with a proliferation of the endoplasmic reticulum, and induction of CYP4A is usually accompanied by peroxisome proliferation.

TABLE 2.4 *Induction of cytochrome P450s*

Family	Individual forms	Induced by	Typical substrates
P450I (P448S)	IA1 IA2	PAHs (e.g. 3MC) TCDD (dioxin) Coplanar PCBs	PAHs Ethoxy-resorufin
P450II	IIB, IIB2 IIC1 → IIC4 } IIE1	Phenobarbital Ethanol	Wide range Ethanol
P450III	IIIA1, IIIA2	Steroids	Testosterone
P450IV (P452)	IVA1	Clofibrate	Lauric acid

The induction process can operate at different levels. The most important mechanisms for particular isoforms are summarised below:

Gene transcription
CYPs 1A1, 1A2, 2B1, 2B2, 2C7, 2C11, 2C12, 2D9, 2E1, 2H1, 2H2, 3A1, 3A2, 3A6, 4A1

mRNA stabilisation
CYPs 1A1, 2B1, 2B2, 2C12, 2E1, 2H1, 2H2, 3A1, 3A2, 3A6

Enzyme stabilisation
CYPs 2E1, 3A1, 3A2, 3A6

The mechanism of induction for CYP1A isoforms operates through the aryl hydrocarbon (Ah) receptor, which is located in the cytosol. The inducing agent (e.g. TCDD, coplanar PCB) binds to the Ah receptor, and then the complex so formed moves into the nucleus. Transcriptional activation of the *CYP1A* gene follows. Interaction of polyhalogenated compounds with the Ah receptor is associated with a variety of toxic responses (Ah receptor-mediated toxicity) which will be discussed later in sections 2.4.4 and 7.2.4. The regulatory elements on the P450 genes are discussed elsewhere (see Lewis, 1996).

Apart from P450s, other enzymes concerned with xenobiotic metabolism may also be induced. Some examples are given in Table 2.5. Induction of glucuronyl transferases is a common response, and is associated with phenobarbital-type induction of CYP family 2. Glutathione transferase induction is also associated with this. A variety of compounds, including epoxides such as stilbene oxide and dieldrin, can induce epoxide hydrolases. Finally, carboxyl esterases can be induced by aminopyrene, phenobarbital and clofibrate, but only to a limited extent (Hosokawa *et al.*, 1987). Generally speaking, these other inductions are of lesser magnitude than many inductions of the P450 system.

Although the inductions described here are conceived as having, in an evolutionary sense, a protective function, they sometimes have the opposite effect upon certain

TABLE 2.5 *Induction of other enzymes*

Type of enzyme	Induced by	Typical substrates
Carboxyl esterases	Phenobarbital, clofibrate amino pyrine	Carboxyl esters
Epoxide hydrolases	Phenobarbital, *trans*-stilbene oxide, 2-acetyl amino fluorene	Epoxides
Glucuronyl transferases	Various, including phenobarbital	Many compounds with OH (and SH)
Glutathione-S-transferases	Various, including phenobarbital	Wide range of compounds that are electrophiles

man-made xenobiotics. The P450 system, as has been explained, can activate a number of mutagens and insecticides. Some of the pesticides developed by industry (e.g. phosphorothionate insecticides, see Chapter 10) depend on oxidative activation for their efficacy. As yet, effective protective systems against such active metabolites (e.g. more effective enzyme systems for rapidly destroying them) do not appear to have been evolved (see, however, discussion of resistance in Chapter 4).

2.3.3 STORAGE

A xenobiotic is said to be stored when it is not available to sites of metabolism or action, and it is not available for excretion. It is held in an 'inert' position, from a toxicological point of view, where it is not able to express toxic action or be acted upon by enzymes.

Storage may be in a fat depot (adipose tissue), bound to an inert protein or other cellular macromolecule, or simply located in a membrane that does not have any toxicological function (i.e. it does not contain or represent a site of toxic action, neither does it contain enzymes that can degrade the xenobiotic).

Highly lipophilic compounds such as organochlorine insecticides, PCBs and polychlorinated dibenzodioxins (PCDDs) tend to be stored in fat depots, where they can reach concentrations 10–50 times higher than in brain, liver, muscle or other metabolically active tissues. Although storage in fat can protect the organism in the short term, it may prove damaging in the long term. Rapid mobilisation of fat depots as a consequence of starvation or disease can lead to rapid release of the stored xenobiotic, and to delayed toxic effects. In one well-documented case in The Netherlands (see Chapter 5), wild female eider ducks (*Somateria mollissima*) experienced delayed neurotoxicity caused by dieldrin. The ducks had laid down large reserves of depot fat before egg laying, and these reserves were run down during the course of egg laying. Dieldrin concentrations quickly rose to lethal levels in the brain. Male eider ducks did not lay down and mobilise body fat in this way, and did not show delayed neurotoxicity due to dieldrin.

Binding to proteins can also represent storage. In the first place, highly lipophilic compounds, such as organochlorine insecticides, associate with lipoproteins. They are circulated in blood in association with lipoproteins. Indeed, their water solubility is so low that only an extremely small proportion of the total concentration present in body fluids is in true solution. Their association with lipoproteins is due, at least in part, to the hydrophobic effect of water. They are excluded from water by the mutual attraction of water molecules and pushed into the hydrophobic zones of the body, including lipoproteins. Also, van der Waals interactions between the compounds, and components of lipoproteins, can cause binding. More polar compounds (including ionic compounds) also interact with proteins, but in different ways. The formation of ionic bonds or hydrogen bonds leads to binding of more polar xenobiotics to functional groups of certain proteins. Albumin, for example, is abundant in mammalian plasma, and can bind a number of relatively polar xenobiotics. Hydroxy metabolites of PCBs can bind to certain plasma proteins. One particular case, the binding of 3,3′,4,4′-tetrachlorobiphenyl to transthyretin, has been closely studied because it is associated with toxicity rather than storage (see section 6.2 and Brouwer *et al.*, 1990).

Storage of lipophilic pollutants in the eggs of invertebrates, birds, amphibians and reptiles is of importance in ecotoxicology. Organochlorine insecticides are carried by lipoproteins from females into eggs. At first they are stored, mainly within the yolk of birds eggs. When the eggs develop they are mobilised, and can cause delayed toxicity in the developing embryo. Such effects have been observed with dieldrin and DDT.

2.3.4 Excretion

As explained in Chapter 1, there is strong evidence for the rapid evolution of enzyme systems concerned with the metabolism of xenobiotics coinciding in time with the movement of animals from water to land. Thus, radiation of the CYP2 family of P450s corresponds closely with the colonization of land at the start of the Devonian period *c.* 400 million years ago. The CYP2 family is particularly concerned with the metabolism of xenobiotics, and is represented by a considerable number of different forms in terrestrial mammals (see 'Monooxygenases'). On the other hand, this family of P450s, and indeed xenobiotic metabolising enzymes more generally, is less well developed in fish, as will be explained later. Fish can 'excrete' many lipophilic xenobiotics by diffusion across the gills into ambient water. This excretion mechanism is not, however, available to terrestrial animals. They depend on the conversion of lipophilic compounds into water-soluble and readily excretable metabolites and conjugates, which are eliminated in urine and bile. In the following account, aquatic animals and terrestrial animals will be treated separately.

Excretion by aquatic animals

Lipophilic xenobiotics are subject to exchange diffusion across the respiratory surfaces that aquatic animals present to ambient water. Thus, excretion can occur across the gills of fish and the permeable skins of certain amphibia, e.g. frogs. Exchange diffusion tends towards an equilibrium or steady state in which the ratio of the concentration of the xenobiotic inside the organism to the concentration in ambient water represents the bioconcentration factor (BCF; see Chapter 4). Thus, loss by diffusion is limited because it is a two-way process. Uptake from the ambient medium works against loss by outward diffusion. However, the dilution volume of the ambient water in relation to that of the organism is very large indeed. Where a pollutant is at low concentration in water and is absorbed mainly from food, or directly from sediment, the system should be very effective. On the other hand, if the pollutant is absorbed mainly from water, effective elimination by the diffusion process will depend upon the concentration in the water falling. A fall in pollutant concentration in ambient water may be due to its removal by degradation, to uptake by other organisms, to adsorption to sediments, etc., or to the aquatic animal moving into cleaner water.

Loss by diffusion tends to be very effective for compounds of K_{ow} close to 1, but less so for compounds of very high K_{ow} (1×10^5 and above). Thus, many anthropogenic compounds of high K_{ow} (e.g. higher chlorinated PCBs, most organochlorine insecticides) can be strongly bioaccumulated by aquatic animals. The relatively weak development in fish of enzyme systems that can metabolise xenobiotics suggests that diffusion has provided reasonably effective protection against naturally occurring xenobiotics during the course of evolution, with only a limited requirement for back up by a metabolic detoxication system. Some anthropogenic pollutants, however, appear to be asking new questions, and there is some evidence that strains of fish with greater metabolic capability have evolved/will evolve in some polluted areas (see section 5.3.5).

Excretion has also been studied in marine invertebrates, such as the bivalve mollusc *Mytilus edulis* (edible mussel). Although the principal excretion mechanism appears again to be passive diffusion, there is recent evidence for the active excretion of some organic nitrogen compounds by *Mytilus* spp. utilising a 'multixenobiotic resistance' mechanism (Kurelec *et al.*, 2000). Metabolism of xenobiotics is generally slower in marine invertebrates than in fish, with crustaceans (e.g. crabs) showing higher metabolic capability than molluscs.

Excretion by terrestrial animals

Excretory processes for xenobiotics are best understood for mammals, as far less work on birds, reptiles and amphibians has been carried out. Highly lipophilic compounds show little tendency to be excreted unchanged. In the absence of effective metabolism, they tend to have very long biological half-lives in depot fat. Thus, half-lives of *c.* 1 year have been reported for p,p'-DDE in birds, whereas higher chlorinated PCBs and higher brominated polybrominated biphenyls (PBBs) have half-lives running into years in certain mammals, including man! Related compounds that are easily metabolised, e.g. lower chlorinated PCBs or 'biodegradable' cyclodienes such as the dieldrin analogues endrin and HCE (hexachloro-octahydra-7,8-epoxy-methano-naphthalene) have half-lives of a few hours or days. The effective elimination of strongly lipophilic compounds depends upon their conversion to water-soluble conjugates and metabolites, leading to excretion in bile and urine. A small qualification to this generalisation is that some lipophilic compounds are excreted to a limited extent with lipoproteins. Thus, organochlorine insecticides such as dieldrin and DDT are transported from female birds, reptiles and insects into eggs. Also, female mammals excrete such compounds in milk.

The conversion of lipophilic xenobiotics into water-soluble metabolites and conjugates by vertebrates occurs principally in the liver, less in the kidney and certain other metabolically active tissues. Most excretion occurs in the form of conjugates. When conjugates are formed in the liver, they may be excreted in bile or in urine. The excretory route depends upon molecular weight. As mentioned earlier (p. 22), most conjugates are anions. Anionic conjugates below a molecular weight of 300 tend to be excreted in urine, whereas above 600 they tend to be excreted in bile. Between these two extremes the preferred excretory route depends upon the species. It has been proposed that there are 'threshold' molecular weights for different species, above which 10% or more of the anionic conjugate is excreted in bile (Figure 2.16). The following values are proposed for the named species: the rat 325 (+50), the guinea-pig 425 (+50) and the rabbit 475 (+50). Thus, rats show a greater tendency to excrete anionic conjugates in bile than do the other two species when considering xenobiotics over a range of molecular weights (see Walker, 1975).

The reason for this pattern has yet to be definitely established. The most likely explanation is that there is selective reabsorption of the conjugates from bile into plasma. Conjugates are moved from hepatocytes into biliary canaliculi during the formation of primary bile. With some conjugates, e.g. glucuronides, this movement appears to be driven by active transport. During subsequent passage through the biliary system, lower-molecular-weight conjugates are absorbed into blood, whereas larger ones remain in bile, eventually to be discharged into the alimentary tract. Such selective absorption may occur through pores. This could explain the reason for the differing thresholds between species. It is possible that rats have tighter gap junctions between hepatocytes than do rabbits, so that larger conjugates are retained in the biliary system, whereas in rabbits they can escape into plasma. The species differences in preferred excretory route have implications for the later course of excretion, and for toxicity. Conjugates that remain within the biliary system may be broken down in the gut to release the conjugated molecule (often a phase 1 metabolite). This molecule, being less polar than the conjugate, can then be reabsorbed into the blood and

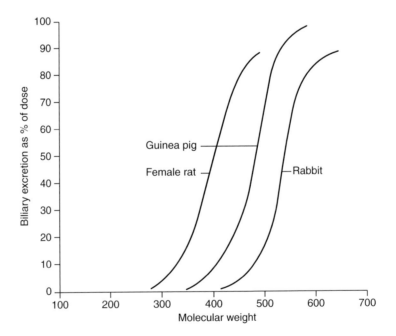

FIGURE 2.16 *Excretion routes of xenobiotic anionic conjugates.*

recirculated to the liver. The process is termed 'enterohepatic circulation', and some compounds may cycle 30 times or more before being finally voided with faeces. Enterohepatic circulation can lead to toxic effects. For example, the drug chloramphenicol is metabolised to a conjugate, which is excreted in bile by the rat. Once in the gut, the conjugate is broken down to release a phase 1 metabolite, which undergoes further metabolism to yield toxic products. When these are reabsorbed, they can cause toxicity. The rabbit, by contrast, excretes chloramphenicol conjugates in urine, and there are no toxic effects at the dose rates in question.

Excretion via the kidney can be a straightforward question of glomerular filtration followed by passage down the kidney tubules into the bladder. However, there can also be excretion and reabsorption across the tubular wall. Reabsorption may occur if an ionised form within the tubule is converted into its non-polar non-ionised form because of a change in pH. The non-ionised form can then diffuse across the tubular wall into plasma. In addition, there are active transport systems for the excretion of lipophilic acids and bases across the wall of the proximal tubule. The antibiotic penicillin can be excreted in this way.

Far less is known about excretion by terrestrial insects than by terrestrial mammals. Metabolism can take place in the midgut and in the fat body. Excretion can occur via the malpighian tubules.

2.4 *Toxicodynamics*

The discussion until now has been concerned with toxicokinetics – the uptake, distribution, metabolism, storage and excretion of organic pollutants within living organisms. These processes determine how much of a toxic molecule – original pollutant or active metabolite – reaches the site(s) of action. At this stage we enter the field of *toxicodynamics*, which is concerned with the interaction of toxic molecules with their site(s) of action and the consequences thereof. Conceptually, it is useful to make a distinction between toxicokinetics and toxicodynamics, because these two factors determine, in contrasting and largely independent ways, how toxic any particular chemical will be to living organisms. Selectivity is due to the differential operation of toxicokinetic and toxicodynamic processes between different species, strains, sexes or age groups (see section 2.5). In the study of the basis of selective toxicity (e.g. resistance of insects to pesticides), it is advantageous to design experiments that permit the distinction between toxicokinetic and toxicodynamic factors (for further discussion, see Walker, 1994a,b).

Most of the organic pollutants described in the present text act at relatively low concentrations because they, or their active metabolites, have high affinity for their sites of action. If there is interaction with more than a critical proportion of active sites, disturbances will be caused to cellular processes, which will eventually be manifest as overt toxic symptoms in the animal or plant. Differences between species or strains in the affinity of a toxic molecule for the site of action are a common reason for selective toxicity.

It should also be mentioned that some compounds of relatively low toxicity act as physical poisons, although such pollutants are seldom important in ecotoxicology. They have no known specific mode of action, but if they reach relatively high concentrations in cellular structures, e.g. membranes, they can disturb cellular processes. Examples include certain ethers and esters, and other simple organic compounds.

Tables 2.6 and 2.7 give examples of the modes of action of pollutants in animals and in plants/fungi respectively. It is noteworthy that many of the chemicals represented are pesticides. Pesticides are designed to be toxic to non-target species. Consequently, manufacturers of them must usually provide evidence of the mode of toxic action before registration is agreed by the regulatory authorities. Other industrial chemicals are not subject to such strict regulatory requirements, and their mode of action is frequently unknown.

The examples given in Tables 2.6 and 2.7 illustrate the wide range of different mechanisms by which pollutants cause toxic effects. The following account will focus on certain broad issues concerning the mode of action. A more detailed description of individual examples will be given in later chapters devoted to particular types of pollutants.

Many of the pollutants expressing high toxicity to animals are lipophilic in character. This appears to be a consequence of toxicokinetic factors. After absorption, lipophilic compounds tend to remain within the organism, and effective excretion depends on their enzymic conversion to water-soluble and readily excretable products (see 'General considerations'). They tend to associate with membranes and hydrophobic macromolecules within the body, thus facilitating interaction with enzymes, receptors for chemical messengers and pore channels. Water-soluble organic compounds, on the other hand, tend to be rapidly excreted unchanged, and they do not tend to associate with lipophilic structures within the organism.

Broadly speaking, toxic interactions between chemicals and cellular sites of action are of two kinds.

1 The pollutant (xenobiotic) forms a stable covalent bond with its target. Examples include the phosphorylation of cholinesterases by the oxon forms of OPs ,the formation of DNA adducts by the reactive epoxides of benzo(*a*)pyrene and other PAHs and the binding of organomercury compounds to the sulphydryl groups of proteins. It should be added that the reactive PAH epoxides and most of the reactive oxons of OPs are not the original pollutants absorbed by animals; they are unstable metabolites generated by monooxygenase attack (see 'Monooxygenases').

2 The pollutant binds to the target without forming a covalent bond. Such interactions are usually reversible, and examples include the binding of the stable organochlorine insecticides *p,p'*-DDT and dieldrin to Na$^+$ channels and GABA

receptors, respectively, the binding of planar PCBs and dioxins to the Ah receptor, the binding of 4-OH'-3,3',4,5'-tetrachlorobiphenyl to the thyroxine binding site of transthyretin and the binding of ergosterol biosynthesis inhibitor (EBI) fungicides to a fungal form of cytochrome P450. The reduction of DDT toxicity to insects with increasing temperature has been attributed to the desorption of p,p'-DDT from the Na^+ channel. (The same phenomenon has been reported with pyrethroid insecticides, which also interact with the Na^+ channel.)

The interaction of toxic molecules with their sites of action leads to physiological effects at the cellular and whole-organism level. The nature of these effects, and their significance in ecotoxicology, depends upon the site of action. It is noteworthy that the four most important groups of insecticides are all neurotoxins, albeit working through different modes of action (Table 2.6). Disturbances of the nervous system can have far-reaching consequences, especially when they arise in the central nervous system (e.g. the effects of dieldrin and HCH on GABA receptors of vertebrates). At an early stage of intoxication there can be behavioural effects (evidence of behavioural effects of organochlorine and organophosphorous insecticides will be discussed in Chapters 5 and 10 respectively). In another example, the action of endocrine disruptors may lead to deleterious effects upon reproduction, as in the case of imposex caused by TBT in the dog whelk. Also, the reproduction of some raptors can be adversely affected by eggshell thinning caused by p,p'-DDE, apparently due to inhibition of Ca^{2+}-ATPase. In a further example, anticoagulant rodenticides, such as warfarin, brodifacoum and flocoumafen, act as antagonists of vitamin K, which has a critical role in the synthesis of clotting proteins in the liver of vertebrates. After exposure to these compounds, synthesis of clotting proteins in the liver is incomplete, and precursors of them are released into blood. After a period of days, the level of clotting proteins falls in the blood and haemorrhaging occurs. Thus, different modes of action lead to different toxic effects, with differing ecotoxicological consequences.

The mode of action can be an important selectivity factor. In the extreme case, one group of organisms has a site of action that is not present in another group. Thus, most of the insecticides that are neurotoxic have very little phytotoxicity; indeed, some of them (e.g. the OPs dimethoate, disyston and demeton-S-methyl) are good systemic insecticides. Most herbicides that act upon photosynthesis (e.g. triazines and substituted ureas) have very low toxicity to animals (Table 2.7). The resistance of certain strains of insects to insecticides is due to their possessing a mutant form of the site of action that is insensitive to the pesticide. Examples include certain strains of housefly with 'Kdr' resistance (mutant form of Na^+ channel that is insensitive to DDT and pyrethroids) and strains of several species of insects that are resistant to OPs because they have mutant forms of acetylcholinesterase. These examples will be discussed in more detail when considering individual groups of pollutants. In contrast to the foregoing examples, some pollutants have wide-ranging toxicity to nearly all forms of life and earn the doubtful accolade of being termed 'biocides'! Probably the best examples of these are uncouplers of oxidative phosphorylation such as

TABLE 2.6 *Mode of action of pollutants in animals*

Pollutant(s)	Type of toxic action	Site of action	Consequence	Comments and references
OP and carbamate insecticides	Neurotoxicity	Acetylcholinesterase	Disruption of synaptic transmission	OPs include insecticides and chemical warfare agents (Eto, 1974; Ballantyne and Marrs, 1992; Kuhr and Dorough, 1976)
DDT and related compounds	Neurotoxicity	Na$^+$ channels of axonal membrane	Disturbance of nerve action potential	Persistent insecticides (Brooks, 1974; Eldefrawi and Eldefrawi, 1990)
Dieldrin and other cyclodiene insecticides	Neurotoxicity	GABA receptors	Can cause CNS disturbances	Dieldrin is persistent and highly toxic to vertebrates (Brooks, 1974, 1992; Eldefrawi and Eldefrawi, 1990; Salgado, 1999)
Pyrethroids	Neurotoxicity	Na$^+$ channels of axonal membrane	Similar action to DDT	Selective insecticides (Eldefrawi and Leahey, 1985; Eldefrawi, 1990; Salgado, 1999)
Benzo(a)pyrene and other PAHs	Mutagens and carcinogens	DNA	Mutagenesis can lead to carcinogenesis	Little known about consequences of mutation other than carcinogenesis (Hodgson and Levi, 1994)
p,p'-DDE	Affects Ca^{2+} transport in avian shell gland	Probably Ca^{2+}-ATPase	Thinning of avian eggshells	Persistent metabolite of p,p'-DDT implicated in the decline of certain predatory birds (Lundholm, 1987; Peakall, 1992)
Dioxins and coplanar PCBs	Ah receptor-mediated toxicity	Ah receptor	A range of toxic effects including endocrine disturbances	Mechanism by which binding to Ah receptor causes toxic effects still unclear (Safe, 1990)

4-OH-3,3',4,5'-tetrachlorobiphenyl	Thyroxine antagonism	Thyroxine binding site of transthyretin	Vitamin A deficiency	Effect caused by this and certain other PCB metabolites (Brouwer et al., 1990)
Anticoagulant rodenticides	Vitamin K antagonism	Vitamin K binding sites of hepatic carboxylase	Haemorrhaging	Warfarin, diphenacoum, bromodiolone, flocoumafen, and brodiphacoum are familiar examples (Thijssen, 1995)
Organomercury fungicides	Neurotoxicity	Sulphydryl groups of proteins	Damage to CNS	Both alkyl and aryl mercury compounds can bind to sulphydryl groups (EHC 86, 1989)
Tributyltin fungicides	Endocrine disruptors in molluscs	Thought to disturb testosterone metabolism by binding to cytochrome P450	Cause 'imposex' in dogwhelks and a variety of other effects on other molluscs	Have caused population declines in certain molluscs (Matthiessen and Gibbs, 1998)

TABLE 2.7 *Mode of action of pollutants in plants and fungi*

Pollutant (s)	Type of toxic action	Site of action	Consequence	Comments and references
Triazine and substituted urea herbicides	Photosynthesis inhibitors	Q_B^* binding site of the D1 protein in photosystem II	Inhibition of electron transport in chloroplast, and consequent reduction of photosynthesis	These herbicides have generally very low toxicity to animals (Hassall, 1990; Sjut, 1997)
Phenoxyalkanoic herbicides	Disruption of growth regulation of plants	Unknown, but apparently the same as for the natural auxins whose action they mimic	Distorted growth patterns including malformed leaves and epinasty of stems	2,4-D, MCPA, CMPP and MCPB are well-known examples. Generally of low toxicity to animals (Hassall, 1990)
Dinitroaniline herbicides	Mitotic disrupters	Polymerase responsible for microtubule formation	Disruption of cell division	Examples include trifluralin and oryzalin (Hassall, 1990)
Paraquat	Photosynthesis inhibitor	Electron transport system of photosystem I	Paraquat diverts electrons from the electron transport system, and passes them on to oxygen to form $O_2^{\cdot -}$, which causes cellular damage	Paraquat and other bipyridylium herbicides are also toxic to animals because of their tendency to generate reactive oxyradicals (Hassall, 1990; Sjut, 1997; Chapter 14)
DNOC	Mitochondrial poison	Inner membrane of mitochondrion	Uncouples oxidative phosphorylation by running down the proton gradient across the inner mitochondrial membrane	DNOC and certain other dinitrophenols are general biocides because their site of action, the mitochondrion, is present in most living organisms (Nicholls, 1982; Hassall, 1990; Chapter 14)
Ergosterol biosynthesis inhibitors (EBIs)	Inhibitors of ergosterol biosynthesis in fungi	Fungal cytochrome P450 involved in sterol metabolism	Destabilising of fungal membranes	Some EBIs, e.g. prochloraz, can act as inducers and/or inhibitors of P450 forms of animals, and thus potentiate the toxicity of other pesticides (Kato, 1986; section 2.6)

*See Sjut (1997) for further information on Q_B.

dinitrophenol and the herbicide dinitroorthocresol (DNOC). These act upon the mitochondrial membrane of animals and plants, and run down the proton gradient across the membrane, which drives the process of ATP formation (see Chapter 14).

2.5 *Selective toxicity*

Selective toxicity (henceforward simply 'selectivity') is of fundamental interest in ecotoxicology. For any pollutant that is, or may become, widely distributed in the environment, it is desirable to know which species or groups will be most sensitive to its toxic action. Sensitive species should be given particular attention in biological monitoring (e.g. in use of biotic indices) and, ideally, should have appropriate representatives in ecotoxicity testing. Selectivity needs to be taken into account when estimating 'environmental toxicity', e.g. PNEC (predicted no effect concentration), from ecotoxicity data obtained with surrogates, where emphasis is placed upon 'the most sensitive species'. A large safety factor needs to be incorporated in such calculations if the test organism is expected to be much less sensitive than key species in the natural environment. In the design of new pesticides, a major objective is to design compounds that are selective between pest species and beneficial organisms, thereby ensuring compliance with the requirements of regulatory bodies for environmental safety. Selective toxicity is also important in relation to the development of resistance or tolerance to pollutants from two distinct points of view. On the one hand, there is interest among scientists concerned with crop protection and disease control in the mechanisms by which crop pests, vectors of disease, plant pathogens and weeds develop resistance to pesticides. Understanding the mechanism should lead to the discovery of a remedy. On the other hand, ecotoxicologists are interested in the development of resistance as an indicator of the environmental impact of pollutants.

As discussed earlier (p. 18), selectivity is the consequence of the interplay between toxicokinetic and toxicodynamic factors. Some examples are given in Table 2.8 and will now be briefly discussed (data from Walker and Oesch, 1983; Walker, 1994a). These and other examples will be described in more detail under specific pollutants later in the text. In Table 2.8 comparisons are made between the median lethal doses (LD_{50}) or median lethal concentrations (LC_{50}) for different species or strains. Comparisons are made of data obtained in lethal toxicity tests where the same route of administration was used for both species. The degree of selectivity is expressed as a selectivity ratio:

$$\frac{LD_{50} \text{ (or } LC_{50}) \text{ of species or strain A}}{LD_{50} \text{ (or } LC_{50}) \text{ of species or strain B}}$$

The OPs dimethoate and diazinon are much more toxic to insects, e.g. housefly, than they are to the rat or other mammals. A major factor responsible for this is rapid detoxication of the active oxon forms of these insecticides by A esterases of mammals.

TABLE 2.8 *Selective toxicity*

Pollutant	Toxicity test	Toxicity to species or strain A	Toxicity to species or strain B	Toxicity to A / Toxicity to B
Dimethoate (OP)	Topical LD_{50}	Rat 925	Housefly 0.20*	4.6×10^3
Diazinon (OP)	Topical LD_{50}	Rat 850	Housefly 1.9*	447
Diazinon	Acute oral LD_{50}	Rat 450	Birds 4.5 (mean four species)	100
Fenvalerate (pyrethroid)	Topical LD_{50}	Rat 4000	Honeybee 0.21*	2.3×10^4
Dimethoate (OP)	Topical LD_{50}	*Myzus persicae** 'R' clone	*Myzus persicae** 'S' clone	500†
cis-cypermethrin (pyrethroid)	Topical LD_{50}	*Heliothis virescens** 'R' strain (PEG 87 d/3 third instar larva)	*H. virescens** 'S' strain (BRCb/1c third instar larva)	7×10^5‡

LD_{50} values expressed as mg/kg (μg/g) except where marked with an asterisk (*), which are given as μg/insect.
†Devonshire and Sawicki (1979).
‡McCaffery *et al.* (1991).
Other data from Walker (1994a) and Walker and Oesch (1983).

Insects in general appear to have no A esterase activity or, at best, low A esterase activity (some earlier studies confused A esterase activity with B esterase activity) (Walker, 1994a,b). Diazinon also shows marked selectivity between birds and mammals, which has been explained on the grounds of rapid detoxication by A esterase in mammals, an activity that is absent from the blood of most species of birds (see p. 33, 'Esterases and hydrolases'). The related OPs pirimiphos methyl and pirimiphos ethyl show similar selectivity between birds and mammals. Pyrethroid insecticides are highly selective between insects and mammals, and this has been attributed to faster metabolic detoxication by mammals and greater sensitivity of target (Na^+ channel) in insects.

Some examples are also given of insect resistance. Here, the selectivity ratio is usually termed the 'resistance factor'. Thus, a strain of housefly resistant to the organophosphorous insecticide malathion can detoxify the insecticide as a result of metabolism by a carboxylesterase (form of B esterase) not found in susceptible strains of this species. Some strains of housefly (*Musca domestica*) developed resistance to DDT as a consequence of possessing a mutant form of the target (Na^+ channel). Clones of the peach potato aphid (*M. persicae*) were generally resistant to OPs because of very high levels of a carboxylesterase. The elevated levels of the enzyme were the consequence of multiple copies of the gene coding for it. Resistant clones had 2, 4, 8, 16 or 32 copies of the gene, and the level of resistance was related to the number of copies in any particular aphid clone (Devonshire, 1991). The clone showing a 500-fold resistance factor possessed 32 copies of the carboxylesterase gene, which conferred resistance. In another example, a resistant strain of the tobacco bud worm (*Heliothis virescens*) showed 70 000-fold resistance to the pyrethroid cypermethrin as a consequence of two major resistance factors: (1) enhanced oxidative detoxication by a P450-based monooxygenase and (2) insensitivity of mutant form of target (Na$^+$ channel). Thus, resistance here was related to both toxicodynamic and toxicokinetic factors.

2.6 *Potentiation and synergism*

Ecotoxicity testing of industrial chemicals and pesticides during the course of environmental risk assessment is usually only performed upon individual compounds. In the real world, however, living organisms are frequently exposed to complex mixtures of chemicals, inorganic as well as organic, and questions arise about the toxicity and ecological effects of mixtures (Chapter 14; Walker *et al*., 1996; Moriarty, 1999). Examples include mixtures of hydrocarbons and mixtures of polyhalogenated aromatic compounds such as PCBs and 'dioxins'. The problem is, of course, that even testing single compounds is a difficult and expensive exercise; the resources do not exist to test more than a very small proportion of the combinations of chemicals found in the living environment. It is usually assumed that the toxicity of mixtures is roughly additive (in other words the toxicity of the mixture is approximately the sum of the toxicities of the individual components). Although this assumption is usually borne out in practice, there are important exceptions; there can be *potentiation*, where the sum of the individual toxicities is considerably greater than additive. *Synergism* is a particular type of potentiation, where the toxicity of one compound is enhanced by another compound (*synergist*), which does not express toxicity by itself at the levels of exposure in question. Thus, potentiation is a broader term than synergism, and includes cases where two or more compounds, which both express toxicity, cause greater than additive toxicity when brought together in a mixture.

A basic problem, then, is recognising where there is a serious risk of potentiation of toxicity when animals are exposed to mixtures of chemicals in the field. Whilst it is clearly impossible to test more than a small proportion of the combinations that exist (or may exist in future as new chemicals are released into the environment), a more logical approach than the random testing of mixtures suggests itself. As knowledge grows of the mechanisms which determine toxicity, it becomes increasingly possible to foresee where potentiation is likely to occur, and to limit any testing of the ecotoxicity of mixtures to a small number of combinations that give cause for concern.

From a mechanistic point of view, potentiation can be seen to arise from interactions occurring within the model given in Figure 2.1. Taking the example of two interacting compounds, potentiation may be due to one compound causing changes in either the toxicokinetics or the toxicodynamics of the other compound – or both of these things.

TABLE 2.9 *Synergism due to inhibition of detoxication*

Organism	Insecticide	Detoxifying enzyme	Inhibitor (synergist)	Synergistic ratio*
Insect strains resistant to pyrethroids, e.g. *H. virescens*	Cypermethrin	Monooxygenase	Piperonyl butoxide (PBO)	< 200
Insects	Carbaryl	Monooxygenase	PBO and other methylene-dioxyphenyls	< 400
Mammals and some malathion-resistant insects	Malathion	Carboxylesterase	Some OPs other than malathion	< 200

*Toxicity without synergist/toxicity with synergist.

Data from Kuhr and Dorough (1976), McCaffery *et al.* (1991), Walker and Oesch (1983).

The first compound may alter the toxicokinetics of the other, so that more of the active form (original compound or metabolite) reaches the site of action. Alternatively, the first compound may alter the toxicodynamics of the other, so that it becomes more effective at the same tissue concentration. In practice, most of the well-understood cases of potentiation are due to one compound affecting the toxicokinetics of the other, so that more of the active form reaches the site of action. In many cases this is due to inhibition of detoxication, but enhanced metabolic activation, increased rate of absorption and reduced storage have all been implicated. Table 2.9 gives a number of examples of synergism due to one compound (the synergist) inhibiting the metabolic detoxication of the other. There can be an increase of toxicity as great as 400-fold, when enzymic detoxication is inhibited. It is noteworthy that two of the examples cited above involve the inhibition of microsomal monooxygenases, enzymes that have a very wide-ranging detoxifying function in the animal kingdom. Understandably, there is concern about the presence of monooxygenase inhibitors among environmental chemicals because of their potential to increase the toxicity of other compounds that are substrates of the enzyme. Inhibitors of monooxygenases, esterases and other enzymes having a detoxifying function are sometimes useful for identifying the cause of resistance to pesticides. The inhibitor reduces the level of resistance by blocking an enzyme that is responsible for enhanced detoxication in the resistant strain.

Other cases of potentiation will be discussed in later chapters dealing with individual pollutants. These will include the synergism of pyrethroid toxicity to honeybees by EBI fungicides due to inhibition of detoxication by monooxygenases (Chapter 12); the increased toxicity of the carbamate insecticide carbaryl to red-legged partridges (*Alectoris rufa*) by the organophosphorous insecticide malathion due to inhibition of detoxication by monooxygenases (Chapter 10); the enhanced toxicity of malathion to birds by EBI fungicides as a consequence of induction of monooxygenase and increased activation (Chapter 10); and the enhanced activation of mutagenic PAHs due to induction of monooxygenases by coplanar PCBs and dioxins (Chapter 9).

2.7 *Summary*

The toxicity of an organic chemical depends upon the operation of toxicokinetic and toxicodynamic processes within living organisms, and selectivity between species, strains, sexes and age groups is the outcome of the differential operation of these processes. A model is presented that describes the fate of lipophilic organic compounds within living organisms, seen from a toxicological point of view. After uptake, the chemical is distributed to sites of action, metabolism, storage and excretion. 'Toxicokinetics' encompasses all of those processes that occur before the arrival of a toxic compound (original compound or active metabolite) at the site of action, processes that determine how much of the toxic compound reaches the site of action. 'Toxicodynamics' is concerned with the interaction of the toxic compound with the site of action and consequent toxic effects.

Emphasis is given to the critical role of metabolism, both detoxication and activation, in determining toxicity. The principal enzymes involved are described, including monooxygenases, esterases, epoxide hydrolases, glutathione-S-transferases and glucuronyl transferases. Attention is given to the influence of enzyme induction and enzyme inhibition upon toxicity.

The mechanism of toxic action of some important organic pollutants is described and discussed and related, where possible, to ecotoxicological effects.

2.8 *Further reading*

There are many texts dealing with the biochemical toxicology of organic compounds, but most of them are principally concerned with human toxicology. The following texts are suitable, in differing ways, as supplementary reading to the present chapter.

Hassall, K. A. (1990) *The Biochemistry and Use of Pesticides*. A readable student text, particularly useful for the metabolism and mode of action of herbicides and fungicides.

Hodgson, E. and Kuhr, R. J. (1990) *Safer Insecticides: Development and Use*. A multiauthor work with a wealth of information on the mechanism of action of insecticides.

Hodgson, E. and Levi, P. (1994) *Introduction to Biochemical Toxicology*. A multiauthor text dealing in depth with some aspects of biochemical toxicology highly relevant to the present topic. Examples of metabolism and mode of action of insecticides and PAHs.

Jakoby, W. B. (ed.) (1980) *Enzymatic Basis of Detoxication*. Although a little out of date, it is still a useful source of the enzymes concerned with xenobiotic metabolism.

Lewis, D. F. V. (1996) *Cytochromes P450*. A detailed account of cytochrome P450, including evolutionary aspects.

CHAPTER 3

Influence of the properties of chemicals on their environmental fate

3.1 Introduction

The previous chapter was concerned with the processes that determine the distribution of chemicals within living organisms and with the relationship between distribution and toxicity. The importance of the properties of chemicals in determining their fate within living organisms was given emphasis. Polarity, molecular size, the presence of functional groups and molecular stability were all seen to influence toxicokinetic processes. The types of enzymes responsible for biotransformations were related to the structures of the chemicals undergoing biotransformation (e.g. esterases for esters, reductases for nitro compounds, monooxygenases for aromatic hydrocarbons). The present chapter will consider the wider question of how the properties of chemicals determine their fate in the gross environment, and how these properties can be incorporated into descriptive and predictive models relating to the movement and distribution of environmental chemicals.

First, there will be a description of chemical and physical properties that determine the fate of organic pollutants in the gross environment, restricting the discussion to chemical and physical processes. Next, there will be an account of how these data are incorporated into models that attempt to describe or predict environmental fate. Finally, the relationship will be discussed between the properties of chemicals *and the operation of biological processes that determine environmental fate.* Drawing on the background given in the previous chapter, emphasis will be given to the relationship between the properties of environmental chemicals and their uptake, excretion and metabolism. The overall aim will be to lay a proper foundation for consideration of more complex issues in Chapter 4, which will address the question of the movement and distribution of chemicals in ecosystems where both biological and abiotic processes come into play. Key issues will be the fate of chemicals in soils and sediments, and the movement of chemicals along terrestrial and aquatic food chains, situations in which their fate is determined by both biological and chemical processes.

3.2 *Properties of chemicals which influence their fate in the gross environment*

Polarity is one of the most important determinants of the environmental fate of organic chemicals. In general, the more polar a compound is, the higher its water solubility and the lower its octanol–water partition coefficient (K_{ow}). Conversely, the less polar a compound is, the lower its water solubility and the higher its K_{ow} (see earlier discussion in section 2.3). K_{ow} is a measure of 'hydrophobicity', and it is regarded as one of the most valuable indicators of the environmental behaviour of organic chemicals. Many examples will be given in the chapters dealing with individual pollutants in the second part of this text. (For more detailed and critical discussion about the determination and utilisation of K_{ow} values, readers are referred to the chapters by Connell and Donkin in Calow, 1994.)

Hydrophobic compounds tend to be excluded from the aqueous phase of the environment because of what has been termed 'the hydrophobic effect' (Tanford, 1980). Polar water molecules are drawn together because of the attraction of d$^+$ and d$^-$ charges on adjacent molecules. Hydrophobic compounds present in the water, which have little or no charge, are thereby 'squeezed out'. In soils and sediments, they tend to become adsorbed to the surface of colloidal material – organic matter and clay – where they are retained by van der Waals forces. In waters they tend to move to the surface (into surface oil films, where present), into sediments or into the waxy cuticle of aquatic macrophytes. The tendency of hydrophobic compounds to bind to colloid surfaces has two important consequences: (1) they are not very mobile, and (2) they are often very persistent because they are not freely available to the enzyme systems (e.g. of microorganisms) that can degrade them. In soils, as in terrestrial animals, refractory hydrophobic molecules tend to be markedly persistent. Of the examples given in Table 2.1, dieldrin and *p,p'*-DDT with K_{ow} values of 5.48 and 6.36 respectively have very long half-lives in soils and sediments.

Another very important factor determining the distribution of chemicals in the gross environment is **vapour pressure**. Vapour pressure is defined as the pressure exerted by a chemical in the vapour state upon its own liquid or solid surface at equilibrium. It may be expressed as millimetres of mercury (torr), as a fraction of normal atmospheric pressure (760 mmHg), or as pascals (Pa) (1 Pa = 0.0075 mmHg). Vapour pressure is related to temperature, and liquids boil at temperatures that raise their own vapour pressures to 760 mmHg. Water, for example, boils at 100 °C. Some solids sublime, that is they pass directly into the vapour state without liquefying. Many pesticides that exist in the solid state under normal temperature and pressure will sublime under field conditions. Volatilisation represents a major source of loss of pesticides from the surface of crops and from the soil. With pollutants generally, high vapour pressure indicates that chemicals will tend to move into the atmosphere, with the attendant risk that they will be transported over large distances if they are stable enough. Chlorofluorocarbons (CFCs) present a striking example of the problem. They are highly volatile and stable enough to reach the ozone layer within the stratosphere. Once there, however, they can reduce the concentration of ozone by interacting with it, an environmental process that is thought to have significantly reduced the extent of the ozone layer. The long-range movement of certain PCBs and organochlorine insecticides into polar regions has been attributed to aerial transport in the vapour state. In the second part of the book examples will be given of pollutants that move readily into the air because of relatively high vapour pressure.

Much of the earth's surface is water, and an important aspect of the volatilisation of pollutants is their movement from water into air. The movement of solutes between water and air is governed by Henry's law, which states that, at equilibrium, the concentration of a chemical in the vapour state bears a constant relationship to the concentration in aqueous solution.

$$H' = \text{concentration air/concentration water} = (n/V)\,(1/S) = P/RTS$$

where H' is a partition coefficient, n/V is the molar concentration of the vapour, S is the saturation solubility of the solute in water, P is vapour pressure, R is the universal gas constant and T is the absolute temperature.

This equation can be simplified to give $H = P/S$, where H is now the Henry's law constant, which has dimensions of atm m^3/mol. Values for H may be calculated or measured (Mackay *et al.*, 1979) and are now widely used in **fugacity modelling** (see next section).

Another important determinant of the environmental fate of pollutants is their **chemical stability**. Environmental chemicals that are highly resistant to chemical degradation have been described as being 'refractory' or 'recalcitrant' (see Crosby, 1998). Many environmental chemicals are not very photochemically stable. When exposed to solar radiation, they may be oxidised or undergo molecular rearrangement. Some chemicals, e.g. OPs, carbamates and pyrethroids, are susceptible to hydrolysis, especially when exposed to water of high pH. By contrast, other compounds only undergo very slow degradation and can, therefore, be transported over large distances in air or water without substantial loss, unless biodegradation is significant (see section

3.4). Typical examples are highly halogenated compounds such as p,p'-DDE, dieldrin, higher chlorinated PCBs and TCDD (dioxin). Highly halogenated compounds tend to be resistant to oxidation and other mechanisms of chemical degradation.

In most cases the chemical instability of pollutants limits risk to the environment because it usually represents a reduction in toxicity. There are, however, exceptions and the devil may be found in the detail. When dieldrin residues are exposed to solar radiation, there is some conversion to the persistent and highly toxic photodieldrin. When the organophosphorous insecticide malathion is stored under hot conditions over long periods, it is converted to highly toxic iso-malathion. When PAHs are exposed to radiation they are converted into products that are highly toxic to fish. Examples such as these illustrate the dangers of hasty or superficial judgement in environmental risk assessment and the importance of rigorous analysis case by case.

Another factor that can influence the environmental distribution of a chemical is the presence of **charged groups**. Some pollutants, such as the sodium or potassium salts of the phenoxyalkanoic herbicides, the dinitrophenols and tetra- or penta-chlorophenol, exist as anions in solution. Others, such as the bipyridyl herbicides diquat and paraquat, are present as cations. In either case the ions may become bound to organic macromolecules or minerals of soils or sediments that bear the opposite charge. Thus, the paraquat cation can be strongly bound (adsorbed) to negatively charged surfaces within clay minerals in competition with exchangeable cations such as Ca^{2+}, K^+ and Na^+. It can also be bound to ionised carboxyl or phenolic groups of soil organic matter. Organic anions may bind to certain positively charged groups within soils and sediments.

The influence of chemical properties on environmental fate is dealt with in much more detail in specialised texts on environmental chemistry, such as Schwarzenbach *et al.* (1993) and Crosby (1998).

3.3 *Models of environmental fate*

Confronted with the complexity of pollution of the environment, and the great expense of actually measuring levels of organic pollutants, the attractions of developing models for environmental fate are obvious. Because of constraints of cost and of time, the determination of actual levels of pollutants in different compartments of the environment can only be carried out to a very limited degree. Many different models have been developed that attempt to describe or predict the fate of chemicals in the gross environment (for a detailed treatment of the subject readers are referred to Jorgensen, 1990; Mackay, 1991; Bacci, 1994; Mackay, 1994, in Calow, 1994). They range from limited single-medium models that are concerned with the fate of a chemical in a phase such as air, water or soil to wide-ranging multimedia models that attempt to describe the movement of chemicals through different phases of the environment over very large areas. The former may be expected to provide reasonably close estimates of environmental concentrations if the database is good enough, whereas the latter can only give a broad and imprecise view of distribution through air, soil, water, sediment and biota on what is approaching a global scale. Some multimedia models are only useful for ranking chemicals according to their tendency to move into particular environmental compartments (e.g. air), and they cannot give reliable estimates of concentrations that will be found in the real world following defined releases of chemicals. There are too many uncontrolled variables such as temperature, wind speed, solar radiation, precipitation of rain and snow. The term 'evaluative models' is sometimes used to describe such systems (Bacci, 1994). In some cases, the multimedia approach has been applied to global pollution problems, for example the geochemical cycling of CO_2 and the distribution of CFCs, that may affect the ozone layer.

Multimedia models can describe the distribution of a chemical between environmental compartments in a state of equilibrium. Equilibrium concentrations in different environmental compartments following the release of defined quantities of pollutant may be estimated by using distribution coefficients such as K_{ow} and H' (see section 3.2). An alternative approach is to use fugacity (f) as a descriptor of chemical quantity (Mackay, 1991). Fugacity has been defined as the tendency of a chemical to escape from one phase to another and has the same units as pressure. When a chemical reaches equilibrium in a multimedia system, all phases should have the same fugacity. Fugacity is usually linearly related to concentration (C) as follows:

$$f = C/Z$$

where Z is a constant, sometimes termed the fugacity capacity constant. Values for Z depend on the chemical, the nature of the absorbing medium and the temperature.

Examples of models of the environmental fate of chemicals utilising partition coefficients and fugacities are given in the texts cited above, also in Chapter 3 of Walker *et al.* (2000).

3.4 *Influence of the properties of chemicals on their metabolism and disposition*

Up to this point, the discussion has been limited to the influence of chemical properties on the physical processes that influence distribution through the environment without taking into account effects upon biological processes which determine movement. Although it is true that biological factors have little or no influence on the fate of chemicals transported in the air, this is not the case with transport by surface waters, or in sediments or soils. Here, living organisms are involved in the uptake, metabolism and transfer of organic pollutants. Stable pollutants are transported over considerable distances by migrating animals and birds, and move through terrestrial and aquatic food chains. Such complexities are not readily accommodated by the relatively simple 'chemical' models described above.

In Chapter 2, the fate of organic pollutants in living organisms was described, drawing attention to the processes involved. Here the relationship between the properties of chemicals and the operation of these processes will be briefly reviewed. Many examples of the influence of the properties of a chemical upon its own metabolic fate will be encountered in the second part of the text.

Once again, the hydrophobicity of compounds is a critical determinant of their fate. Compounds with high K_{ow} values tend to undergo bioaccumulation by animals because they move from the aqueous phase into the hydrophobic environment of fat depots and biological membranes. Water-soluble compounds do not have this tendency. Aquatic organisms do not tend to bioconcentrate them. When terrestrial animals absorb them from food or water, they are usually rapidly excreted in their unchanged forms. With lipophilic compounds, however, rapid elimination from the body depends upon efficient metabolism to water-soluble and readily excretable metabolites and conjugates. This trend is particularly marked in terrestrial animals; aquatic species can lose lipophilic compounds to some extent by 'exchange diffusion'. Interestingly, aquatic animals generally have relatively low levels of detoxifying enzymes such as cytochrome P450s of family 2 in comparison with terrestrial ones, presumably because they have less need of them. Fish, for example, have considerably lower levels of such enzymes than omnivorous terrestrial animals and birds (Chapter 2; Walker, 1978, 1980).

Many of the lipophilic pollutants described here are not persistent along food chains because they are readily biodegradable by terrestrial vertebrates. The main exceptions are polyhalogenated compounds such as some organochlorine insecticides, higher halogenated PCBs and PBBs, PCDDs and polychlorinated dibenzofurans (PCDFs). As mentioned earlier, the main mechanism of primary metabolism of these compounds is by P450-catalysed attack. The problem is that highly halogenated compounds are resistant to such oxidative attack. Thus, they present a general problem of persistence and have long biological half-lives in most animals, terrestrial or aquatic. They are, therefore, liable to undergo biomagnification with movement along food chains, and also to survive transportation over long distances in migrating animals (e.g. fish or whales) and birds. Some organometallic compounds, e.g. methyl mercury, are also very persistent in terrestrial vertebrates and are only slowly metabolised.

Some compounds that are readily and rapidly metabolised and excreted by terrestrial vertebrates only degrade very slowly in other species, e.g. marine invertebrates such as bivalve molluscs. The general question of major differences between species and groups in metabolic capacity will be discussed in section 4.2. At this stage the critical point is that there are compounds, such as PAHs, that tend to bioconcentrate/ bioaccumulate in certain species low in aquatic food chains but are rapidly metabolised by fish, mammals and birds occupying higher trophic levels. Thus, unlike many polyhalogenated compounds, they are not biomagnified as they move to the top of the food chain. This illustrates the limitations of using standard laboratory species such as rats, mice or Japanese quail as metabolic models in ecotoxicology. There are many chemicals that are rapidly metabolised by these species which are recalcitrant and consequently persistent in certain aquatic invertebrates (Livingstone, 1991; Walker and Livingstone, 1992).

The biotransformation of lipophilic pollutants into water-soluble and readily excretable products represents the main mechanism for their elimination by terrestrial animals. Its effectiveness, however, depends on the availability of the pollutants to the relevant enzymes. Where pollutants are stored in fat reserves, they usually become available to the liver in the course of time with the mobilisation of lipids. Rates of elimination depend on rates of mobilisation of fat reserves. The very strong binding of pollutants to proteins can severely limit the availability of the chemicals to enzyme systems or for direct excretion and can result in very long biological half-lives. Examples include the binding of superwarfarins to liver proteins (Chapter 11) and the binding of hydroxymetabolites of PCBs to plasma proteins (see section 6.2.4). Such long-term binding is limited by the number of available binding sites, and in the examples given only relates to the persistence of low concentrations of chemical. Higher concentrations are freely available for metabolism/excretion.

The refractory nature of some pollutants, notably persistent polyhalogenated compounds, has raised problems of bioremediation of contaminated sites, e.g. sediments, dumping sites. There has been interest in the identification, or the production by genetic manipulation, of strains of microorganisms that can metabolically degrade recalcitrant molecules. For example, there are bacterial strains that can reductively dechlorinate PCBs under anaerobic conditions. The chemical industry has moved towards the design of more readily biodegradable products in the interests of environmental quality, e.g. the production of 'soft' detergents and readily biodegradable pesticides (see Chapters 10 and 12).

3.5 *Summary*

The environmental fate of chemicals is determined by both chemical/physical and biological processes; in turn, the operation of these processes is dependent upon the properties of the environmental chemicals themselves. Polarity, vapour pressure, partition coefficients and chemical stability are all determinants of movement and distribution in the physical environment. Here, constants such as vapour pressure, partition coefficients (e.g. K_{ow}) and fugacity values have been incorporated into models that describe or predict environmental fate. In these relatively simple situations biological factors are largely ignored. With movement along food chains, or fate in soils or sediments, biological factors such as uptake, metabolism and excretion are critically important and need to be taken into account if developing models. These complications will be dealt with in the next chapter.

3.6 *Further reading*

Crosby, D.G (1998) *Environmental Toxicology and Chemistry*. A very readable teaching text with many useful examples, which has the advantage of linking chemistry to toxicology.

Schwarzenbach, R. P. *et al.* (1993) *Environmental Organic Chemistry*. An authoritative text on basic principles.

CHAPTER 4

Distribution and effects of chemicals in communities and ecosystems

4.1 Introduction

In Chapter 3 the distribution of environmental chemicals through compartments of the gross environment was related to the chemical factors and processes involved. In the present chapter, the discussion moves on to the more complex question of movement and distribution in communities and ecosystems where biological as well as physical and chemical factors are involved. The movement of chemicals along food chains and the fate of chemicals in the complex communities of sediments and soils are basic issues here.

Ecotoxicology deals with the study of the harmful effects of chemicals in ecosystems. This includes harmful effects upon individuals, although the ultimate concern is about how effects on individuals are translated into changes at the levels of population, community and ecosystem. In the concluding sections of the chapter, emphasis will move from the distribution and environmental concentrations of pollutants to the effects that they have at the levels of individual, population, community and ecosystem. Dose–response relationships are of fundamental interest in ecotoxicology, and **biomarker assays** represent a practical development of the concept that can provide a measure of harmful effects under field conditions.

4.2 *Movement of pollutants along food chains*

Persistent organic pollutants are compounds that have sufficiently long half-lives in living organisms for them to pass along food chains and to undergo biomagnification in higher trophic levels. Some compounds such as PAHs (Chapter 9) can be bioconcentrated/bioaccumulated in lower trophic levels but are rapidly metabolised by vertebrates in higher levels. These will not be discussed further here, where the issue is biomagnification with movement along the entire food chain. The best-studied examples of this are the organochlorine insecticides dieldrin and p,p'-DDE (see Chapter 5) and PCBs (see Chapter 6), where concentrations in the tissues of predators of the highest trophic levels can be 10^4- to 10^5-fold higher than in organisms in the lowest trophic levels. Other examples include PCDDs, PCDFs and some organometallic compounds (e.g. methyl mercury).

Biomagnification along terrestrial food chains is principally due to biaccumulation from food, the principal source of most pollutants (Walker, 1990b). In a few instances, the major route of uptake may be from air, from contact with contaminated surfaces or from drinking water.

The bioaccumulation factor (BAF) of a chemical is given by the following equation:

Concentration in organism/concentration in food = BAF

Biomagnification along aquatic food chains may be the consequence of bioconcentration as well as bioaccumulation. Aquatic vertebrates and invertebrates can absorb pollutants from ambient water; bottom feeders can take up pollutants from sediments. The bioconcentration factor (BCF) of a chemical absorbed directly from water is defined as:

Concentration in organism/concentration in ambient water = BCF

One of the challenges when studying biomagnification along aquatic food chains is establishing the relative importance of bioaccumulation versus bioconcentration.

The processes that lead to biomagnification have been investigated with a view to developing predictive toxicokinetic models (Walker, 1990b). When organisms are continuously exposed to pollutants maintained at a fairly constant level in food and/or in ambient water/air, tissue concentrations will increase with time until either (1) a lethal concentration is reached and the organism dies *or* (2) a steady state is reached when the rate of uptake of the pollutant is balanced by the rate of loss. The BCF or BAF at the steady state is of particular interest and importance because (1) it represents the highest value that can be reached, and therefore indicates the maximum risk; (2) it is not time dependent; *and* (3) the rates of uptake and loss are equal, thereby facilitating the calculation of the rate constants involved.

BCFs and BAFs measured before the steady state is reached have little value because they are dependent on the period of exposure of the organism to the chemical and may greatly underestimate the degree of biomagnification that is possible. This statement should be qualified by the reservation that there may be situations in which the duration of exposure cannot be long enough for the steady state to be reached, e.g. where the life stage of an insect is very short. The principal processes of uptake and loss by different types of organisms are indicated in Table 4.1.

A rough indication of the relative importance of different mechanisms of uptake and loss is indicated by a scoring system on the scale $+ \rightarrow ++++$. Within each category of organism there will be differences between compounds in the relative importance of different mechanisms, e.g. because of differences in polarity and biodegradability. The main points to bring out are:

1 that uptake and loss by exchange diffusion from ambient water is important for aquatic organisms but not for terrestrial ones;
2 that metabolism is the main mechanism of loss in terrestrial vertebrates but is less important in fish, which can achieve excretion by diffusion into ambient water;
3 that most aquatic invertebrates have very little capacity for metabolism; this is particularly true for molluscs. Crustaceans, e.g. crabs and lobsters, appear to have greater metabolic capability than molluscs (see Livingstone and Stegeman, 1998; Walker and Livingstone, 1992).

The balance between competing mechanisms of loss in the same organism depends on the compound and the species in question. In fish, for example, some compounds that are good substrates for monooxygenases, hydrolases, etc., can be metabolised relatively rapidly, even though fish, as a group, have relatively low metabolic capacity (Chapter 2). Thus, in this case, metabolism as well as diffusion is an important factor determining the rate of loss. By contrast, many polyhalogenated compounds are only metabolised very slowly by fish, so metabolism does not make a significant contribution to detoxication, and loss by diffusion is the dominant mechanism of elimination.

TABLE 4.1 *Principal mechanisms of uptake and loss for lipophilic compounds*

Habitat/type of organism	Mechanisms of uptake			Mechanisms of loss	
	Diffusion	From food	From ingested water	Diffusion	Metabolism
Aquatic					
Molluscs	++++	+		++++	
Fish	++++	$+ \rightarrow +++$		++++	$+ \rightarrow +++$
Terrestrial					
Vertebrates		++++	< ++		++++

Some further aspects of detoxication by fish need to be briefly mentioned. When fish inhabit polluted waters, exchange diffusion occurs until a steady state is reached where no net loss will occur until the concentration in water falls. When a recalcitrant pollutant is acquired from prey, digestion can lead to the tissue levels of that pollutant temporally exceeding those originally existing while in the steady state. Here, diffusion into the ambient water may provide an effective excretion mechanism in the absence of effective metabolic detoxication. Seen from an evolutionary point of view, the requirements of fish for metabolic detoxication would appear to have been limited, because loss by diffusion would often have prevented tissue levels becoming too high. The poor metabolic detoxication systems of fish relative to those of terrestrial omnivores and herbivores are explicable on these grounds (Chapter 2). However, the advent of refractory organic pollutants that combine high toxicity with high lipophilicity has exposed the limitations of existing detoxication systems of fish. The very high toxicity of compounds such as dieldrin and other cyclodiene insecticides to fish was soon apparent, with fish deaths occurring at very low concentrations in water (see Chapter 5), and metabolically resistant strains of fish being reported in polluted rivers such as the Mississippi. More rapid elimination was needed than could be provided by passive diffusion in order to prevent tissue concentrations reaching toxic levels.

Some models for predicting bioconcentration and biomagnification are presented in Box 4.1.

4.3 *Fate of pollutants in soils and sediments*

Regarding soils, a central issue is the persistence and movement of pesticides that are widely used in agriculture. Many different insecticides, fungicides, herbicides and molluscicides are applied to agricultural soils, and there is concern not only about the effects that they may have on non-target species residing in soil, but also on the possibility of the chemicals finding their way into adjacent water courses.

Soils are complex associations between living organisms and mineral particles. Decomposition of organic residues by soil microorganisms generates complex organic polymers ('humic substances' or simply 'soil organic matter'), which bind together mineral particles to form aggregates that give the soil its structure. Soil organic matter and clay minerals constitute the colloidal fraction of soil; because of their small size they present a large surface area in relation to their volume. Consequently, they have a large capacity to adsorb the organic pollutants that contaminate soil. Within a freely draining soil there are air channels and soil water, the latter being closely associated with solid surfaces. Depending on their physical properties, organic compounds become differentially distributed between the three phases of soil, soil water and soil air.

Hydrophobic compounds of high K_{ow} become very strongly adsorbed to soil colloids (section 3.2) and consequently tend to be immobile and persistent. Organochlorine

BOX 4.1 *Models for bioconcentration and bioaccumulation.*

As indicated in Table 4.1 aquatic molluscs present a relatively simple picture because they have little capacity for biotransformation of organic pollutants, the principal mechanism of uptake being diffusion. It is not surprising, therefore, that BCFs for diverse lipophilic compounds, measured at the steady state, are related linearly to log K_{ow} values (Figure 4.1). Thus, the more hydrophobic a compound is, the more it tends to partition from water into the lipids of the mollusc. The relationship shown in Figure 4.1 has been demonstrated in several species of aquatic molluscs, including the edible mussel (*M. edulis*), the oyster (*Crassostrea virginica*) and soft clams (*Mya arenaria*) (Ernst, 1977). A similar relationship has also been found with rainbow trout and other fish for some pollutants. On the other hand, some organic pollutants do not fit the model well (Connor, 1983). It seems probable that some compounds, which are metabolised relatively rapidly by fish, will be eliminated faster than would be expected if diffusion were the only process involved (Walker, 1987). Such compounds would not be expected to follow closely a model for BCFs based on K_{ow} values alone. This point aside, K_{ow} values can give a useful prediction of BCF values at the steady state for lipophilic pollutants in aquatic invertebrates. A great virtue of the approach is that K_{ow} values are easy and inexpensive to measure or predict (Connell, 1994). Other more complex and sophisticated models have been developed for fish (see, for example, Norstrom *et al.*, 1976) but are too time-consuming/expensive to be used widely in environmental risk assessment where cost-effectiveness is critically important.

Modelling for bioaccumulation by terrestrial animals presents greater problems, and BAFs cannot be reliably predicted from K_{ow} values (Walker, 1987). For example, benzo(*a*)pyrene and dieldrin have log K_{ow} values of 6.50 and 5.48, respectively, but their biological half-lives range from a few hours in the case of the former to 10–369 days for the latter! Endrin is a stereoisomer of dieldrin with a similar K_{ow}, but it has a half-life of only 1 day in man compared with 369 days for dieldrin. These large differences in persistence are attributed to differences in the rate of metabolism by P450-based monooxygenases (Walker, 1981). Effective predictive models for bioaccumulation of strongly lipophilic compounds by terrestrial animals need to take account of rates of metabolic degradation. This is not a straightforward task, and would require the sophisticated use of enzyme kinetics to be successful. In one model, it has been suggested that Lineweaver–Burke plots for microsomal metabolism might be used to predict BAF values in the steady state (Walker, 1987) (Figure 4.2). In principle, when an animal ingests a lipophilic compound at a constant rate in its food, a steady state will eventually be reached where the rate of intake of the compound is balanced by the rate of its metabolism. It is assumed that the rate of loss of the unchanged compound by direct excretion is negligible. Primary metabolic attack upon many highly lipophilic compounds takes place predominantly in the endoplasmic reticulum, particularly that of the liver in vertebrates. Thus, microsomes (especially hepatic microsomes of vertebrates) can serve as model systems for measuring rates of enzymic detoxication. Lineweaver–Burke and similar metabolic plots can relate concentrations of pollutants in microsomal membranes to rates of metabolism. In the steady state, the rate of intake of a chemical should equal the rate of metabolism in the membranes of the

BOX 4.1 *Continued*

endoplasmic reticulum. The concentration of the chemical required in the membranes to give this balancing metabolic rate can be estimated from the Lineweaver–Burke plot. The necessary balancing metabolic rate can be calculated from the defined rate of intake in food, and then the microsomal concentration that will give this rate can be read from the plot. Thus, the concentration in the endoplasmic reticulum can be compared with the dietary concentration to give an estimate of BAF. Estimates can also be made of BAF for the liver, or BAF for the whole body; to do this one needs the approximate ratios of concentrations of the chemical in different compartments of the body when at the steady state.

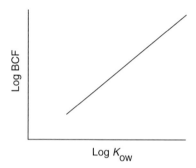

FIGURE 4.1 *Relationship of BCF to log* K_{ow}.

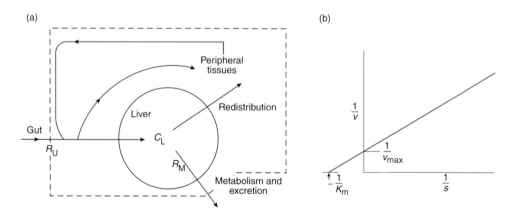

FIGURE 4.2 *(a) A bioaccumulation model for terrestrial organisms. A kinetic model for liver.* R_U, *rate of uptake from the gut;* R_M, *rate of metabolism in liver;* C_L, *concentration of pollutant in liver. The arrows indicate the routes of transfer of pollutant within the animal. The rates of uptake and metabolism are expressed in terms of kilograms of body weight. The final elimination of water-soluble products (metabolites and conjugates) is in the urine. (b) Lineweaver–Burke plot to estimate the bioaccumulation factor.* V_{max} *and* v *are expressed as milligrams of pollutant metabolised per kilogram of body weight per day. S is expressed as the concentration of pollutant, ppm by weight (either in terms of grams of liver or milligrams of hepatic microsomal protein). After Walker (1987).*

insecticides such as DDT and dieldrin are good examples of hydrophobic compounds of rather low vapour pressure that have long half-lives – sometimes running into years – in temperate soils (Chapter 5). Because of their low water solubility and their refractory nature, the main mechanism of loss from most soils is by volatilisation. Metabolism is limited by two factors: (1) being tightly bound, they are not freely available to enzymes of soil organisms that can degrade them; and (2) they are, at best, only slowly metabolised by enzyme systems. Because of strong adsorption and low water solubility, there is little tendency for them to be leached down the soil profile by percolating water. The degree of adsorption, and consequently the persistence and mobility, is also dependent on soil type. Heavy soils, high in organic matter and/ or clay, adsorb hydrophobic compounds more strongly than light sandy soils, which are low in organic matter. Strongly lipophilic compounds are most persistent in heavy soils. When organochlorine insecticides are first incorporated into soil, before they become extensively adsorbed to soil colloids, they are lost relatively rapidly, mainly as a result of volatilisation (Figure 4.3). With time, however, most residual organochlorine insecticide becomes adsorbed, and subsequently there is a period of very slow exponential loss.

In marked contrast to hydrophobic compounds, more polar ones tend to be less adsorbed and to reach relatively high concentrations in soil water. Phenoxyalkanoic acids such as 2,4-D and 2-methyl 4-chloro-phenoxyacetic acid (MCPA) are good examples (Figure 4.3). Their half-lives in soil are measured in weeks rather than in years, and they are more mobile in soils than are organochlorine insecticides. When first applied they are lost only slowly. After a lag period of a few days, however, they disappear very rapidly as a consequence of metabolism by soil microorganisms. This has been explained on the grounds that it takes time for a build-up in numbers of strains of microorganisms that can metabolise them; these microorganisms use the herbicides as an energy source. It has also been suggested that the lag period relates to the time it takes for enzyme induction to occur. Whatever the explanation, soils treated with these compounds stay enriched for a period, and further additions of the original compounds will be followed by rapid metabolism without a lag phase. If, however, the soils are untreated for a long period they will revert to their original state and not show any enhanced capacity for degrading the herbicides. An important difference from organochlorine insecticides and related hydrophobic pollutants is that, because of their polarity and water solubility, they are freely available to the microorganisms that can degrade them. Interestingly, the phenoxyalkanoic acid 2,4,5-T (2,4,5-trichlorophenoxyacetic acid) is more persistent than either 2,4-D or MCPA. With three substituted chlorines in its phenyl ring, it is metabolised less readily than the other two compounds; it would appear that metabolism is a rate-limiting factor.

It was long assumed that there is little tendency for pesticides or other organic pollutants to move through soil into drainage water. Indeed, this is to be expected with intact soil profiles. Hydrophobic compounds will be held back by adsorption, whereas water-soluble ones will be degraded by soil organisms. Some soils, however, depart from this simple model. Soils high in clay can crack and develop deep fissures

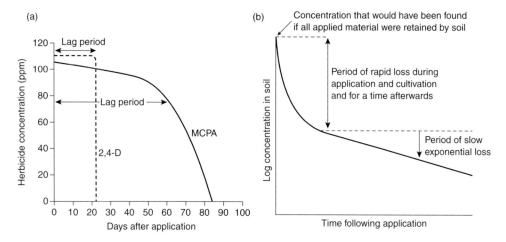

FIGURE 4.3 *Loss of pesticides from soil. (a) Breakdown of herbicides in soil. (b) Disappearance of persistent organochlorine insecticides from soils. From Walker et al. (2000).*

during dry weather. If rain then follows, pesticides, in solution or adsorbed to mobile colloids, can be washed down through the fissures and appear in neighbouring drainage ditches and streams. This was found to happen with pesticides such as carbofuran, isoproturon and chlorpyriphos in the Rosemaund experiment, conducted in England during the period 1987–93 (Williams *et al.*, 1996). The influence of polarity on movement of chemicals down through the soil profile has been exploited in the selective control of weeds using soil herbicides (Hassall, 1990). In general, the more polar and water soluble the herbicide, the further it will be taken down into the soil by percolating water. Insoluble herbicides such as the triazine compound simazine (water solubility 3.5 ppm) remain in the first few centimetres of soil when applied to the surface. More water-soluble compounds, such as the urea herbicides diuron and monuron (water solubilities 42 ppm and 230 ppm respectively), are more mobile and can move further down the soil profile. Selective weed control can be achieved by judicious use of this range of herbicides ('depth selection'). When applied to the soil surface, simazine should only be toxic to shallow-rooting weeds and should not affect crops that root further down. The other herbicides can give weed control to greater depths in situations where the rooting systems of the crops are far enough down. When attempting depth selection in weed control, account needs to be taken of soil type. Herbicides will move further down the profile in light sandy soils than in heavy clay or organic soils.

Although the major concern about fate of organic pollutants in soil has been about pesticides in agricultural soils, other scenarios are also important. The disposal of wastes on land, e.g. at landfill sites, has raised questions about movement of pollutants contained in the wastes into the air or neighbouring rivers or water courses. The presence of PCBs or PAHs in such wastes can be a significant source of pollution. Likewise, the disposal of some industrial wastes in landfill sites, e.g. by Chemical Industry, raises questions about movement into air or water and needs to be carefully controlled and monitored. Currently, there is debate about the most appropriate means of disposal of chemical weapons on land.

In certain respects **sediments** resemble soils. Sediments also represent an association between mineral particles, organic matter and resident organisms. The main difference is that they are situated underwater and are, in varying degrees, anaerobic. The oxygen level influences the type of organisms and the nature of biotransformations that occur in sediments. A feature with sediments, as of soils, is the limited availability of chemicals that are strongly adsorbed. Again, compounds with high K_{ow} values tend to be strongly adsorbed, relatively unavailable and highly persistent. There is much interest in the question of sediment toxicity (Burton, 1992), and the availability to bottom-dwelling organisms of compounds adsorbed by sediments (Hill *et al.*, 1993). One case in point is pyrethroid insecticides (see Chapter 12), which are strongly retained in sediments on account of their high K_{ow} values. Although, because of their ready biodegradability, they are not biomagnified in aquatic food chains, they are available to bottom-dwelling organisms low in the food chain. Questions are asked about the possible long-term build-up of pyrethroids in sediments and their effects on organisms in lower trophic levels.

4.4 *Effects of chemicals at the population level: population dynamics*

In environmental risk assessment, the objective is to establish the likelihood of a chemical (or chemicals) expressing toxicity in the natural environment. Assessment is based on a comparison of ecotoxicity data from laboratory tests with estimated or measured exposure in the field. The question of effects at the level of population or above that may be the consequence of such toxicity is not addressed. This issue will now be discussed.

Toxic effects upon individuals in the field may be established and quantified in a number of ways. Concerning lethal effects, a simple approach is to collect and count corpses after the application of a chemical, as in field trials with new pesticides. This is an imprecise technique because many individual casualties will escape detection – especially with mobile species such as birds. With very stable pollutants such as dieldrin, heptachlor epoxide, p,p'-DDT and p,p'-DDE the determination of residues in carcasses found in the field can provide evidence of lethal toxicity in the field. Such data may also be used to obtain estimates of the effects of chemicals upon mortality rates of field populations, which can then be incorporated into population models (see section 5.3.5).

Biomarker assays for toxic effect may also be used to quantify toxic effects in the field. When carcasses are in a good enough condition, relatively stable biomarker assays such as cholinesterase inhibition may be used to establish the cause of death. Of greater potential is the use of biomarker assays to monitor the effects of chemicals on living organisms in the field. Here, biomarker assays can provide measures of sublethal toxic effects. Ideally, these should be non-destructive, to allow serial sampling of individuals.

The present section will be limited to the question of effects on populations that are a direct consequence of toxicity to individuals, be it lethal or sublethal. The problem of indirect effects will be touched upon, briefly, when discussing effects at the level of community or ecosystem (section 4.6). It should be emphasised, however, that the reduction in numbers of one species caused by toxic chemicals will have knock-on effects in an ecosystem. The state of a population can be expressed in terms of numbers (population density) or genetic composition (population genetics). This section will deal with questions relating to population dynamics; the next will deal with population genetics.

The population density of animals in the natural environment is far from constant. Seasonal fluctuations are normal, with an increase in numbers during and immediately after breeding, but a decline between this time and the next breeding occasion. In temperate climates, most breeding occurs during the warm season, when food is most readily available. Numbers fall during the cold season, when there is a shortage of food.

Given these complications, it is understandable that ecologists and ecotoxicologists are particularly interested in the growth rate of populations (r). Population growth rate is defined as the population increase per unit time, divided by the number of individuals in the population (see Chapter 12; Gotelli, 1998; Walker *et al.*, 2000) Population growth rate may be positive, negative or zero. When $r = 0$, the rates of recruitment and of mortality are equal, and the population density in the field represents 'carrying capacity'. In the field, population numbers are determined by, among other things, density-dependent factors such as availability of food, water or breeding sites. Population density in the field cannot exceed, for any appreciable period, the carrying capacity. When investigating the effects of factors such as pollutants on population numbers, an important question is whether they bring population numbers below carrying capacity. A pollutant may reduce survivorship or reproductive success, but this does not necessarily reduce population numbers below those which would normally be maintained by density-dependent factors.

In a general way, it can be stated that the population density of an animal depends on the balance between the rate of recruitment and the rate of mortality. In the context of ecotoxicology, the influence of pollutants upon either of these factors is of fundamental interest and importance. When a population is at or near its carrying capacity these two factors are in balance, and the critical question about the effects of pollutants is whether they can adversely affect this balance, and bring a population decline. The population growth rate (r) of an organism can be calculated from the Euler–Lotka equation:

$$1 = \tfrac{1}{2} n_1 l_1 e^{-rt_1} + \tfrac{1}{2} n_2 l_2 e^{-rt_2} + \tfrac{1}{2} n_3 l_3 e^{-rt_3} + \ldots$$

where t_1 is the age of first breeding, t_2 is the age of second breeding, t_3 is the age of third breeding, etc., l_1 is the probability of the female surviving to the age of first breeding, l_2 the probability of reaching age of second breeding, etc., n_1 is the number of offspring produced at the first breeding, n_2 the number of offspring produced at the second breeding, etc.

In predicting the effects of a pollutant on population growth rate the effects of the chemical on values for t, l and n are of central interest. Chemical residue data and biomarker assays that provide measures of toxic effects are relevant here because they can, in concept, be used to relate the effects of a chemical upon the individual organism to a population parameter such as survivorship or fecundity (Figure 4.4). Examples of this are discussed in the second part of the text, including the reduction of survivorship of sparrowhawks caused by dieldrin (Chapter 5), the reduction in breeding success of raptors caused by p,p'-DDE (Chapter 5) and the reduction in fecundity of dog whelks caused by TBT (Chapter 8). The measurement of responses of free-living organisms to pollutants in the field using appropriate biomarker assays can be a prelude to estimating the consequent effects on t, l or n from graphs of the type shown in Figure 4.4. The estimated values can then be incorporated into equations, such as the one shown above, to predict the effects of the chemical on r. Such models can be tested in field studies or in mesocosm studies to establish their validity. It is also possible, in concept, to utilise biomarker responses determined in the laboratory to predict the effects at the population level of known or predicted exposures to particular chemicals in the field. In the simplest case, such an approach would assume that the relationship between the biomarker response and the consequent change in the population parameter would be similar in the field and the laboratory. However, this approach would need to be rigorously tested in the field, because the relationship might be substantially different between the laboratory and the field.

The development of models incorporating biomarker assays to predict the effects of chemicals upon parameters related to r has obvious attractions from a scientific

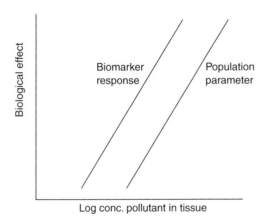

FIGURE 4.4 *Schematic diagram of the relationship between biomarker response and the change in population parameter.*

point of view, and it is preferable, in theory, to the crude use of ecotoxicity data currently practised in environmental risk assessment. However, the development of this approach would involve considerable investment in research, and might prove too complex and costly to be widely used in environmental risk assessment.

4.5 *Effects of pollutants on population genetics*

As explained in Chapter 1, the toxicity of 'natural' xenobiotics has exerted a selection pressure upon living organisms since very early in evolutionary history. There is abundant evidence of compounds produced by plants and animals that are toxic to species other than their own and which are used as chemical warfare agents (Chapter 1). It is only very recently that organic compounds synthesised by humans have begun to exert a selection pressure upon natural populations with the consequent emergence of resistant strains. Pesticides are a prime example, and they will be the principal subject of the present section. It should be mentioned, however, that other types of biocides, e.g. antibiotics and disinfectants, can have a similar effect, albeit largely restricted to organisms that cause disease in humans.

The large-scale use of pesticides commenced in developed countries after the Second World War. In due course they came to be more widely used in developing countries, notably for the large-scale control of important vectors of disease such as malarial mosquitoes and tsetse flies. With the continuing use of pesticides, problems of resistance began to emerge. The emergence of strains of pest species possessing genes that confer resistance were an inevitable consequence of the continuing selection pressure of the pesticides. Indeed, this can be seen to mirror the development of defence systems towards natural toxins much earlier in evolutionary history. It is highly probable that many of the defence systems now emerging in resistant strains were originally developed to combat natural toxins (e.g. P450 forms of some resistant strains of insects).

The development of resistant strains of pest species of insects has been intensively studied (for sound economic reasons!), and there are many good examples. For further information see Brown (1971), Georghiou and Sato (1983), McCaffery (1998) and Oppenoorth and Welling (1985). Examples of mechanisms of insect resistance are given in Table 4.2.

The levels of resistance (RF values) developed by insects towards insecticides in the field can be as high as several hundred-fold relative to susceptible strains of the same species. Such high levels of resistance usually lead to loss of effective control of the pest. Broadly speaking, resistance mechanisms are of two kinds: (1) those that depend on toxicokinetic factors such as reduced uptake, increased metabolism or increased storage; and (2) those that depend on toxicodynamic factors, principally alterations in the site of action that lead to decreased sensitivity to the insecticide. Taking these two kinds of mechanism in turn.

TABLE 4.2 *Examples of resistance of insects to insecticides*

Insecticide	Species	Strain (RF)	Mechanism	Comment
Cypermethrin (*cis* isomers)	*Heliothis virescens*	PEG 87 (70 000+)	P450 (major) Altered NaCh (minor)	Resistance sensitive to PBO
Cypermethrin (*cis* isomers)	*H. virescens*	Field strains (85–315)	Altered NaCh (principal)	Little metabolic resistance
Parathion Methyl parathion	*H. virescens*	Field strains	Altered ChE	
OPs	*Myzus persicae*	Resistant clones	Enhanced 'B' esterase	Multiple copies of gene
Cyclodienes	*H. virescens*	Field strains	Altered GABA	
DDT	*Musca domestica*	Many strains	Enhanced DDT-ase	DDT-ase inhibitors reduce resistance
Carbamates	Several	Various (< 200)	P450	PBO reduces resistance

NaCh, sodium channel; RF, resistance factor, which is LD_{50} R strain/LD_{50} S strain; GABA, gamma-aminobutyric acid receptor; PBO, piperonyl butoxide.

Data from McCaffery (1998), Oppenoorth and Welling (1976) and Devonshire (1991).

1 Resistance mechanisms associated with changes in toxicokinetics are very largely cases of enhanced metabolic detoxication. With readily biodegradable insecticides such as pyrethroids and carbamates, enhanced detoxication by P450-based monooxygenase is a common resistance mechanism (Table 4.2). The existence of this type of resistance can often be established by toxicity studies with synergists. Inhibitors of P450 forms such as piperonyl butoxide (PBO) can substantially reduce the level of resistance shown by the resistant strains. OP resistance is sometimes due to enhanced esterase activity, as in the case of a series of clones of the peach potato aphid that possess multiple copies of a gene encoding for a carboxyl esterase (see Devonshire, 1991; and section 10.2.2 of this text) A specialised example of metabolic resistance is that shown by houseflies (*M. domestica*) and other insects to DDT. Resistant strains have elevated levels of DDT-dehydrochlorinase, a form of glutathione transferase that is able to dehydrochlorinate some organochlorine insecticides. Metabolic resistance has been attributed to elevated levels of glutathione-S-transferases in the case of diazinon and some other OPs (Brooks, 1972; Oppenoorth and Welling, 1976).

Metabolic resistance may be the consequence of the appearance of a novel gene on the resistant strain which is not present in the general population; it may also be due to the presence of multiple copies of a gene in different strains or clones as in the example of OP resistance in the peach potato aphid mentioned above.

2 Many cases of resistance are due to the existence of insensitive forms of the target site in the resistant strain (Table 4.2). The change of a single amino acid residue due to mutation can be enough to radically alter the affinity of an insecticide for its active site. For example, the replacement of a single leucine residue by phenylalanine in the sodium channel of the housefly can radically reduce the effectiveness of DDT or pyrethroids (Devonshire *et al.*, 1998; Salgado, 1999) The same substitution on the sodium channel protein is also found in resistant strains of the diamondback moth (*Plutella xylostella*), the German cockroach (*Blatella germanica*) and peach potato aphid. Resistance to OPs and carbamates is sometimes due to the presence of altered forms of AChE in the resistant strains. Again, the substitution of a single amino acid residue by another can bring resistance. This type of resistance has been reported in several species of insects as well as in some red spider mites and cattle ticks (Oppenoorth and Welling, 1976). The forms of AChE in resistant insects are discussed in section 10.2.5. Resistance to cyclodiene insecticides such as dieldrin and endosulfan has been related to the presence of an altered form of the GABA receptor in resistant strains of insects. Dieldrin, heptachlor epoxide and other active forms of cyclodiene insecticides are refractory, and it may well be that metabolic resistance is unlikely to arise – leaving alteration of the target site as the main, if not the only, viable type of resistance mechanism.

In some resistant strains both types of resistance mechanism have been shown to operate against the same insecticide. Thus, the PEG 87 strain of the tobacco bud worm (*H. virescens*) is resistant to pyrethroids on account of both a highly active form of cytochrome P450 and an insensitive form of the sodium channel (Table 4.2 and McCaffery, 1998).

Apart from the resistance of insects to insecticides, resistance has been developed by plants to herbicides, fungi to fungicides and rodents to rodenticides. Rodenticide resistance is discussed in section 11.6.

4.6 *Effects of pollutants upon communities and ecosystems: the natural world and model systems*

The state of communities and ecosystems can be established according to both structural and functional parameters (see Chapter 14 in Walker *et al.*, 2000). Functional analyses include the measurement of nutrient cycling, turnover of organic residues, energy flow and 'niche-metrics'. Structural analyses include the assessment of species present, their population densities and their genetic composition.

Changes in the composition of some communities and ecosystems are relatively easy to measure and to monitor using biotic indices (ecological profiling). One well-established system is the River Invertebrate Prediction and Classification System (RIVPACS) used in Britain to assess the quality of rivers (Wright, 1995). Macroinvertebrate profiles have been established for normal unpolluted rivers of diverse kinds, which are used as standards. Pollution can cause departures from these profiles, e.g. because of the removal of sensitive species by direct toxicity. Another system of this kind is the Invertebrate Community Index, now receiving much attention in the USA, which gives a quantitative index of the structure and function of aquatic invertebrate communities. A further system, used in some US states as part of a regulatory mechanism, is the Index of Biotic Integrity (IBI) for aquatic communities (Karr, 1981).

Biotic indices that are relatively simple and inexpensive to apply can be very useful for identifying environmental problems caused by pollutants. The problem can be, however, identifying which pollutants – or other environmental factors! – are responsible for significant departures from normality. This dilemma illustrates well the importance of having both a 'top down' and a 'bottom up' approach to pollution problems in the field. Chemical analysis, and the use of biomarker assays, can be used to identify chemicals responsible for adverse changes in communities detected by the use of biotic indices.

When new pesticides are developed, their effects upon soil communities are tested. Typically these tests use functional parameters, e.g. generation of CO_2 or nitrification (Somerville and Greaves, 1987). Many effects shown on soil communities are of short duration and are thought to lie within the range of normal fluctuations in soil processes.

In the quest for better methods of establishing the environmental safety (or otherwise!) of chemicals, interest has grown in the use of microcosms and mesocosms: artificial systems in which the effects of chemicals on populations and communities can be tested in a controlled way, with replication of treatments. Mesocosms have been defined as 'bounded and partially enclosed outdoor units that closely resemble the natural environment, especially the aquatic environment' (Crossland, 1994). Microcosms are smaller, and less complex, multispecies systems. They are less similar to 'the real world' than are mesocosms. Experimental ponds and model streams are examples of mesocosms. The effects of chemicals at the levels of population and community can be tested in mesocosms, although the extent to which such effects can be related to events in the natural environment is questionable. Although mesocosms have been developed both by industrial and government laboratories, there is uncertainty about the interpretation of the results obtained with them. Results coming from mesocosm tests are not yet of much use in environmental risk assessment. However, refinement of techniques could make mesocosms more valuable for risk assessment (see next section).

4.7 *New approaches to predicting ecological risks presented by chemicals*

As discussed above, current risk assessment practices do not deal with the fundamental question about possible effects at the level of population and above. In the foregoing sections consideration was given to ways in which effects at these higher levels might be identified – and even predicted – by using data from field studies, laboratory studies and mesocosms. At the fundamental level, the use of population models that can predict population growth rate (r) has obvious attractions. The incorporation of data from field and laboratory studies into such models should, in principal, allow reasonable predictions to be made of the effects of defined environmental levels of chemicals upon populations. The critical roles of **biomarker assays** and residue data in establishing (1) the relationship between dose and toxic effect and (2) the relationship between toxic effect and population change has already been emphasised here and in the wider literature (Huggett *et al.*, 1992; Peakall, 1992; Peakall and Shugart, 1993; Fossi and Leonzio, 1994; Walker *et al.*, 1998). In the forthcoming sections, examples will be given where this approach has already been successful in the retrospective investigation of pollution problems. The effects of TBT on dog whelks (section 8.3.4), dieldrin on sparrowhawks (section 5.3.5) and p,p'-DDE on peregrines and bald eagles (section 5.2.5) are all cases in point.

The more difficult thing is to develop models that can, with reasonable confidence, be used to predict ecological effects. A detailed discussion of ecological approaches to risk assessment lies outside the scope of the present text. For further information see Suter (1993), Landis *et al.* (1998) and Peakall and Fairbrother (1998). There is also the question of refining the testing protocols used in mesocosms. Mesocosms could have a more important role in environmental risk assessment if the data coming from them could be better interpreted. The use of biomarker assays to establish toxic effects and, where necessary, to relate them to effects produced by chemicals in the field might be a way forward. The issues raised in this section will be returned to in Chapter 15, after consideration of the individual examples given in Part 2.

4.8 *Summary*

The movement of organic pollutants along food chains and their fate in soils and sediments depend on biological as well as chemical factors. The chemical and biochemical properties of pollutants determine the rates at which they move between compartments of the environment, cross membranous barriers or undergo chemical or biochemical degradation. Highly lipophilic compounds with high K_{ow} values are of particular concern because they tend to be immobile and persistent in soils and sediments. Where they are chemically stable and resist metabolic degradation they tend to be biomagnified in food chains, reaching relatively high concentrations in top predators. Examples include persistent organochlorine insecticides and PCBs, PCDDs, PCDFs and methyl mercury.

In ecotoxicology, the largest concern is with effects of organic pollutants at the levels of population, community and ecosystem. Population effects may be on numbers (population dynamics) or on gene frequencies (population genetics). In communities and ecosystems there may be effects on structure and function. The potential use of population models incorporating biomarker data for studying pollutant effects is discussed.

4.9 *Further reading*

Newman, M. C. (1996) *Fundamentals of Ecotoxicology*. A valuable account of ecological effects of pollutants.

Schuürmann, G. and Markert B. (eds) (1998) *Ecotoxicology*. A multiauthor work giving a very detailed account of the environmental fate of certain chemicals.

PART 2

Major organic pollutants

CHAPTER 5

The organochlorine insecticides

5.1 Background

The organochlorine insecticides can be divided into three main groups, each of which will be discussed separately in the sections that follow. These are DDT and related compounds, the cyclodiene insecticides and isomers of HCH (Brooks, 1974; Figure 5.1).

The first organochlorine insecticide to become widely used was DDT. Although first synthesised by Zeidler in 1874, its insecticidal properties were not discovered until 1939 by Paul Mueller of the Swiss company J. R. Geigy. DDT production commenced in the Second World War, during the course of which it was mainly used for the control of insects that are vectors of disease, including malarial mosquitoes and the ectoparasites that transmit typhus (e.g. lice and fleas). DDT was used to control malaria and typhus both in the military personnel and in the civilian population. After the war it came to be used widely to control agricultural and forest pests. After the introduction of DDT, the related compounds rhothane (DDD) and methoxychlor were also marketed as insecticides, but they were only used to a very limited extent. Restrictions began to be placed on the use of DDT from the late 1960s on, with the discovery of its persistence in the environment, and with growing evidence of its ability to cause harmful side-effects.

HCH, sometimes misleadingly termed benzene hexachloride (BHC), exists in a number of different isomeric forms, of which the γ-isomer has valuable insecticidal properties. The insecticidal properties of HCH were discovered during the 1940s, and it came to be widely used as an insecticide to control crop pests and certain ectoparasites of farm animals after the Second World War. Crude technical BHC, a mixture of isomers, was the first form of HCH to be marketed. In the course of time it was largely replaced by a refined product called lindane, containing 99% or more of the insecticidal γ-isomer.

The cyclodiene insecticides aldrin, dieldrin, endrin, heptachlor, endosulfan and others started to be introduced in the early 1950s. They were used to control a variety of pests, parasites and, in developing countries, certain vectors of disease such as the tsetse fly. However, some of them (e.g. dieldrin) combined high toxicity to vertebrates with marked persistence, and were soon found to have serious side-effects when used in the field, notably in Western European countries, where they were extensively used. During the 1960s severe restrictions were placed on the cyclodienes, so that few uses remained by the 1980s.

The organochlorine insecticides that came to be widely marketed were stable solids that acted as neurotoxins. Some organochlorine insecticides, or their stable metabolites, proved to have very long biological half-lives and marked persistence in the living environment. Where persistence was combined with high toxicity, as in the case of dieldrin and heptachlor epoxide (stable metabolite of heptachlor), there were sometimes serious side effects.

5.2 *DDT {1,1,1-trichloro-2,2-bis(p-chlorophenyl) ethane}*

5.2.1 CHEMICAL PROPERTIES

The principal insecticidal ingredient of technical DDT is *p,p'*-DDT (Table 5.1 and Figure 5.1). The composition of a typical sample of technical DDT is given in Table 5.2. The composition of the technical insecticide differs between batches, but the *p,p'* isomer usually accounts for 70% or more of the total. The *o,p'* isomer is the other major constituent, accounting for some 20% of the whole. *o,p'*-DDT is more readily degradable and less toxic to insects and vertebrates than the *o,p'* isomer. The presence of small quantities of *p,p'*-DDD deserves mention, because it has been marketed as an insecticide in its own right (rhothane) as well as being a reductive metabolite of *p,p'*-DDT.

p,p'-DDT is a stable white crystalline solid with a melting point of 108°C. It is of very low solubility in water and is highly lipophilic ($\log K_{ow} = 6.36$); thus there is a high potential for bioconcentration and bioaccumulation. It has a low vapour pressure, and is consequently relatively slow to sublimate when applied to surfaces (e.g. leaves, walls or surface waters).

p,p'-DDT is not very chemically reactive. One important chemical reaction is dehydrochlorination to form *p,p'*-DDE, which takes place in the presence of KOH

TABLE 5.1 *Chemical properties of organochlorine insecticides*

Chemical	Description	Water solubility (mg/L)	Log K_{ow}	Vapour pressure (mmHg) (25°C)
p,p'-DDT	Solid m.p. 108°C	< 0.1	6.36	1.9×10^{-7} (20°C)
p,p'-DDT	Solid m.p. 109°C	< 0.1		
Aldrin	Solid m.p. 104°C	< 0.1		6.5×10^{-5}
Dieldrin	Solid m.p. 178°C	0.2	5.48	3.2×10^{-6}
Heptachlor	Solid m.p. 93°C	0.056	5.44	4.0×10^{-4}
Endrin	Solid m.p. 226–230°C	0.2	5.34	2.0×10^{-7}
Endosulfan	Solid m.p. 79–100°C	0.06–0.15		1.0×10^{-5}
γ-HCH	Solid m.p. 112°C	7.0	3.78	9.4×10^{-6}

m.p., melting point.

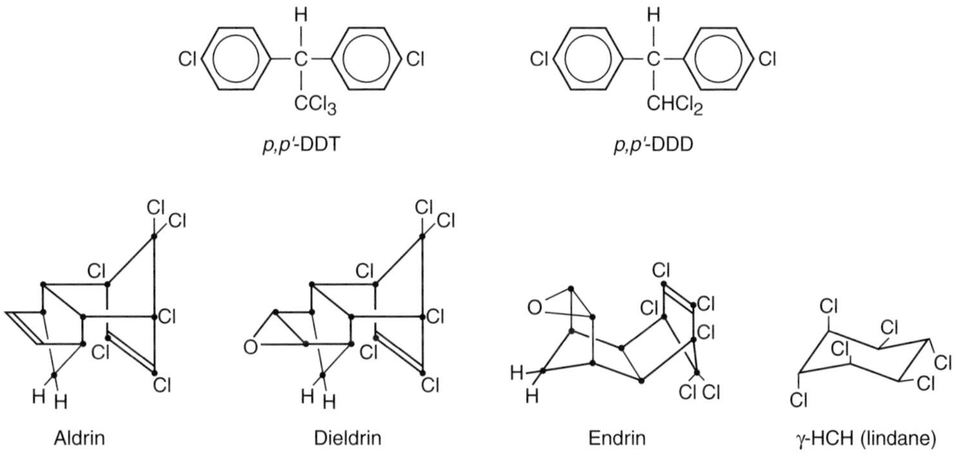

FIGURE 5.1 *Organochlorine insecticides.*

TABLE 5.2

Compound	Percentage of technical product
p,p'-DDT	72
o,p'-DDT	20
p,p'-DDD	3.0
o,o'-DDT	0.5
Other	4.5

(potassium hydroxide), NaOH (sodium hydroxide) and other strong alkalis. Dehydrochlorination is also a very important biotransformation and will be discussed further in section 5.22. *p,p'*-DDT undergoes reductive dechlorination by reduced iron porphyrins and slow photochemical decomposition.

5.2.2 METABOLISM OF DDT

p,p'-DDT is rather stable biochemically as well as chemically. Thus, it is markedly persistent in many species on account of its slow biotransformation. Metabolism of *p,p'*-DDT is complex, and there is still some controversy over the details of it. The most important metabolic pathways are shown in Figure 5.2.

A major route of biotransformation in animals is dehydrochlorination to the stable lipophilic and highly persistent metabolite *p,p'*-DDE. *p,p'*-DDE is far more persistent than *p,p'*-DDE, so dehydrochlorination does not promote excretion, although it usually results in detoxication because the metabolite is less acutely toxic than the parent compound. However, as will be seen, *p,p'*-DDE does cause sublethal effects. The metabolic conversion of parent compounds into persistent lipophilic metabolites also occurs with other organochlorine insecticides (see later sections), and may be regarded as a malfunction of detoxication systems that originally evolved to promote the elimination of naturally occurring lipophilic xenobiotics through the rapid excretion of their water-soluble metabolites and conjugates (Chapter 1). The dehydrochlorination of *p,p'*-DDT is catalysed by a form of glutathione-S-transferase, and involves the formation of a glutathione conjugate as an intermediate.

Under anaerobic conditions *p,p'*-DDT is converted to *p,p'*-DDD by reductive dechlorination, a biotransformation that occurs post mortem in vertebrate tissues such as liver muscle, and in certain anaerobic microorganisms (Walker and Jefferies, 1978). Reductive dechlorination is carried out by reduced iron porphyrins, and can be

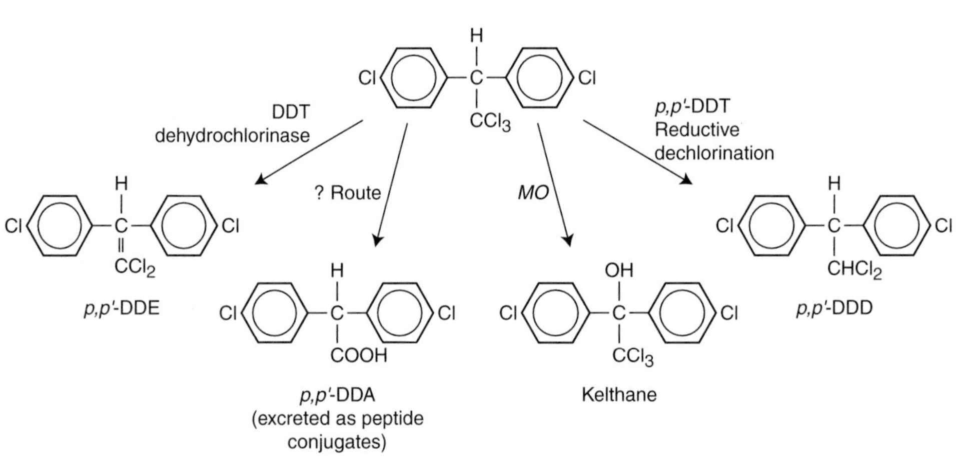

FIGURE 5.2 *Metabolism of* p,p'*-DDT.*

accomplished by cytochrome P450 of vertebrate liver microsomes supplied with NADPH in the absence of oxygen (Walker, 1969; Walker and Jefferies, 1978). The latter process can account for the relatively rapid conversion of p,p'-DDT to p,p'-DDD in avian liver immediately after death, but not during life, and mirrors the reductive dechlorination of other organochlorine substrates (e.g. CCl_4 and halothane) under anaerobic conditions. It is uncertain to what extent, if at all, the reductive dechlorination of DDT occurs *in vivo* in vertebrates (Walker, 1974).

A major, albeit slow, route of detoxication in animals is conversion to the water-soluble acid p,p'-DDA, which is excreted unchanged or as a conjugate. In one study the major urinary metabolites of p,p'-DDT in two rodent species were p,p'-DDA-glycine, p,p'-DDA-alanine and p,p'-DDA-glucuronic acid (Gingell, 1976). The route by which p,p'-DDA is formed remains uncertain. Early studies suggested that conversion might be via p,p'-DDD, but the later observation that this is a post-mortem process casts doubt on these findings. Some or all of the p,p'-DDD found in livers in these studies would have been generated after death because analysis was carried out after a period of storage. Another possibility is that this process, like dehydro-chlorination, takes place via glutathione conjugation. After conjugation and loss of HCl, the DDE moiety still bound to glutathione may undergo hydrolysis, leading eventually to deconjugation and formation of p,p'-DDA. A mechanism of this type has been proposed for the conversion of dichloromethane to HCHO (see Figure 2.15; Schwarzenbach *et al.*, 1993: p. 514).

One other biotransformation deserving mention is the oxidation of p,p'-DDT to kelthane, a molecule that has been used as an acaricide. This biotransformation occurs in certain DDT-resistant arthropods, but does not appear to be important in vertebrates.

In the absence of biotransformation, p,p'-DDT is normally lost very slowly by land vertebrates. There can, however, be a small amount of 'excretion' of the unchanged insecticide by females in milk or across the placenta into the developing embryo (mammals) or into eggs (fish, birds, reptiles and amphibia).

5.2.3 ENVIRONMENTAL FATE OF DDT

In discussing the environmental fate of technical DDT, the main issue is the persistence of p,p'-DDT and its stable metabolites, although it should be borne in mind that certain other compounds – notably o,p'-DDT and p,p'-DDD – also occur in the technical material and are released into the environment when it is used. The o,p' isomer of DDT is neither very persistent nor acutely toxic; it does, however, have oestrogenic properties (see section 5.2.4). A factor favouring more rapid metabolism of the o,p' isomer than of the p,p' isomer is the presence, on one of the benzene rings, of an unchlorinated *para* position which is available for oxidative attack. The other major impurity of technical DDT, p,p'-DDD, has been used as an insecticide in its own right (rhothane), and is also generated in the environment as a metabolite of DDT. In practice, the most abundant and widespread residues of DDT found in the environment have been p,p'-DDT, p,p'-DDD and p,p'-DDE.

When DDT was widely used, it was released into the environment in a number of ways. The spraying of crops and the spraying of water surfaces and land to control insect vectors of disease were major sources of environmental contamination. Waterways were sometimes contaminated with effluents from factories where DDT was used. Sheep dips containing DDT were discharged into water courses. Thus, it is not surprising that DDT residues became so widespread in the years after the war. It should also be remembered that, because of their stability, DDT residues can be circulated by air masses, and ocean currents, to reach remote parts of the globe. Very low levels have been detected in Antarctic snow!

Some data on the half-lives of these three compounds are given in Table 5.3. All of them are highly persistent in soils, with half-lives running into years, once they have become adsorbed by soil colloids (especially organic matter) (see Chapter 3). The degree of persistence varies considerably between soils, depending on soil type and temperature. The longest half-lives have been found in temperate soils with high levels of organic matter. Of particular significance are the very long half-lives for p,p'-DDE in terrestrial animals, approaching 1 year in some species, and greatly exceeding the comparable values for the other two compounds. This appears to be the main reason for the much higher levels of p,p'-DDE than the other two compounds in food chains even when technical DDT was still widely used. After the wide-ranging bans on the use of DDT in the 1960s and 1970s, levels of p,p'-DDT have fallen to very low levels in biota, but significant residues of p,p'-DDE are still found; for example in terrestrial food chains such as earthworms–thrushes–sparrowhawks in Britain (Newton, 1986), and in aquatic food chains.

When considering the fate of residues arising from technical DDT in ecosystems, p,p'-DDE is more stable and persistent (i.e. refractory) than either p,p'-DDT or p,p'-DDD and undergoes strong biomagnification with transfer along food chains. Studies

TABLE 5.3 *Half-lives of* p,p′*-DDT and related compounds*

Compound	Material/ organism	t_{50} (years)	Compound	Material/ organism	t_{50} (days)
p,p'-DDT	Soil	2.8	p,p'-DDT	Feral pigeon (*Columba livia*)	28
p,p'-DDE	Soil	10+ (British soils)	p,p'-DDD	Feral pigeon	24
			p,p'-DDE	Feral pigeon	250
			p,p'-DDT	Bengalese finch (*Lonchura striata*)	10
			p,p'-DDT	Hens (*Gallus domesticus*)	36–56 (in fat)
			p,p'-DDT	Rat	57–107
			p,p'-DDT	Rhesus monkey	32 and 1520

Data from Edwards (1973) and Moriarty (1975).

on the marine ecosystem of the Farne islands in 1962–4 showed that p,p'-DDE reached concentrations in fish-eating birds at the top of the food chain that were over 1000-fold higher than were present in macrophytes at the bottom of the food chain (Figure 5.3). The fish-eating shag (*Phalocrocorax aristotelis*) contained residues some 50-fold higher than those in its main prey species, the sand eel (*Ammodytes lanceolatus*). The sand eel was evidently the principal source of p,p'-DDE for the shag, so there had apparently been very efficient bioaccumulation over a considerable period (Robinson *et al.*, 1967a). However, as explained in Chapter 4, the biomagnification of highly lipophilic chemicals along food chains is not only a consequence of bioaccumulation through the different stages of the food chain. In aquatic systems an important source of residues to organisms at lower trophic levels is direct uptake from water or sediment, and this is more important than uptake from food in particular cases.

A nationwide investigation of organochlorine residues in bird tissues and bird eggs was conducted in great Britain in the early 1960s (Moore and Walker, 1964). The most abundant residue was p,p'-DDE – levels of p,p'-DDT and p,p'-DDD were considerably lower. Levels in depot fat were some 10- to 30-fold higher than in tissues such as liver or muscle. The magnitude of residues was related to the position in the food chain, with low levels in omnivores and herbivores, and the highest levels in predators at the top of both terrestrial and aquatic food chains (see Walker *et al.*, 1996: Chapter 4). Similar results were obtained with bird tissues and eggs. The highest p,p'-DDE levels (9–12 ppm) were found in the eggs of sparrowhawks, which are bird eaters, and in herons (*Ardea cinerea*), which are fish eaters.

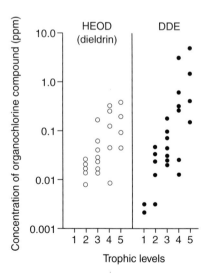

FIGURE 5.3 *Organochlorine insecticides in the Farne Island ecosystem. From Walker et al. (2000). Trophic levels: 1. Serrated wrack, oar weed. 2. Sea urchin, mussel, limpet. 3. Lobster, shore crab, herring, sand eel. 4. Cod, whiting, shag, eider duck, herring gull. 5. Cormorant, gannet, grey seal.*

p,p′-DDE can undergo bioaccumulation in terrestrial food chains. Studies with earthworms and slugs indicate that there can be a bioconcentration of total DDT residues (*p,p′*-DDT + *p,p′*-DDE + *p,p′*-DDD) relative to soil levels of one- to fourfold by earthworms, and greater than this by slugs (Edwards, 1973; Bailey *et al.*, 1974). When DDT was still widely used in orchards in Britain, blackbirds (*Turdus merula*) and song thrushes (*Turdus philomelis*) that had been found dead contained very high levels of DDT residues compared with those in the earthworms upon which they were feeding. Some results from a study on one orchard sprayed with DDT are given in Table 5.4. Interpretation of field data, involving such small numbers of individual specimens, needs to be done with caution. However, the principal source of DDT residues for the two *Turdus* species appears to have been earthworms and other invertebrates (including slugs and snails). The birds found dead were probably poisoned by DDT residues, and the levels found in them were some 20-fold higher than in the earthworms upon which they were feeding, suggesting marked bioaccumulation. Birds that were shot also contained DDT residue levels well above those recorded for earthworms. It should be added that the relatively high levels of *p,p′*-DDE found in sparrowhawks in the 1980s from areas where DDT was once used may be due, in part, to transfer from soil sinks via soil invertebrates to the insectivorous birds upon which these raptors feed (Newton, 1986). The virtual absence of *p,p′*-DDT coupled with the high levels of *p,p′*-DDD in tissues other than fat from the dead birds strongly suggests post-mortem conversion of the former to the latter by reductive dechlorination. By contrast, relatively high levels of *p,p′*-DDT were present in the earthworms upon which the birds were feeding and in eggs of both species sampled in the same area. It has been shown that little or no conversion of *p,p′*-DDT to *p,p′*-DDD occurs in birds eggs until embryo development commences (Walker and Jefferies, 1978); thus the relative levels of the two compounds in eggs should reflect what is present in the birds' food, and in the tissues of the birds during life.

To summarise, *p,p′*-DDE can undergo strong bioaccumulation to reach particularly high levels at the top of both aquatic and terrestrial food chains food chains; this is also true to a lesser extent of *p,p′*-DDT and *p,p′*-DDD, which are more readily biodegradable than *p,p′*-DDE.

TABLE 5.4 *DDT residues from samples taken in orchard near Norwich 1971–2*

Species/ sample	Sampling procedure	*p,p′*-DDT	*p,p′*-DDE	*p,p′*-DDD	Total DDT residues
Soil	Random	1.2–3.5	0.5–1.1	0.22–0.72	2.1–5.3
Earthworm	Random	1.1–6.8	1.4–4.2	0.46–5.5	3.9–11.5
Blackbird	Two birds found dead	0/6.8	130/180	58/195	195/249
Blackbird	Two birds shot	0/2.4	24/33	14/30	49/53
Song thrush	Two birds found dead	0	164/192	81/128	273/292
Song thrush	One bird shot	0	30	42	72

Residue concentrations expressed as ppm by weight.

Data from Bailey *et al.* (1974).

Although DDT has been banned in most countries for many years, residues of p,p'-DDE occur widely in ecosystems. The loss of p,p'-DDE from contaminated soils and sediments is so slow that they act as sinks, ensuring the contamination of terrestrial and aquatic ecosystems for many decades to come.

5.2.4 TOXICITY OF DDT

The acute toxicity of p,p'-DDT to both vertebrates and invertebrates is attributed mainly to action upon axonal Na^+ channels that are voltage dependent (see Figure 5.4) (Eldefrawi and Eldefrawi, 1990). The molecule binds reversibly to a site on the channel, thereby altering its function. Normally, when a Na^+ current is generated during the passage of a nerve action potential, the signal is rapidly terminated by the closure of the sodium channel. In DDT-poisoned nerves the closure of the channel is delayed, an event which can cause disruption of the regulation of action potential and can lead to repetitive discharges. p,p'-DDT can also act upon the K^+ channel, which is concerned with the repolarisation of the axonal membrane after passage of the action potential.

Apart from the action upon Na^+ channels, p,p'-DDT and/or its metabolites can have certain other toxic effects. It has been reported that p,p'-DDT can inhibit certain ATPases (see EHC 83, 1989) In fish, the inhibition of ATPases can affect osmoregulation. p,p'-DDE can inhibit the Ca^{2+}-ATPase of avian shell gland, which is thought to be the main reason for the severe eggshell thinning caused by this compound in certain species of birds, including the American kestrel (*Falco sparverius*), the sparrowhawk (*Accipiter nisus*), the peregrine falcon (*Falco peregrinus*) and the gannet (*Sula bassana*). Dietary levels of as low as 3 ppm have been shown to cause shell thinning in the American kestrel (Wiemeyer and Porter, 1970; Peakall *et al.*, 1993). Finally, there is evidence that constituents of technical DDT can have a feminising effect upon avian embryos. In a study of the California gull, o,p'-DDT was found to be considerably more potent than p,p'-DDE (Fry and Toone, 1981). It should be

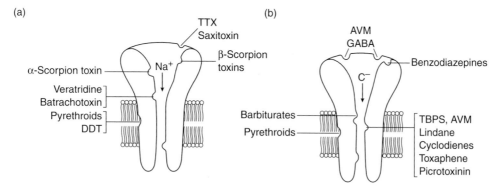

FIGURE 5.4 *Sites of action of organochlorine insecticides. (a) Sodium channel. (b) GABA receptor. From Eldefrawi and Eldefrawi (1990) with permission.*

remembered that the foregoing effects are all upon lipophilic membrane-bound proteins. *p,p'*-DDT, *p,p'*-DDE and other lipophilic organochlorine compounds can reach very high concentrations at or near such hydrophobic domains, relative to those in tissue fluids and blood.

From an ecotoxicological point of view it has often been suspected that sublethal effects, such as those described above, can be more important than lethal ones. Both *p,p'*-DDT and *p,p'*-DDD are persistent neurotoxins, and may very well have caused behavioural effects in the field. This issue was not resolved when DDT was widely used and remains a matter for speculation. More is known, however, about eggshell thinning caused by *p,p'*-DDE and the effects of this upon reproduction, which will be discussed in the next section.

It can be seen in Table 5.5 that *p,p'*-DDT is toxic to a wide range of vertebrates and invertebrates. That said, it is considerably less toxic than dieldrin, heptachlor or endrin to most species. If applied topically, it is 180-fold more toxic to the housefly than to the rat. In general, it appears to be selective between insects and mammals. Some aquatic invertebrates are very sensitive to it, but there is a very wide range of susceptibility among freshwater invertebrates. The rather wide range of values for mammals is partly the consequence of the use of different vehicles. Oil solutions tend to be appreciably more toxic than solid formulations, presumably because of more rapid and/or efficient absorption from the gut. Thus, the lower part of the range should be more representative of the toxicity that will be shown when the compound is passed through the food chain, when it is 'dissolved' in the fatty tissues of prey species. In general *p,p'*-DDE is less toxic than *p,p'*-DDT, especially in insects, in which the reductive dechlorination of *p,p'*-DDT represents a detoxication mechanism despite the greater persistence of the metabolite than of the parent compound.

TABLE 5.5 *Ecotoxicity of* p,p'-DDT *and related compounds*

Compound	Organism	Test	Median lethal dose or concentration
p,p'-DDT	Marine Invertebrates	LC_{50} (48 or 96 h)	0.45–2.4 µg/L
p,p'-DDE	Marine invertebrate (brown shrimp)	LC_{50} (48 or 96 h)	28 µg/L
p,p'-DDT	Freshwater invertebrates	LC_{50} (48 or 96 h)	0.4–1800 µg/L
p,p'-DDT	Fish (smaller fish most susceptible)	LC_{50} (96 h)	1.5–5.6 µg/L
p'-DDT	Mammals	LD_{50} (acute oral)	100–2500 mg/kg
p,p'-DDE	Rodents	LD_{50} (acute oral)	880–1240 mg/kg
p,p'-DDD	Rat	LD_{50} (acute oral)	400–3400 mg/kg
p,p'-DDT	Birds	LD_{50} (acute oral)	> 500 mg/kg

Data from EHC 9 (1979), EHC 83 (1989) and Edson *et al.* (1966).

5.2.5 ECOLOGICAL EFFECTS OF DDT

As explained in section 5.2.3, p,p'-DDE is much more persistent in food chains than either p,p'-DDT or p,p'-DDD, and during the 1960s when DDT was still extensively used it was often the most abundant of the three compounds in birds and mammals found in the field. Since the large-scale banning of DDT, very little is now released into the environment, and p,p'-DDE is by far the most abundant DDT residue found in biota. In considering ecological effects of DDT and related compounds, effects upon population numbers will be considered before effects upon population genetics (gene frequencies).

Effects upon population numbers

An early indication of the damage that organochlorine compounds can cause in the higher levels of food chains came from a study on the East Lansing Campus of the Michigan State University in 1961 and 1962 (Bernard, 1966). Over several years, leading up to and including 1962, American robins (*Turdus migratororius*) and several other species of birds were virtually eliminated from a 75-ha study area in the spring after the application of high levels of DDT (< 25 lb/acre) to elm trees. The purpose of the exercise was to control Dutch elm disease. Subsequent investigation established that American robins dying in this way all contained above 50 ppm of total DDT in the brain. Comparison with experimentally poisoned birds led to the conclusion that these levels were high enough to have caused lethal DDT poisoning. In these early days, before the development of gas chromatography, it was difficult to distinguish between the different compounds derived from DDT, and a limitation of the study was that deductions were based on estimates of 'total' DDT. As has been pointed out, the various impurities and metabolites arising from the technical material differ considerably in their toxicity, so an estimate of total DDT residues is of limited usefulness when attempting to establish the cause of death. However, with the benefit of hindsight, it seems clear that many birds did die of DDT poisoning after these very high levels of application, and that transfer through earthworms and other invertebrates made a major contribution to the residues in the birds. It also appeared that the effects were localised, seasonal and transitory.

In another widely quoted early study, Hunt and Bischoff (1960) reported the decline of western grebe (*Aechmophorus occidentalis*) populations on Clear Lake, CA, after the application of rhothane (DDD) over several years. There was evidence of a progressive build-up of p,p'-DDD residues in sediments over the period, and an analytical study of biota from the lake yielded the results shown in Table 5.6.

The levels of DDD found in dying or dead grebes were high enough to have caused acute lethal poisoning. As with the study on the American robin, there was strong evidence for a local decline in a species occupying a high trophic level of an ecosystem, a decline consequent upon the toxicity of a persistent organochlorine compound obtained via its food. At first there was a tendency to explain the very large differences

TABLE 5.6 *Results from an analytical study of biota from Clear Lake, California*

Species sample	*p,p'*-DDD (ppm, wet weight)
Lake water	0.02
Plankton	5.0
Non-predatory fish (fat)	40–1000
Predatory fish (fat)	80–2500
Predatory fish (flesh)	1–200
Western grebe (fat)	1600

Data from Hunt and Bischoff (1960).

in DDD concentrations between the top and the bottom of the food chain in terms of progressive bioaccumulation with movement up the food chain. On closer examination, however, much of this increase is explicable on the grounds of strong bioconcentration with direct uptake from water, both by plankton and by fish. The difference in concentration in body fat between predatory and non-predatory fish is not very large, and there is no clear evidence of strong bioaccumulation of *p,p'*-DDD by the predators from their prey. The grebes, however, contained levels in their depot fat nearly equal to the highest found in fish (above the top of the range for non-predatory fish), pointing to bioaccumulation in the last step of the food chain. Indeed, it seems very probable that the birds died from DDD poisoning while the tissue levels of the insecticide were still increasing, i.e. some time before a steady state was reached, so that the level of bioaccumulation found was below that which would have been achieved at a lower level of exposure.

The two examples just given are of localised effects associated with the acute toxicity of DDT and DDD to organisms in high trophic levels. A much wider-ranging toxic effect associated with population decline was eggshell thinning caused by the relatively high levels of *p,p'*-DDE in some predatory birds (Table 5.7).

In North America, the decline of several species of birds of prey was associated with eggshell thinning caused by *p,p'*-DDE. Peregrine populations in North America declined or were extirpated with eggshell thinning of 18–25% associated with DDE residues in excess of 10 ppm in eggs (Peakall, 1993). The bald eagle showed a marked decline in many area of North America, a decline that was first reported in Florida in 1946 (Broley, 1958). Shell thinning of 15% was associated with residues of 16 ppm *p,p'*-DDE in this species (Wiemeyer *et al.*, 1993), and at the time of the initial decline (1946–57) shell thinning of 15–19% was associated with diminished breeding success. In field studies carried out during 1969–84 the picture was complicated by the fact that, although breeding success was negatively correlated with *p,p'*-DDE levels in the eggs, the correlation between breeding success and eggshell thinning was poor. This raises the possibility that *p,p'*-DDE may have had other toxic effects in addition to eggshell thinning (Nisbet, 1989). There is the further complication that other

TABLE 5.7 *Population declines associated with eggshell thinning caused by* p,p′-DDE

Species/area	Years	p,p′-DDE residues (ppm)	Degree of thinning (%)	Population effect
Peregrine (New Jersey, Massachusetts, S. California, Belgium)	1950–73	Mostly > 20 ppm in eggs	18–25	Extirpated
Bald eagle (USA)	1970	Mean 18 ppm in carcasses		Declining
Gannet (Bonaventura Island, Quebec)	1969	19–30 ppm in eggs	17–20	Declining; low reproductive success

Data from Peakall (1993), Elliott *et al*. (1988) and Kaiser *et al*. (1980).

organochlorine insecticides such as dieldrin and heptachlor epoxide were present in the same samples and might have had toxic effects.

Clearly, caution is needed when attempting to relate levels of eggshell thinning caused by *p,p′*-DDE to population effects. In Britain, eggshell thinning related to the presence of *p,p′*-DDE occurred in the peregrine falcon and sparrowhawk from 1946/7 but population declines did not occur until some 8 years later. These declines coincided with the introduction of the cyclodiene insecticides and will be discussed in section 5.3.3. It is important to emphasise, however, that levels of DDT were higher and levels of cyclodienes lower in North America than in Britain and other Western European countries. The weight of evidence suggests that declines in the bald eagle, the peregrine and the osprey (*Pandion haliaetus*) in the USA were mainly due to the effects of DDT, and especially to eggshell thinning caused by *p,p′*-DDE.

In another study, on Bonaventura Island, Quebec, Canada, during the 1960s and early 1970s, gannets (*Sula bassanus*) showed a sharp population decline that was associated with poor breeding success. In 1969 there was clear evidence of severe shell thinning caused by *p,p′*-DDE residues of 19–30 ppm in eggs. Subsequently, pollution of the St Lawrence river by DDT was reduced. The *p,p′*-DDE in the gannets fell, and by the mid- to late 1970s shells became thicker, reproductive success increased and the population recovered (Elliot *et al*., 1988). Taken overall, these findings illustrate very clearly the ecological risks associated with the wide dispersal of a highly persistent pollutant that can cause sublethal toxicity.

Before leaving the question of effects of organic pollutants upon populations, brief mention should be made of indirect effects. Sometimes insect populations increase in size because an insecticide reduces the numbers of a predator or parasite that keeps the numbers in check. Such an effect was found in a controlled experiment where

DDT was applied to a brassica crop infested with the caterpillars of the cabbage white butterfly (*Pieris brassicae*) (see Dempster in Moriarty, 1975). Field applications of DDT severely reduced the population of carabid beetles, which prey upon and control the numbers of *P. brassicae* larvae. The infestation of the crop was initially controlled by DDT but, as the residues declined on the crop, the caterpillars returned, eventually to reach much higher numbers than on plots untreated by DDT, where the natural predators maintained control of the pest. Thus, the long-term indirect effect of DDT was to increase the numbers of the pest species! When DDT was used as an orchard spray, it was implicated, together with certain other insecticides, in the triggering of an epidemic of red spider mites (Mellanby, 1967). The insecticides successfully controlled the black-kneed capsid bug (*Blepharidopterus angulatus*), which normally keep down the numbers of red spider mites, and this led to a population explosion of the latter – and to a new pest problem! These examples illustrate well a fundamental difference between ecotoxicology and normal medical toxicology. The well-established test procedures of the latter may tell us very little about what will happen when toxic chemicals are released into ecosystems.

Effects upon population genetics (gene frequencies)

DDT had not been in general use for very long before there were reports of resistance in insect populations that were being controlled by the insecticide. Examples included resistant strains of houseflies (*M. domestica*) and mosquitoes (Oppenoorth and Welling, 1976; Georghiou and Saito, 1983). For further discussion see Brown (1971). Two contrasting resistance mechanisms have been found in resistant strains of housefly. First, metabolic resistance, usually due to enhanced levels of DDT dehydrochlorinase. In one resistant strain of housefly, enhanced monooxygenase activity was found that might cause increased rates of detoxication to kelthane and other oxidative metabolites (Oppenoorth and Welling, 1976). By contrast, some houseflies showed knock-down (kdr or super-kdr) resistance as a result of nerve insensitivity. It now seems clear that this is the consequence of the appearance of a mutant form (or forms) of the Na^+ channel in resistant strains that is insensitive to DDT (Salgado, 1999). Axonal Na^+ channels represent the normal target for p,p'-DDT (and, incidentally, for pyrethroid insecticides), as explained earlier (pp. 55 and 85). The appearance of resistant strains such as these can give valuable retrospective evidence of the environmental impact of pollutants. Assays for resistance mechanisms, and the genes that operate them, are valuable tools in ecotoxicology.

5.3 *The cyclodiene insecticides*

The insecticides belonging to this group are derivatives of hexachlorocyclopentadiene, synthesised by the Diels–Alder reaction (see Brooks, 1974). They did not come into use until the early 1950s and were not used, to any important extent, in Europe or North America before the mid-1950s. Thus, they did not begin to produce environmental side-effects until at least 6 years after the onset of DDE-induced eggshell thinning in sparrowhawks, peregrines and bald eagles, as described in section 5.2.5. The cyclodienes include dieldrin, aldrin, heptachlor, endrin, chlordane, endosulfan, telodrin, chlordecone and mirex. Some of them have only been used to a limited extent, and the following account will be restricted to dieldrin, aldrin, heptachlor, endrin, telodrin and chlordane, the compounds that have caused most concern about environmental side-effects.

5.3.1 CHEMICAL PROPERTIES

The cyclodienes are stable solids of low water solubility and marked lipophilicity (Table 5.1), and their active ingredients have cage structures (Figure 5.5). The technical insecticides aldrin and dieldrin contain the molecules HHDN and HEOD (abbreviations of the full formal chemical names) as their active insecticidal ingredients (see Figure 5.5). Here, the term 'aldrin' will be used synonymously with HHDN, and the term 'dieldrin' will be used synonymously with HEOD unless otherwise indicated, thus following a common practice in the literature. Similarly the common names of the other cyclodienes (e.g. heptachlor and endrin) will be used to refer to the chemical structures of the principal insecticidal ingredients of the technical products.

Of the examples given in Table 5.1, aldrin and heptachlor have the lowest water solubilities and the highest vapour pressures. They are readily oxidised, both chemically and biochemically, to their epoxides – dieldrin and heptachlor epoxide respectively (Figure 5.5). (It should be noted that dieldrin is marketed as an insecticide in its own right whereas heptachlor epoxide is not.) The two epoxides have greater polarity, and consequently greater water solubility and lower volatility, than their precursors. Endrin is also an epoxide – in fact a stereoisomer of dieldrin – and again shows greater water solubility and lower vapour pressure than aldrin or heptachlor. These relationships illustrate the importance of the electron-withdrawing power of oxygen atoms in determining the properties of organic compounds.

Apart from the oxidations just mentioned, cyclodienes are rather stable chemically. It should, however, be noted that dieldrin can undergo photochemical rearrangement under the influence of sunlight to the persistent and toxic molecule 'photodieldrin', which occurs as a residue after the application of the insecticide in the field.

5.3.2 THE METABOLISM OF CYCLODIENES

In terrestrial animals, cyclodienes such as dieldrin and other refractory lipophilic pollutants can be 'excreted' in their unchanged forms, notably with exported lipoproteins into milk (mammals), eggs or developing embryos (mammals). In a few cases (e.g. laying hens), this can represent an important mechanism of loss. In general,

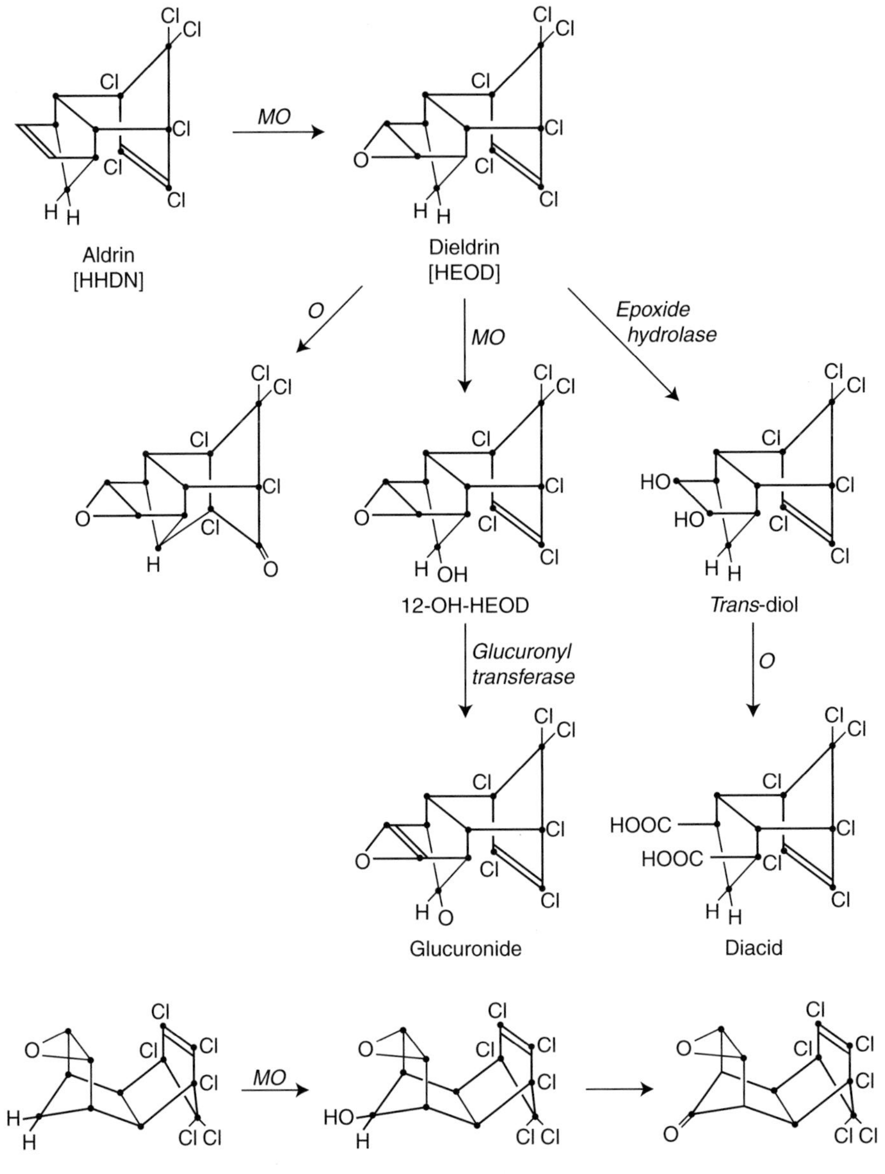

FIGURE 5.5 *Metabolism of cyclodienes.*

however, it is not a sufficiently rapid mechanism to give much protection to the adult organism, and it constitutes a hazard for the next generation! Effective elimination depends upon biotransformation into water-soluble and readily excretable metabolites and conjugates.

The metabolism of aldrin, dieldrin, endrin and heptachlor in vertebrates is shown in Figure 5.5. As with *p,p'*-DDT and related compounds, the high level of chlorination greatly limits the possibility of metabolic attack by forming what is, in effect, a protective shield of halogen atoms. Monooxygenase attack is, in principal, the most effective and rapid primary biotransformation of these compounds, but it does not occur to any important extent on C–Cl bonds – or for that matter on C–Br or C–F bonds. Thus, the most effective attack is on other positions, for example the C=C of the unchlorinated rings of aldrin and heptachlor epoxide, and upon the endomethylene bridges across same in the cases of dieldrin and endrin (Figure 5.5) (Brooks, 1974; Walker, 1975; Chipman and Walker, 1979). The first type of oxidation yields stable epoxides that are toxic and much more persistent than the parent compounds, and represents activation not detoxication! The second line of attack is a typical phase 1 detoxication, yielding monohydroxy metabolites more polar than the parent compounds dieldrin and endrin; moreover, these monohydroxy metabolites readily undergo conjugation to form glucuronides and sulphates, which are usually rapidly excreted. The hydroxylation of endrin occurs relatively rapidly owing to the endomethylene bridge being in an exposed position for monooxygenase attack. It may be deduced that the molecule is bound to one or more forms of P450 belonging to gene family 2, and that the endomethylene group is thereby exposed to an activated form of oxygen generated from molecular oxygen attached to haem iron (see Chapter 2). The endomethylene group of dieldrin is less exposed than is that of endrin, being screened by bulky neighbouring chlorine atoms, and metabolic detoxication is consequently a good deal slower (Hutson, 1976; Chipman and Walker, 1979). Dieldrin is considerably more persistent in vertebrates than endrin, despite the fact that the two compounds are stereoisomers with very similar physical properties, a logical consequence of the differential rates of metabolism. Thus, a stereochemical difference between two compounds having the same empirical formula may be reflected in large differences in toxicokinetics!

Cyclodiene epoxides such as dieldrin and heptachlor epoxide are also detoxified, albeit rather slowly, by epoxide hydrolase attack to form *trans*-dihydrodiols (Figure 5.5). The diols are relatively polar compounds, which may be excreted unchanged or as conjugates. There are very marked species differences in ability to detoxify cyclodienes by epoxide hydrolase attack (Walker *et al.*, 1978; Walker, 1980). Using the readily biodegradable cyclodienes HEOM (hexachloro-octahydro-6,7-epoxy-methano-naphthalene) and HCE, mammals showed much higher microsomal epoxide hydrolase activities than birds or fish. Of the mammals, pigs and rabbits had particularly high epoxide hydrolase activity, and it is noteworthy that the *trans*-diol has been shown to be an important *in vivo* metabolite of dieldrin in the rabbit, but not in the rat or the mouse, and not in birds (Korte and Arent, 1965; Chipman and Walker, 1979; Walker, 1980). In general, the principal primary detoxication of dieldrin, endrin and heptachlor epoxide is by monooxygenase attack.

In mammals, dieldrin and endrin are also converted into keto metabolites (Figure 5.5). In the rat, the keto metabolite is only a minor one that, because of its lipophilicity, tends to be stored in fat. With endrin, a keto metabolite is formed by the dehydrogenation of the primary monohydroxy metabolite. In mammals, the *trans*-diol of dieldrin is converted into a diacid *in vivo* (Oda and Muller, 1972).

5.3.3 ENVIRONMENTAL FATE OF CYCLODIENES

During the 1950s, cyclodiene insecticides came to be widely used for a number of different purposes. They were used to control agricultural pests, insect vectors of disease and ectoparasites of farm animals, to treat wood against wood-boring insects and to moth-proof fabrics. Because of their very low solubility in water, they were usually formulated as emulsifiable concentrates or wettable powders. After the discovery that they were having undesirable environmental side-effects, the use of cyclodienes for many purposes was discontinued during the 1960s and early 1970s in Western Europe and North America. The following account will focus on the environmental fate of aldrin, dieldrin and heptachlor, three insecticides that gave rise to persistent residues and have been shown to cause serious and widespread environmental side-effects. Other cyclodienes were less widely used, and some (e.g. endrin and endosulfan) were far less persistent.

As mentioned earlier (Figure 5.5), aldrin and heptachlor are rapidly metabolised to their epoxides (i.e. dieldrin and heptachlor epoxide repectively) by most vertebrate species. These two stable toxic compounds are the most important residues of the three insecticides found in terrestrial or aquatic food chains. In soils and sediments, aldrin and heptachlor are epoxidised relatively slowly and, in contrast to the situation in biota, may reach significant levels (note, however, the difference between aldrin and dieldrin half-lives in soil shown in Table 5.8). The important point is that, after entering the food chain, they are quickly converted to their epoxides, which become the dominant residues.

It should be emphasised that these data are approximate; there is considerable variation in half-lives published in the literature, especially between different soil types (see section 4.3), but also from different studies on the same species. The strains used, the manner of dosing, the diet and the method of analysing data can all influence the result (see Moriarty, 1975) for further discussion of this point). Looking at the estimated half-lives shown in Table 5.8, dieldrin, like p,p'-DDT, is markedly persistent, but less so than p,p'-DDE. Heptachlor epoxide is likely to have a similar half-life to dieldrin. Dieldrin is also highly persistent in sediments, as became clear in a study of British rivers conducted many years after the banning of the insecticide. Eels from all rivers investigated contained substantial dieldrin residues. In vertebrates there are considerable interspecies differences in dieldrin half-lives. Male rats eliminate dieldrin more rapidly than female rats, pigeons or dogs. In humans the estimated dieldrin half-life is 369 days! By contrast, the half-life of endrin in humans is only 1–2 days (EHC 130).

The rate of oxidative detoxication appears to be a critical factor in determining cyclodiene half-lives. As noted earlier, endrin is rapidly detoxified by monooxygenase attack whereas dieldrin is not. Also, male rats tend to have substantially higher monooxygenase activity towards cyclodiene substrates than do pigeons or humans. In general, the rate of elimination of cyclodienes by terrestrial animals is dependent upon the rate at which they are converted into water-soluble and readily excretable

TABLE 5.8 *Half-lives of cyclodienes*

Compound	Material/organism	Half-life
Dieldrin	Soil	2.5 years
Aldrin	Soil	0.3 years
Heptachlor	Soil	0.8 years
Dieldrin	Male rat	12–15 days
Dieldrin	Pigeon	47 days (mean)
Dieldrin	Dog	28–32 days
Dieldrin	Man	369 days (mean)

Soil data from Edwards (1973).
Data for pigeon from Robinson *et al.* (1967b).
Other data from EHC 91.

metabolites (Chipman and Walker, 1979; Walker, 1981). There is little tendency for the original lipophilic molecules to be excreted unchanged in urine or faeces.

Dieldrin, like p,p'-DDE, p,p'-DDD, p,p'-DDT and many other organohalogenated compounds with high K_{OW} values, can undergo very marked bioconcentration by aquatic organisms (see Walker and Livingstone, 1992). BCFs in the steady state exceeding 1000 are usual. Thus, Ernst (1977) gives a BCF value of 1570 for *M. edulis* and Holden (1973) a value of 3700 for the rainbow trout. With aquatic organisms, the rate of dieldrin metabolism is generally very low (especially in molluscs), and these high BCFs are a reflection of the passive exchange equilibrium between the organism and the ambient water (see Chapter 4). With terrestrial organisms metabolism is generally faster than in aquatic organisms but, as mentioned above, there are large species differences, and species which metabolise dieldrin slowly may strongly bioaccumulate the insecticide if exposed to it over long periods (Chipman and Walker, 1979; Walker, 1987, 1990a).

In an incident at London Zoo, 22 owls of diverse species died as the result of dieldrin poisoning (Jones *et al.*, 1978). The source of dieldrin was the mice that they had been fed. Comparison of the dieldrin liver concentrations in mice (geometric mean approximately 2.6 ppm over two batches) with the dieldrin liver concentrations in the 22 owls (geometric mean 28 ppm) suggested a bioaccumulation factor of about 11. This can only be an approximation, because the levels in the mice were not monitored during the actual exposure of the owls. It should be recalled, however, that specialised predatory birds tend to have very low monooxygenase activities (Chapters 2 and 4; Walker, 1980; Ronis and Walker, 1989; Walker, 1998a–c). These data, and the evidence from the Farne Island study (see next paragraph), strongly indicate that low metabolic capability is associated with an ability to strongly bioaccumulate dieldrin (and certain other organochlorine insecticides).

Dieldrin and heptachlor epoxide undergo bioconcentration and bioaccumulation with movement along food chains (see section 5.2.3), reaching their highest concentrations in predators at the apex of food pyramids. Thus, in the study conducted during 1962–4 dieldrin, like p,p'-DDE, was shown to exist at the highest concentration in the fish-eating birds of the Farne Islands ecosystem (Figure 5.3 and Robinson *et al.*, 1967a). The mean concentration of dieldrin in carcasses of the shag (*P. aristotelis*) ($n = 8$) was 1.0 ppm, whereas the mean concentration in the whole bodies of its principal prey in this area, the sand eel (*A. lanceolatus*) ($n = 16$), was 0.016 ppm; this indicated the extraordinarily high BCF of 63! Figures from uncontrolled field studies need to be interpreted with great care. Nevertheless, this and other evidence points to the marked bioaccumulation of dieldrin from prey by specialised predators at the apex of food pyramids. It is interesting to note that the gradient of increasing concentration with movement from trophic level 1 to trophic level 5 is steeper for p,p'-DDE than it is for dieldrin. Existing data suggest that dieldrin is more biodegradable and is eliminated more rapidly than p,p'-DDE, which may explain the difference in the degree of biomagnification.

Dieldrin also shows persistence in terrestrial ecosystems, but the situation is complicated here because of the very high concentrations of the chemical that existed at the beginning of the food chain, when it was widely used in agriculture. Dieldrin and heptachlor were once commonly used for dressing cereal and other crop seeds to give protection against soil pests. Newly treated seed had cyclodiene residues of approximately 800 ppm by weight associated with it (Turtle *et al.*, 1963). Here bioaccumulation was impossible because animals and birds are lethally poisoned by tissue levels of dieldrin and heptachlor epoxide of 10–100 ppm by weight, which is far below the levels on the freshly dressed seed.

In the study of organochlorine residues in wild birds referred to earlier (section 5.2.3), dieldrin showed much the same distribution pattern as p,p'-DDE, although the levels were usually lower. Once again the highest levels were found in predators, in both terrestrial and aquatic ecosystems (Moore and Walker, 1964).

5.3.4 TOXICITY OF CYCLODIENES

There is strong evidence that the primary target for dieldrin, heptachlor epoxide, endrin and other cyclodienes in the mammalian brain is the GABA receptor (Figure 5.4), against which they act as inhibitors (Eldefrawi and Eldefrawi, 1990). In insects too, cyclodiene toxicity is attributed, largely or entirely, to interaction with GABA receptors of the nervous system. Toxaphene and γ-HCH also act on this receptor. GABA receptors are found in the brains of both vertebrates and invertebrates, as well as in insect muscle; they possess chloride channels that, when open, permit the flow of Cl⁻, with consequent repolarisation of nerves and reduction of excitability. They are particularly associated with inhibitory synapses. In vertebrates, the action of cyclodienes upon them can lead to convulsions.

Given this mode of action upon the central nervous system, it is not surprising that cyclodienes can have a range of sublethal effects. These have been observed in humans occupationally exposed to aldrin or dieldrin (EHC 91; Jager, 1970). The symptoms observed included headache, dizziness, drowsiness, hyperirritability, general malaise, nausea and anorexia. Sublethal effects included characteristic changes in EEG (electroencephalogram) patterns, and were observed over a wide range of blood concentrations. At this early stage of intoxication muscle twitching and convulsions sometimes occurred. According to various authors, patients showing these symptoms had blood dieldrin levels in the range 8–530 μg/L (EHC 91). The relationship between blood levels and toxic effects of dieldrin in humans is shown in Figure 5.6. With increasing tissue levels severe convulsions occurred leading eventually to death. In cases of lethal poisoning, blood levels were found to exceed 600 μg/L (850 μg/L in one suicide case). Individuals showing sublethal effects made complete recoveries after discontinuation of exposure to insecticide. Experimental animals exposed to sublethal doses of cyclodienes show a similar picture, with changes in EEG patterns, disorientation, loss of muscular coordination and vomiting, as well as convulsions, which become more severe with increasing dose (Hayes and Laws, 1991). It is clear from these wide-ranging studies that a number of neurotoxic effects can be caused by cyclodienes at levels well below those that are lethal. In the human studies described above, subclinical symptoms were regularly reported when dieldrin blood levels were in the range 50–100 μg/L, an order of magnitude below those associated with lethal intoxication.

Sublethal neurotoxic effects such as these have been associated with changes in behaviour. In one study with dieldrin, squirrel monkeys were reported to show changes in learning ability and in EEG pattern after receiving doses of 0.01 or 0.1 mg/kg over 54 days (Van Gelder and Cunningham, 1975). In a study with goldfish (*Carassias auratus*), 96-h exposure to 0.44 μg/L toxaphene caused alterations in a number of

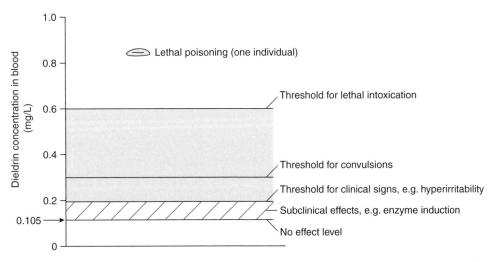

FIGURE 5.6 *Dieldrin intoxication in man – relationship to blood levels. The hatched area represents the sublethal effects seen at 15–30% of lethal concentration in blood. After Jager (1970).*

behavioural parameters (Warner *et al.*, 1966). The changes in response were more marked after 264 h, although the fish remained outwardly healthy. As noted above, toxaphene has the same mode of action as cyclodienes. The question of possible behavioural effects is an important one when considering the impact of cyclodienes in the field (see section 5.3.5).

Some data on cyclodiene toxicity are presented in Table 5.9. Aldrin and dieldrin have similar levels of acute toxicity; indeed the toxicity of aldrin has been largely attributed to its stable metabolite, dieldrin. Dieldrin is highly toxic to fish, mammals and birds. Heptachlor has a similar order of toxicity. Endrin is rather more toxic than aldrin or dieldrin to rodents and to rabbits (data not shown), despite the fact that it is more readily metabolised and therefore less persistent than dieldrin. Thus, in general, the principal cyclodiene insecticides are more toxic than DDT to land vertebrates but of a similar order of toxicity to fish. Lethal poisoning of vertebrates by cyclodiene insecticides in the field was frequently reported when the insecticides were widely used, and will be discussed in the next section (section 5.3.5).

Because of its lipophilicity and refractory character, the toxic effects of dieldrin may be carried through to the next generation. In one example, dosing of female small tortoiseshell butterflies (*Aglais urticae*) with dieldrin led to an increased number of deformed adults emerging from pupae (Moriarty, 1968).

A further feature of dieldrin toxicity, shared with other persistent organochlorine insecticides, is delayed neurotoxicity. In one example from the field in The Netherlands, female, but not male, eider ducks (*S. mollissima*) with neurotoxic symptoms died during the breeding season. They had stored dieldrin residues, which had built up in depot fat during the period before going into lay. These fat reserves, together with the insecticide stored in them, were rapidly mobilised during egg laying, and blood dieldrin levels rose sharply as a consequence. Eventually, lethal concentrations were reached in the nervous system causing convulsions and other symptoms of dieldrin toxicity (Koeman and van Genderen, 1970; Koeman, 1972).

TABLE 5.9 *Toxicity of cyclodienes*

Compound	Species	Toxicity
Dieldrin	*Daphnia magna*	96 h LC_{50} 330 µg/L
Dieldrin	Fathead minnow	96 h LC_{50} 4–18 µg/L
Dieldrin	Rainbow trout	96 h LC_{50} 1.2–9.9 µg/L
Heptachlor	Rainbow trout	96 h LC_{50} 7 µg/L
Dieldrin	Rat	Acute oral LD_{50} 37–87 mg/kg
Dieldrin	Rabbit	Acute oral LD_{50} 45–50 mg/kg
Heptachlor	Rat	Acute oral LD_{50} 40–162 mg/kg
Dieldrin	Pigeon (*Columba livia*)	Acute oral LD_{50} 67 mg/kg
Endrin	Rat	Acute oral LD_{50} 4–43 mg/kg

Data from EHC 38, 91 and 130.

5.3.5 ECOLOGICAL EFFECTS OF CYCLODIENES

The following account will be largely restricted to the effects of aldrin, dieldrin and heptachlor, compounds that were once widely used and for which there is the clearest record of ecological effects.

Effects on population numbers

Within a short time after the introduction of aldrin, dieldrin and heptachlor seed dressings into Western Europe in the mid-1950s, there were reports of large-scale kills of birds on farmland, including predatory birds. Individuals contained high enough residues of dieldrin and or heptachlor epoxide in their tissues to cause death (little or no aldrin or heptachlor were found in tissues because of rapid biotransformation into their respective epoxides). Also, characteristic symptoms, e.g. convulsions, were observed in individuals, which were later found to contain lethal levels of the compounds. Thus, there was early evidence of secondary poisoning due to transfer of the compounds to predators via their prey. Foxes were also victims of secondary cyclodiene poisoning. The toxic effects were largely attributed to dieldrin residues, although heptachlor epoxide was also implicated in certain areas. As mentioned earlier, dressed seed would typically contain 800 ppm of cyclodiene. Thus, grain-eating birds or mammals did not need to consume very large amounts to acquire lethal doses. The question remained, however, whether lethal poisoning was on a large enough scale to cause the decline of vertebrate populations. One of the earliest reports of cyclodiene poisoning in the field came from a coastal area of The Netherlands (Koeman *et al.*, 1967; Koeman, 1972), where sandwich terns (*Sterna sandvicensis*) died showing symptoms of cyclodiene poisoning. Both adults and chicks were affected. Upon analysis, they were found to contain residues of dieldrin, endrin and telodrin at high enough levels, singly or in combination, to cause lethal toxicity. The source turned out to be a neighbouring factory that was synthesising the compounds. These mortalities were linked to a local decline in the sandwich tern population that began around 1962. In agricultural areas of The Netherlands, a decline of the buzzard (*Buteo buteo*), which brought the species to near extinction locally during 1965–6, was attributed very largely to lethal poisoning by dieldrin (Fuchs, 1967). Analysis of birds found in the field that had died showing symptoms of cyclodiene poisoning showed a mean residue of dieldrin in the liver of 18 ppm. From Britain came evidence of widespread declines of the sparrowhawk (*A. nisus*) and the peregrine (*F. peregrinus*). The declines coincided in time with the introduction of aldrin, dieldrin and heptachlor into the country in 1956, and occurred in areas where the chemicals were widely used (Newton, 1986; Ratcliffe, 1993) (see Figures 5.7 and 5.8). Both species are bird-eating raptors, which were exposed to high levels of dieldrin and heptachlor epoxide in the tissues of grain-eating birds upon which they preyed. Cyclodiene poisoning was confirmed in individual cases on the grounds of lethal tissue levels of dieldrin and/or heptachlor epoxide and on toxic symptoms shown before death. Later evidence suggested that British merlins

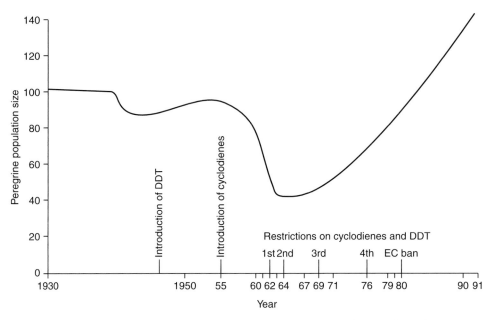

(*Falco columbarious*) had also declined because of cyclodiene toxicity (Newton *et al.*, 1978).

Sparrowhawks declined very sharply in the agricultural areas of eastern England, where cyclodienes were widely used, becoming virtually extinct in parts of East Anglia where they had once been common. Such declines were not seen, however, in more western and northern areas, where there was much less use of cyclodienes (Newton and Haas, 1984; Newton, 1986). Reasonable numbers were also maintained in the New Forest in southern England, where there was little or no use of these insecticides. Peregrines declined sharply in coastal areas of southern England and Wales, disappearing completely from the south-east coast. They were less affected in northern England and Scotland, maintaining numbers well in the Scottish Highlands (Ratcliffe, 1993). It is important to emphasise that these declines came in the mid-1950s, long after the marked eggshell thinning caused by p,p'-DDE, which became clearly manifest by 1947 (Ratcliffe, 1967). Sparrowhawks commence breeding at 1–2 years, peregrines at 2–3 years; thus, any population effects caused by a reduction in breeding success due to eggshell thinning would have been evident long before the population crashes of the mid- to late 1950s which are associated with the introduction of cyclodienes.

There are detailed records of the residues of dieldrin, heptachlor epoxide and other organochlorine insecticides in British sparrowhawks and kestrels (*F. tinnunculus*) from 1963 to the 1990s (Newton, 1986; Newton and Wyllie, 1992) The kestrel was another predatory species showing a sharp decline in agricultural areas over the period under

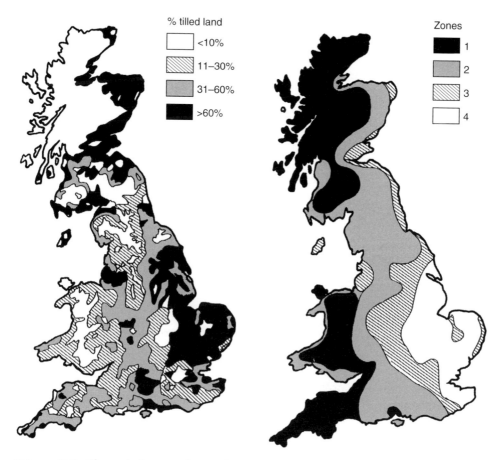

% tilled land

□ <10%

▨ 11–30%

▨ 31–60%

■ >60%

Zones

■ 1

▨ 2

▨ 3

□ 4

FIGURE 5.8 *Changes in the status of sparrowhawks in relation to agricultural land use and organochlorine use. The agricultural map (left) indicates the proportion of tilled land, where most pesticide is used. The sparrowhawk map (right) shows the status of the species in different regions and time periods. Zone 1, sparrowhawks survived in greater numbers through the height of the 'organochlorine era' around 1960; the population decline is judged to be less than 50% and recovery is effectively complete before 1970. Zone 2, the population decline is more marked than in zone 1, but it had recovered to more than 50% by 1970. Zone 3, the population decline is more marked than in zone 2, but it had recovered to more than 50% by 1980. In general, the population decline was more marked, and the recovery later, in areas with a greater proportion of tilled land (based on agricultural statistics for 1966). From Newton and Haas (1984), reproduced in Newton (1986).*

consideration, although less marked than in the case of the sparrowhawk. Residues in the livers of sparrowhawks and kestrels are compared in Figure 5.8 (Walker and Newton, 1998, 1999). Looking at the distribution of liver dieldrin levels in birds found dead from different areas, both species show peaks centring on approximately 20 ppm of dieldrin, which account for a high proportion of the total samples coming from the 'high-cyclodiene' area of eastern England during 1963–75. On the evidence of the magnitude of the residues, and the toxic symptoms sometimes observed, it is concluded that these peaks contain individuals acutely poisoned by cyclodienes. In the samples taken during 1963–75 from the area of low cyclodiene use, the peaks centring on approximately 20 ppm are relatively very small, representing no more than a few per cent of individuals collected.

These results provide strong evidence that dieldrin and heptachlor epoxide were primarily responsible for the population declines of both species in eastern England over the period 1963–75. After 1975, use of cyclodienes as seed dressings for autumn-sown cereal was banned. Dieldrin levels began to fall sharply in sparrowhawks and kestrels after the ban came into force. However, populations did not begin to recover until the geometric mean of the dieldrin concentration in liver fell below 1 ppm. As lethal toxicity of dieldrin in experimentally poisoned birds is associated with dieldrin levels of 10 ppm or more, this is a surprising observation. Very few (not more than 5%) of individuals in samples with geometric means of 1 ppm could have residue levels as high as 10 ppm.

Inspection of the distribution of liver dieldrin levels in birds collected between 1976 and 1982 reveals the virtual disappearance of the 'lethal toxicity' peak centring on approximately 20 ppm. On the other hand, another peak appears centring on 4.6 ppm in the kestrel (Figure 5.9). As mentioned earlier (p. 111), dieldrin and other cyclodienes can produce a variety of sublethal neurotoxic effects, including disorientation, lack of coordination and tremors, at tissue concentrations well below those that cause death. It may therefore be suggested that these individuals with liver dieldrin levels between approximately 3 and 9 ppm experienced sublethal neurotoxic effects before death. Such a hypothesis could also provide a solution to the dilemma mentioned earlier – why did mean liver levels need to fall below 1 ppm before populations recovered? The delayed recovery is easier to understand if a substantially larger proportion of birds than< 5% had died because of the effects of dieldrin. Loss of coordination and other sublethal neurotoxic effects would surely impair the hunting skills of these raptors, and failure to catch prey would lead to starvation (Walker and Newton, 1998, 1999). Following this reasoning, a higher proportion of the recorded deaths of both species in eastern England during 1976–82 would be attributed to dieldrin than would be predicted from an acute lethal threshold of 10 ppm.

Further examination of Figure 5.9 reveals a difference between sparrowhawks and kestrels. Considering dieldrin levels of 3 ppm and above, the kestrel has proportionally more individuals in excess of 9 ppm than has the sparrowhawk. If the foregoing theory is correct, sparrowhawks showed a higher incidence of deaths due to sublethal toxicity than did kestrels and, conversely, kestrels showed a higher proportion of deaths due to direct acute poisoning. This suggests that sparrowhawks are more susceptible to sublethal effects than kestrels. This could be because the sparrowhawk depends on great manoeuvrability to catch its prey in the air, whereas the kestrel drops on its prey from a hovering or perching position. These data for the sparrowhawk have been incorporated, together with other data for the species, into a population model (Sibly *et al.*, 2000). The population decline of the sparrowhawk in eastern England during 1963–75 was predicted reasonably well by a model that assumed that all birds with

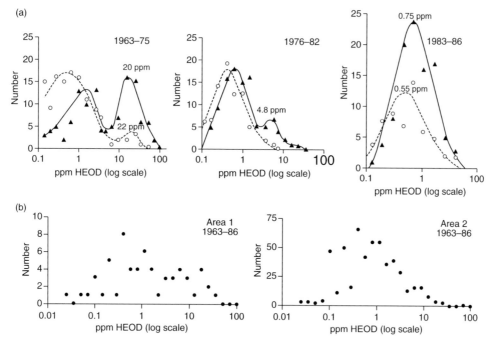

FIGURE 5.9 *(a) Distribution of dieldrin (HEOD) residues in the livers of kestrels from two different areas of Britain. The HEOD residues are expressed as ppm wet weight. They are plotted on a log scale. The numbers of individuals with dieldrin residues falling within the ranges of concentrations represented by 0.15 log units are given on the vertical axis. The concentrations plotted represent the midpoints of each log range. Area 1 (highest cyclodiene use),* ▲*; area 2 (lowest cyclodiene use),* ○*. (b) Distribution of dieldrin (HEOD) residues in the livers of sparrowhawks. Area 1 is zone 4, and area 2 is zones 1 and 2 in Figure 5.8. From Walker and Newton (1999).*

dieldrin levels of 3 ppm or more died because of poisoning by the insecticide. The substantially smaller number of dead birds with 9 ppm or more (i.e. presumed to have died from direct acute toxicity) was insufficient to predict the rate and scale of the population decline that actually occurred.

Dieldrin and endrin were once used for the control of locusts and tsetse flies in Africa (Koeman and Pennings, 1970). Large numbers of birds and mammals were poisoned as a consequence, including many insectivores. The latter point is of interest because insects are usually poor 'vectors' of insecticides, on account of their high sensitivity to the toxic effects. However, in the field programmes in Africa, relatively large quantities of these two highly toxic compounds were sprayed over substantial areas, and were freely available for uptake (cf. the limited availability of cyclodienes used as seed dressings in Western Europe and North America). Under these conditions, individual insects would have picked up, by direct contact, doses of insecticide far in excess of those needed to kill them; they would then pass on high doses to the birds feeding upon them. With the general spraying of the area, the insecticide would have also have been absorbed by birds directly through inhalation, drinking water and leaf surfaces. It is not clear what effects these casualties had at the population level or above.

Development of resistance to cyclodienes

There have been many examples of insects developing resistance to aldrin or dieldrin, sometimes associated with cross-resistance to lindane (Brooks, 1974; Salgado, 1999). Examples include resistant strains of the fruit fly (*Drosophila melanogaster*), the housefly (*M. domestica*), the German cockroach (*B. germanica*), the cabbage root fly (*Erioischia brassicae*) and the onion maggot (*Hylemya antiqua*). There is now good evidence that resistance is due to the presence of a mutant form of the GABA$_A$ receptor (Brooks, 1992), which is relatively insensitive to cyclodienes and gamma-HCH. A single gene conferring 4000-fold resistance to dieldrin has been cloned from a field strain of *D. melanogaster* and found to code for a subunit of the GABA receptor (Ffrench-Constant *et al.*, 1993). Resistance in this case was associated with a single point mutation of alanine to serine in the second transmembrane domain of the protein, which is thought to line the pore of the chloride channel of the GABA receptor. Similar mutations in strains of the housefly and the German cockroach also conferred very high levels of resistance to dieldrin. The levels of resistance of these strains to HCH and certain other cyclodienes were far less (Salgado, 1999).

Some field studies have produced evidence suggesting that resistance is developed by wild vertebrates that have been regularly exposed to cyclodienes. Examples include resistance of pine mice (*Microtus pinetorum*) to endrin (Webb *et al.*, 1972) and the resistance of mosquito fish (*Gambusia affinis*) in the Mississippi River to aldrin (Wells *et al.*, 1973). In both cases, resistance was attributed to enhanced metabolic detoxication by the resistant strain.

5.4 *Hexachlorocyclohexanes*

Hexachlorocyclohexanes (HCHs) have not caused the problems of bioconcentration or bioaccumulation associated with DDT or the cyclodienes, and they have not been implicated in large-scale environmental problems. They will only be discussed briefly here, drawing attention to certain differences from the foregoing groups.

The first commercially available HCH insecticide [sometimes misleadingly called benzene hexachloride (BHC)] was a mixture of isomers, principally α-HCH (65–70%), β-HCH (7–10%) and γ-HCH (14–15%). Most of the insecticidal activity was due to the γ isomer, a purified preparation of which (>99% pure) was marketed as lindane. In Western countries technical HCH was quickly replaced by lindane, but in some other countries (e.g. China) the technical product, which is cheaper and easier to produce, has continued to be used. HCH has been used as a seed dressing, as a crop spray and as a dip to control ectoparasites of farm animals. It has also been used to treat timber against wood-boring insects.

γ-HCH (Figure 5.1) is more polar and more water soluble than most other organochlorine insecticides (Table 5.1), and it is metabolised relatively rapidly to water-soluble products. The metabolism of γ-HCH is complex and involves both dehydrochlorination reactions mediated by GSH and hydroxylations mediated by cytochrome P450. The main excreted metabolites are various chlorophenols, free or in conjugated form; prominent among them are trichlorophenols (EHC 124). γ-HCH appears to be rapidly eliminated by vertebrate species, and residues in free-living vertebrates and invertebrates were found to be low when the compound was widely used in agriculture, e.g. levels in livers of birds of prey were 0.01–0.1 ppm.

γ-HCH acts upon GABA receptors in a similar fashion to cyclodienes (Eldefrawi and Eldefrawi, 1990), and cross-resistance of insects between the two types of insecticide is sometimes due to a mutant form of the receptor, as discussed in section 5.3.5. Acute oral LD_{50} values of γ-HCH to rodents range from 55 to 250 mg/kg. Thus, it is of a similar order of toxicity to DDT, but on the whole it is less toxic than aldrin, dieldrin or endrin. It has not been implicated in field mortalities of birds, in contrast to the cyclodienes. On the other hand, bats can be lethally poisoned when exposed to lindane-treated wood (Boyd *et al.*, 1988). This poses a risk to bats roosting in lofts containing treated timbers. γ-HCH is highly toxic to fish, with LC_{50} values for fish falling into the range 0.02–0.09 mg/L. When used as a sheep dip, there was a hazard to freshwater fish when sheep farmers discharged sheep dips into neighbouring water courses.

Of the other HCH isomers, the alpha and beta forms are less toxic than the gamma form. However, the beta form is more persistent than the gamma form, and unacceptably high residues have sometimes been reported in foods originating from countries where technical HCH is still used (EHC 123).

5.5 *Summary*

Organochlorine insecticides such as aldrin, dieldrin, heptachlor and DDT illustrate well the environmental problems associated with persistent lipophilic compounds of high toxicity, be that toxicity lethal or sublethal. They serve as models for environmental pollutants of this kind, having been studied in much greater depth and detail than most other pollutants. The original insecticides, or their stable metabolites, can undergo biomagnification of several orders of magnitude with movement along food chains. Attention has been drawn to the structural features of these polyhalogenated compounds that make them so resistant to biodegradation.

The organochlorine insecticides have been shown to cause widespread population declines of certain raptorial birds, such as the peregrine falcon, the sparrowhawk and the bald eagle. DDT has caused population declines of predatory birds in North America, as a result of eggshell thinning, brought about by its highly persistent metabolite *p,p'*-DDE. Aldrin, dieldrin and heptachlor were responsible for the population crashes of peregrines and sparrowhawks in Britain and certain other

European countries during the late 1950s. Although acute lethal toxicity was important here, there is some evidence to suggest that sublethal neurotoxic effects were also important.

There are many examples of insects developing resistance to dieldrin. The best known mechanism is the production of mutant forms of the target site (GABA receptor) that are insensitive to the insecticide. DDT resistance has been attributed to enhanced DDT-dehydrochlorinase and aberrant forms of the Na^+ channel (kdr and super-kdr).

5.6 *Further reading*

Brooks, G. T. (1974) *The Chlorinated Insecticides*, vols 1 and 2. A detailed and authoritative standard reference work on the chemistry, biochemistry and toxicology of the organochlorine insecticides.

EHC 91 (1989) *Aldrin and Dieldrin*. A valuable source of information on the environmental toxicology of aldrin and dieldrin.

Moriarty, F. (ed.) (1975) *Organochlorine Insecticides*; *Persistent Organic Pollutants*. A collection of focused chapters on ecotoxicological aspects of the organochlorine insecticides.

CHAPTER 6

Polychlorinated biphenyls and polybrominated biphenyls

6.1 Background

The PCBs and PBBs are industrial chemicals that do not occur naturally in the environment. The properties, uses and toxicology of the PCBs are described in detail in Safe (1984) and EHC 140; PBBs are described in Safe (1984) and EHC 152.

PCBs were first produced commercially around 1930. The commercial products are complex mixtures of congeners, generated by the chlorination of biphenyl. Most of them are very stable viscous liquids, of low electrical conductivity and low vapour pressure. Their principal commercial applications have been (1) as dielectrics in transformers and large capacitors, (2) in heat transfer and hydraulic systems, (3) in the formulation of lubricating and cutting oils and (4) as plasticisers in paints and as ink solvents in carbonless copy paper. With such a diversity of uses they entered the natural environment by many different routes before they were subject to bans and restrictions. The level of chlorination determines the composition and properties, and ultimately the commercial use of PCB mixtures. Depending on reaction conditions, levels of chlorination ranging from 21% to 68% (percentage by weight) have been achieved. The commercial products are known by names such as 'Aroclor', 'Clophen' and 'Kanechlor', usually superseded by a code number, which indicates the quality of the product. Thus, in one series of products, 'Aroclor 1242' and 'Aroclor 1260' contain approximately 42% chlorine and approximately 60% chlorine respectively. The first two numbers of the code indicate that the product is derived from biphenyl, the second two indicate the approximate level of chlorination. Since the discovery of pollution problems during the 1960s, the production of PCBs has greatly declined, and there are few remaining uses at the time of writing.

PBBs have also been marketed as mixtures of congeners, produced in this case by the bromination of biphenyl. The main commercial use of them has been as fire retardants, for which purpose they were introduced in the early 1970s. The most widely known commercial PBB mixture was 'Firemaster', first produced in 1970 in the USA, with production discontinued in 1974 after the recognition of pollution problems.

Many of the components of PCB and PBB mixtures are both lipophilic and stable chemically and biochemically. Like the persistent organochlorine insecticides and their stable metabolites, they can undergo strong bioconcentration and bioaccumulation to reach relatively high concentrations in predators.

6.2 *Polychlorinated biphenyls*

6.2.1 CHEMICAL PROPERTIES

In theory, there are 209 possible congeners of PCB. In practice, only about 130 of these are likely to be found in commercial products. The structures of some congeners are shown in Figure 6.1. The more highly chlorinated a PCB mixture is, the greater the proportion of higher chlorinated congeners in it. Thus, in Aroclor 1242 (42% chlorine), some 60% of the mass is in the form of tri- or tetrachlorobiphenyls, whereas in Aroclor 1260 (60% chlorine) some 80% of the mass is as hexa- and hepta biphenyls. Small amounts of PCDFs are found in commercial products (see Chapter 7).

Individual PCB congeners are often crystalline, but most commercial mixtures exist as viscous liquids, turning into resins with cooling. Highly chlorinated mixtures, such as Aroclor 1260, are resins at room temperature. In general PCBs are very stable compounds of low chemical reactivity; they have rather high density and are fire resistant. They have low electrical resistance which, in combination with their heat stability, makes them very suitable as cooling liquids in electrical equipment. They have low water solubility and high lipophilicity, and they have low vapour pressures (see Table 6.1). With increasing levels of chlorination, vapour pressure and water solubility tend to decrease, and log K_{ow} to increase.

Some PCB congeners have coplanar structures (see, for example, 3,3',4,4'-tetrachlorobiphenyl in Figure 6.1). The coplanar conformation is taken up when there is no chlorine substitution in *ortho* positions. If there is substitution of chlorine in only one *ortho* position, the molecule may still be close to coplanarity, because of only

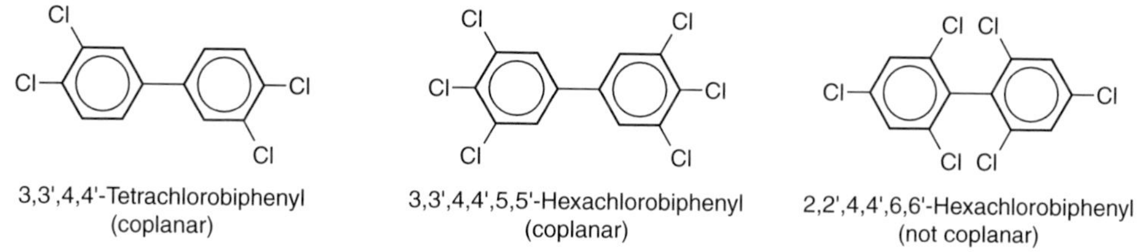

3,3',4,4'-Tetrachlorobiphenyl
(coplanar)   3,3',4,4',5,5'-Hexachlorobiphenyl
(coplanar)   2,2',4,4',6,6'-Hexachlorobiphenyl
(not coplanar)

FIGURE 6.1 *Some PCB congeners.*

limited interaction between Cl and H on adjoining rings. Substitution of chlorines in adjacent *ortho* positions leads to movement of rings away from planarity to accommodate the overlap of the orbitals of the bulky halogen atoms (Figure 6.2). Most PCBs have non-planar structures because of chlorine substitutions in *ortho* positions. There are important biochemical differences between coplanar and non-planar PCB congeners that will be described in later sections.

6.2.2 METABOLISM OF PCBs

In terrestrial animals the elimination of PCBs, like that of organochlorine insecticides, is largely dependent upon metabolism. The rate of excretion of the unchanged congeners is generally very slow, although it should be noted that small amounts are 'excreted' into milk (mammals) or eggs (birds, amphibians, reptiles and insects), presumably transported by lipoproteins (see Chapter 2). In mammals there can also be transport across the placenta into the developing embryo. Although such 'excretions' do not usually account for a very large proportion of the body burden of PCBs, the translocated congeners may still be in sufficient quantity to cause embryo toxicity.

In animals primary metabolism of PCBs is predominantly by ring hydroxylation, mediated by different forms of cytochrome P450, to yield phenolic metabolites. The position of attack is influenced by the substitutions by chlorine. As with other lipophilic polychlorinated compounds, oxidative attack does not usually occur directly on C–Cl positions; it tends to occur where there are adjacent unsubstituted *ortho–meta* or *meta–*

TABLE 6.1 *Properties of PCB congeners*

Compound	Structure	Vapour pressure (atm) $-\text{Log } P_o$	Water solubility (mol/L) $-\text{Log } C$	$\text{Log } K_{ow}$
4,4'-DCB	Coplanar	7.32	6.53	5.33
2',3,4-TCB	Coplanar	6.88	6.52	5.78
2,2',5,5'-TCB	Non-planar	7.60	7.06	6.18
2,2',4,5,5'-PCB	Non-planar	8.02	7.40	6.36
2,2',3,3',4,4'-HCB	Non-planar	9.65	8.72	6.97

All values estimated at 25 °C
Data from Schwarzenbach *et al.* (1993)

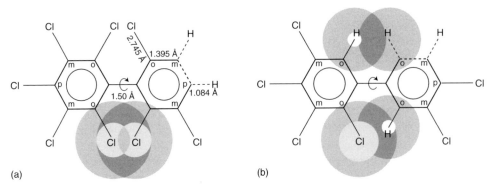

FIGURE 6.2 *Planar and coplanar PCBs. Structural features of CB congeners influencing enzymatic metabolism. Areas where the principal enzymatic reaction occurs are given by broken lines. For atoms in the ortho position, the outer circle represents the area within the van der Waals radius of an atom; the dotted inner circle represents the part of this area which is also within the single bond covalent radius. The van der Waals radius indicates the maximum distance for any possible influence of an atom. The covalent radius represents the minimum distance to which atoms can approach each other. (a) Vicinal atoms in the meta and para positions. Overlapping covalent radii for two ortho Cl show that a planar configuration is highly improbable when three or four ortho Cl are present. (b) Vicinal atoms in the ortho and meta positions. Non-overlapping covalent radii for ortho Cl and ortho H show that a planar configuration causes a much lower energy barrier when chlorine atoms do not oppose each other. Reproduced from Boon et al. (1992).*

para positions on the aromatic ring. Unchlorinated *para* positions are particularly favoured for hydroxylation, a mode of metabolism associated with P450s of family 2 rather than P4501A1/1A2. In the case of aromatic hydroxylations, it has been suggested that primary attack is by an active form of oxygen generated by the haem nucleus of P450 (see Chapter 2) to form an unstable epoxide, which then rearranges to a phenol (for further discussion of mechanism see Trager, 1989; Crosby, 1998). Two examples of hydroxylations of PCBs are shown in Figure 6.3. One PCB is planar, the other coplanar.

Monooxygenase attack upon the coplanar PCB 3,3′,4,4′-tetrachlorobiphenyl (3,3′,4,4′-TCB) is believed to occur at unsubstituted *ortho–meta* (2′,3′) or *meta–para* (3′,4′) positions, yielding one or other of the unstable epoxides (arene oxides) shown in Figure 6.3. Rearrangement leads to the formation of monohydroxy metabolites. In one case, a chlorine atom migrates from the *para* to the *meta* position during this rearrangement (NIH shift), thus producing 4′-OH-3,3′,4,5′-tetrachlorobiphenyl. The mechanism of formation of 2′-OH-3,3′,4,4′-TCB is unclear (Klasson-Wehler, 1989).

In the rabbit, the non-planar PCB 2,2′,5,5′-tetrachlorobiphenyl (2,2′,5,5′-TCB) is converted into the 3′,4′ epoxide by monooxygenase attack upon the *meta–para* position and rearrangement yields two monohydroxymetabolites with substitution in the *meta* and *para* positions respectively (Sundstrom *et al.*, 1976). The epoxide is also transformed into a dihydrodiol by epoxide hydrolase attack (see Chapter 2). This latter conversion is inhibited by 3,3,3-trichloropropene-1,2-oxide (TCPO), thus providing strong confirmatory evidence for the formation of an unstable epoxide in the primary oxidative attack (Forgue *et al.*, 1980).

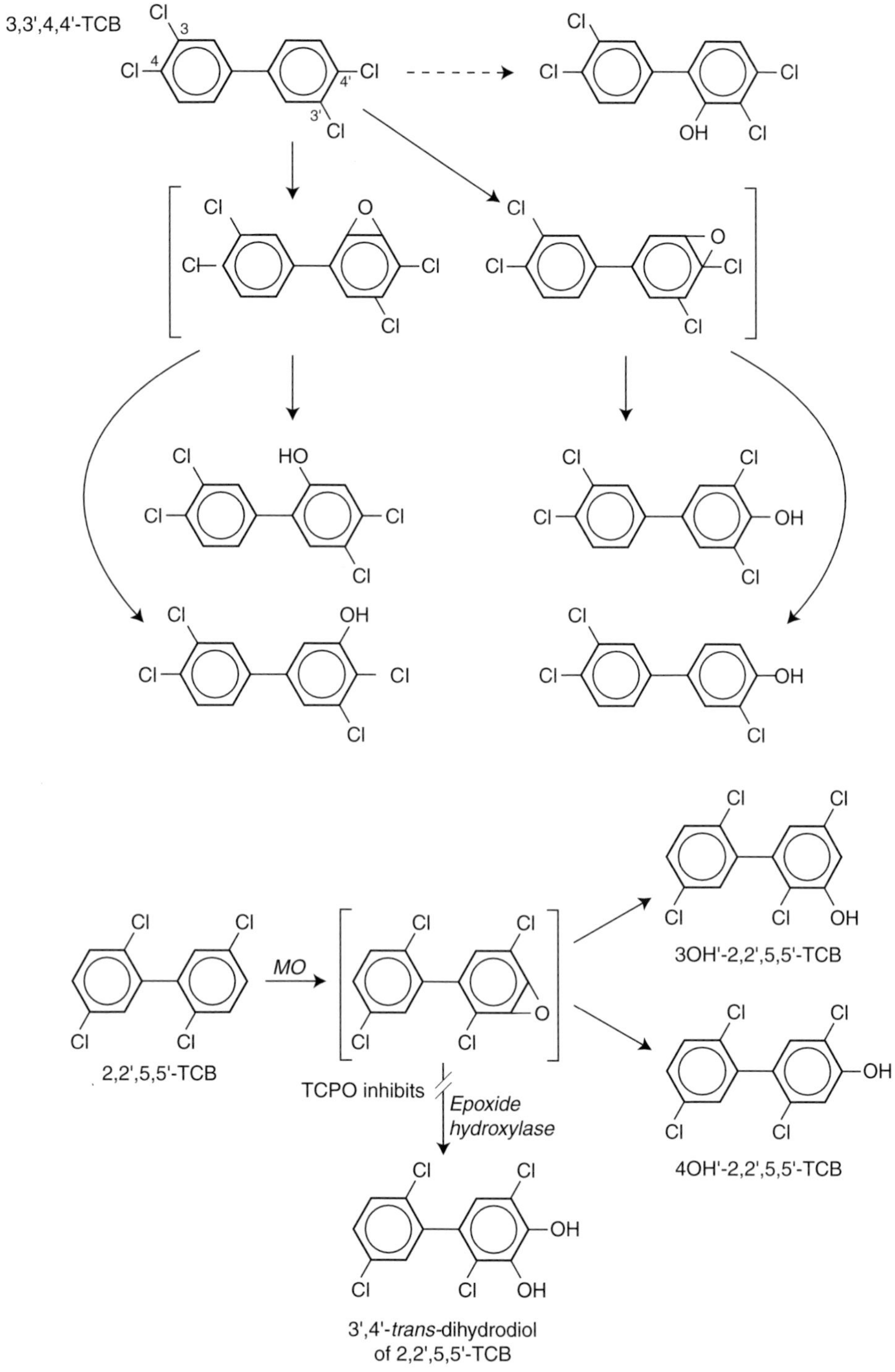

FIGURE 6.3 *Metabolism of PCBs.*

In the examples given there is good evidence for the formation of an unstable epoxide intermediate in the production of monohydroxymetabolites. However, there is an ongoing debate about the possible operation of other mechanisms of primary oxidative attack that do not involve epoxide formation, e.g. in the production of 2′-OH-3,3′,4,4′-TCB (Figure 6.3). As mentioned earlier, P450s of gene family 1 (*CYP1*) tend to be specific for planar substrates including coplanar PCBs; they do not appear to be involved in the metabolism of non-planar PCBs. On the other hand, P450s of gene family 2 (*CYP2*) are more catholic and can metabolise both planar and non-planar PCBs.

Having more unsubstituted ring positions available for metabolic attack, lower chlorinated PCBs are usually more rapidly metabolised than higher chlorinated PCBs. Reflecting this, the pattern of PCB residues changes with movement along food chains (Figure 6.4). Lower chlorinated PCBs decline in relative abundance or disappear altogether at higher trophic levels. The more highly chlorinated compounds, which are refractory to metabolic attack, become dominant in predators (Norstrom, 1988; Boon *et al.*, 1992), which tend have smaller ranges of PCB congeners than do the species below them in the food chain. These trends are readily recognised by comparing capillary gas chromatography (GC) analyses of tissues from organisms representing different trophic levels (Figure 6.4). The early, fast-running peaks representing lower

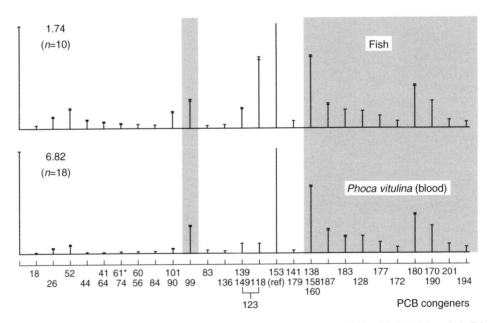

FIGURE 6.4 *Mean concentration of CB 153 in* μg/g *pentane-extractable liquid (PEL) in whole fish from the Dutch Wadden Sea and the cellular fraction of the blood of harbour seals. Numbers of CBs are given in order of elution from the GC column by their systematic numbers according to IUPAC rules as proposed by Ballschmitter and Zell (1980). All concentrations are proportional to the height of the bar. Reproduced from Boon* et al. *(1992).*

chlorinated congeners give way to the slower-moving peaks representing more highly chlorinated compounds with movement along the food chain. Some predatory species such as cetaceans and fish-eating birds metabolise PCBs relatively slowly (Walker and Livingstone, 1992), in keeping with their very low microsomal monooxygenase activities towards lipophilic xenobiotics (Walker, 1980; Walker *et al.*, 1996).

Several studies have related the structures of PCBs to their rates of elimination by mammals. In one study (Mizutani *et al.*, 1977), the elimination of tetrachlorobiphenyl congeners was studied in mice that had been fed diets containing a single isomer for 20 days. The estimated half-lives were as follows:

2,2′,3,3′-TCB	0.9 days
2,2′,4,4′-TCB	9.2 days
2,2′,5,5′-TCB	3.4 days
3,3′,4,4′-TCB	0.9 days
3,3′,5,5′-TCB	2.1 days

In another study (Gage and Holm, 1976), the influence of molecular structure on the rate of excretion by mice for 14 different congeners was studied. The results were as follows:

Most rapidly eliminated
4,4′-DCB, 3,3′,4′,6′-TCB, 2,2′,3,4′,6′-PCB and 2,2′,3,4,4′,5′-HCB
Most slowly eliminated
2,2′,4,4′,5,5′-HCB and 2,2′,3,4,4′,5′-HCB

In the latter example, the most slowly eliminated compounds were non-planar, and lacked vicinal unsubstituted carbons in either the *ortho–meta* or the *ortho–para* positions. The more rapidly eliminated compounds all possessed unsubstituted vicinal *ortho–meta* positions. In the former example, the most persistent compound was non-planar, and lacked unsubstituted carbons in the *meta–para* positions. Interestingly, both of the coplanar compounds were eliminated more rapidly, even though one of them (3,3′,5,5′-TCB) lacked unsubstituted vicinal carbons in either position. This suggests that P4501A1/1A2 was able to hydroxylate the molecule reasonably rapidly without any vicinal unsubstituted carbons, presumably without the formation of an epoxide intermediate.

Working with rats, Lutz *et al.* (1977) compared the rates of loss from blood of 4-CB (rapidly metabolised) with that of 2,2′,4,4′,5′-HCB (slowly metabolised). Both showed biphasic elimination, with the former disappearing much more rapidly than the latter. Estimations were made of the rates of hepatic metabolism *in vitro*, which were then incorporated into toxicokinetic models to predict rates of loss. The predictions for HCB were very close to actual rates of loss for the entire period of elimination. For 1-CB, the prediction was good for the initial rate of loss, but loss was overestimated in the later stages of the experiment.

Looking at the foregoing results overall, the rates of loss *in vivo* are related to the rates of metabolism *in vitro*, measured or estimated. As with the organochlorine insecticides, problems of persistence are associated with compounds that are not readily metabolised, e.g. 2,2′,4,4′,5,5′-HCB in the examples above. For further discussion of the dependence of elimination of lipophilic xenobiotics on metabolism (see Walker, 1981).

Residues of PCBs in animal tissues include not only the original congeners themselves but also hydroxy metabolites that bind to cellular proteins, e.g. transthyretin (TTR) (Brouwer, 1991; Klasson-Wehler *et al.*, 1992; Lans *et al.*, 1993). Small residues are also found of methyl-sulphonyl metabolites of certain PCBs (Bakke *et al.*, 1982, 1983). These appear to originate from the formation of glutathione conjugates of primary epoxide metabolites, thus providing further evidence of the existence of epoxide intermediates. Further biotransformation, including methylation, yields methyl-sulphonyl products that are relatively non-polar and persistent (cf. *p,p′*-DDE metabolism, Figure 2.15).

PCBs can act as inducers of P450, and consequently accelerate the rate of their own metabolism. Coplanar PCBs bind to the Ah receptor and thereby induce P450s 1A1/1A2. Inductions of P4501A1/1A2 by organohalogen compounds are associated with a number of toxic effects (Ah receptor-mediated toxicity), which will be discussed in section 6.2.4. It should also be noted that induction of these P450s can increase the rate of activation of a number of carcinogens and mutagens, e.g. certain PAHs. Non-planar PCBs can induce cytochrome P450s belonging to family 2. The induction of P450 forms by PCBs and other pollutants provides the basis for valuable biomarker assays that are being increasingly used in field studies (Rattner *et al.*, 1993; Walker *et al.*, 1998).

Certain anaerobic bacteria can reductively dechlorinate PCBs in sediments (EHC 140). Higher chlorinated PCBs are degraded more rapidly than lower chlorinated ones, which is in contrast to the trend for oxidative metabolism described above. Genetically engineered strains of bacteria have been developed to degrade PCBs in bioremediation programmes.

6.2.3 ENVIRONMENTAL FATE OF PCBs

PCBs, like persistent organochlorine insecticide residues, have become widely distributed around the globe, including in snow and biota of polar regions (Muir *et al.*, 1992). Long-range aerial transport and subsequent deposition has been the major factor here (Mackay, 1991). At the time of writing, little PCB is being released into the environment, but redistribution is evidently still occurring from 'sinks' such as contaminated sediments and landfall sites, from which the persistent congeners are only slowly being lost. The levels of higher chlorinated PCBs are still undesirably high in predators – notably mammals and fish-eating birds at the top of marine food chains (Walker and Livingstone, 1992; de Voogt, 1996).

Although higher chlorinated PCBs are degraded more rapidly than lower chlorinated ones in anaerobic sediments, the reverse is true in terrestrial and aquatic food chains (section 6.2.2). As explained earlier (pp. 126–127), hydroxylations tend to be very slow in the absence of unchlorinated positions favourable for oxidative attack. Recalcitrant higher chlorinated PCBs tend to be strongly bioaccumulated and bioconcentrated with movement along food chains.

An early indication of the tendency for certain PCB congeners to be biomagnified came from studies on the Great Lakes of North America. The concentration of total PCBs in the food chain was found to be as follows:

Phytoplankton	0.0025 ppm
Zooplankton	0.123 ppm
Rainbow trout smelt	1.04 ppm
Lake trout	4.83 ppm
Herring gull eggs	124 ppm

There have been a number of estimates of bioconcentration factors for total PCBs in aquatic species after long-term exposure to PCB mixtures (EHC 140). Values for both invertebrates and fish have been extremely variable ranging from values below 1 to many thousands. Bioaccumulation factors for birds and mammals for different Aroclors have indicated only limited degrees of bioaccumulation from food, e.g. 6.6 and 14.8 for the whole carcasses of big brown bats (*Eptesicus fuscus*) and white pelican (*Pelecanus erythrorhynchus*) respectively (see EHC 140).

As with organochlorine insecticides (Chapter 5), residue data need to be interpreted with caution. However, it is clear that there can be biomagnification by several orders of magnitude with movement up the food chain. Moreover, the values for total PCBs underestimate the biomagnification of refractory higher chlorinated PCBs. The marked biaccumulation of refractory PCB congeners is illustrated by the data for fish-eating birds given in Table 6.2 (Borlakoglu *et al.*, 1988; Norstrom, 1988; Walker and Livingstone, 1992). In the Canadian study on the herring gull (*Larus argentatus*), a comparison was first made between the concentration of PCB congeners in eggs with those present in a fish, the alewife (*Alosa pseudoharengus*), its principal food in the area of study, Lake Ontario. Some 80% of the total PCBs in the birds were accounted for by about 20 refractory congeners (Norstrom, 1988). One congener, 2,2',4,4',5 (IUPAC code, PCB no. 153, see also Table 6.1) was among the most strongly bioaccumulated and was used as a reference compound. It was assigned a 'bioaccumulation index' of 1.0, and the bioaccumulation factors of other PCBs were expressed relative to this. In another study (Borlakoglu *et al.*, 1988; Walker and Livingstone, 1992) the pattern of PCB congeners found in fish-eating seabirds collected in British and Irish coastal waters were compared with the pattern in the PCB mixture Aroclor 1264. The species studied included the cormorant (*Phalocrocorax carbo*), shag (*P. aristotelis*), guillemot (*Uria aalge*), razorbill (*Alca torda*) and puffin (*Fratercula arctica*).

With both studies, the congeners that were strongly bioaccumulated had one feature in common; they lacked free adjacent *meta–para* positions on the rings. Also, they

TABLE 6.2 *Bioaccumulation of PCB isomers by seabirds*

IUPAC no.	Isomer Cl substitution		No. of unsubstituted adjacent carbons		Relative BF* in herring gull	Enrichment† index seabirds
	Ring 1	Ring 2	*om*	*mp*		
74	4	2,4,5	2	0	0.86	
66	2,4	3,4	2	0	0.43	
99‡	2,4	2,4,5	1	0	0.74	
114	4	2,3,4,5	2	0	< 0.10	
118‡	3,4	2,4,5	1	0	0.83	23.60
146	2,3,5	2,4,5	0	0	1.00	
153‡	2,4,5	2,4,5	0	0	1.00‡	1.44
105‡	3,4	2,3,4	2	0	0.57	
138‡	2,3,4	2,4	0	0	0.93	3.10
178	2,3,5	2,3,4,6,6	0	0	0.52	
187‡	2,4,5	2,3,5,6	0	0	0.94	
128‡	2,3,4	2,3,4	2	0	0.77	
177	2,3,4	2,3,5,6	1	0	0.59	
180‡	2,4,5	2,3,4,5	0	0	1.10	1.28
170‡	2,3,4	2,3,4,5	1	0	1.05	1.35
201	2,3,5,6	2,3,4,5	0	0	0.98	
196‡	2,3,4,5	2,3,4,6	0	0	1.07	
194‡	2,3,4,4	2,3,4,5	0	0	0.90	
202	2,3,5,6	2,3,5,6	0	0		3.52
52	2,5	2,5	0	2	< 0.10	ND
70	2,5	3,4	1	1	0.02	
101	2,5	2,4,5	0	1	0.24	0.40
97	2,3	2,4,5	1	1	< 0.10	0.20
87	2,5	2,3,4	1	1	0.12	
110	3,4	2,3,6	1	1	0.21	
151	2,5	2,3,5,6	0	1	0.02	0.10
149	2,3,6	2,4,5	0	1	0.18	0.17
141	2,5	2,3,4,5	0	1	0.19	
174	2,3,6	2,3,4,5	0	1	0.07	

*Bioaccumulation factor (BF) of herring gull (*Larus argentatus*) eggs/alewife (*Alosa pseudoharengus*) relative to PCB no. 153 (Norstrom, 1988).

†The 'enrichment index' is PCB as a percentage of total PCB in seabird depot fat/ PCB as a percentage of total PCB in Aroclor 1260.

‡These were among the 14 PCB congeners found at the highest concentrations in eggs from the Mediterranean Sea and the Atlantic Ocean (Renzoni *et al.*, 1986). These authors also found nos 156, 172 and 183; the last two were also reported by Borlakoglu *et al.* (1988).

were predominantly non-planar. This suggested that persistence was related to the failure of P450 forms (notably those belonging to family 2) to metabolise such non-planar congeners. Interestingly, coplanar congeners, e.g. 3,3′,4,4′-TCB were not among the most persistent compounds. Their relative abundance was considerably less than in original PCB mixtures. Presumably they had been extensively metabolised by birds, or lower in the food chain by monooxygenase attack. The metabolism of planar compounds such as these is particularly associated with P4501A1/1A2, which are found in birds and in fish, although their characteristics are somewhat different from mammalian forms (Livingstone and Stegeman, 1998; Walker, 1998a,b,c). Cetaceans and seals also have a marked capacity to bioaccumulate a limited number of refractory highly chlorinated PCBs (Boon *et al.*, 1992; Tanabe and Tatsukawa, 1992). Seals appeared to metabolise PCBs lacking free *meta–para* ring positions more rapidly than did cetaceans, irrespective of whether the molecules were planar or non-planar. In a study of total PCB residues in cetaceans from the Pacific Ocean, high levels were found in certain species of whale, with particularly high levels in killer whales (*Orcinus orca*) (approximately 800 ppm in blubber) (Tanabe and Tatsukawa, 1992). Killer whales were found to have very low levels of hepatic microsomal monooxygenase activity towards aldrin and aniline, namely 0.041 and 0.22 nmol/mg protein/min. Comparable values for the rat were 1.58 and 2.24 respectively. Activities towards these substrates have been attributed mainly to P450s of family 2, and these data suggest very low levels of this type of detoxication in the killer whale. This fits in well with the general picture of poor P450-mediated oxidative metabolism of lipophilic xenobiotics by specialised predators (see Chapter 2).

6.2.4 THE TOXICITY OF PCBs

The acute toxicity of PCB mixtures to vertebrates tends to be low, typically 1–10 g/kg to rats. The concern is with sublethal and chronic toxicity. Different PCB congeners show different modes of toxic action, so it is not surprising that mixtures of them can produce a disconcerting range of toxic effects. Some early work on PCB toxicity to birds attempted to deal with this complexity, drawing attention to, among other things, effects upon the thyroid (Jefferies, 1975; Jefferies and Parslow, 1976). Unfortunately, this work was largely overlooked at the time, and it was not until some time later that mechanisms of toxicity of PCBs began to be unravelled. Despite much recent progress, understanding of the toxic effects of PCBs is still very incomplete.

That said, it has come to be recognised that coplanar congeners as a group can express toxicity through a common mechanism; interaction with the cytosolic Ah receptor (Safe, 1990; Ahlborg *et al.*, 1994; EHC 140). Although the full picture has yet to be elucidated, many toxic end points, including porphyria, indices of hepatotoxicity, and mortality of embryos of birds are correlated with the degree to which planar polychlorinated compounds bind to the Ah receptor. As described earlier (Chapter 2, and section 6.2.2), interaction with the Ah receptor also leads to induction of P4501A1/1A2, a response that is closely linked to Ah receptor-mediated toxicity.

The complex formed between coplanar PCBs and the Ah receptor migrates to the nucleus, where it triggers the induction of the P450s and certain other enzymes, events associated with the initiation of toxic responses by mechanisms as yet unknown. It should be added that P4501A1/1A2 is involved in the activation of some mutagens/carcinogens. Thus, if organisms are exposed to coplanar PCBs in combination with, for example, mutagenic PAHs there may be an increased rate of DNA adduct formation because the former induce P4501A1/2 (see Walker and Johnston, 1989).

Ah receptor-mediated toxicity is particularly associated with the highly toxic compound 2,3,7,8-tetrachlorodibenzo-*p*-dioxin (TCDD), commonly referred to as 'dioxin'. TCDD toxicology, and the concept of toxicity equivalency factors (TEFs) based upon TCDD, is dealt with in Chapter 7. The main point to make at this juncture is that the toxicity of each individual coplanar congener in a mixture can be expressed in terms of a toxic equivalent calculated relative to the toxicity of dioxin. Summation of the toxic equivalents of the individual coplanar PCBs gives a measure of the toxicity of the whole mixture, as expressed through the Ah receptor mechanism.

A number of field studies have linked the induction of P450s with the presence of PCBs in fish and birds. In one study of pipping embryos of the black-crowned night heron (*Nycticorax nycticorax*), induction of P450s was related to total PCB levels (Rattner *et al.*, 1993). There were close correlations between total PCB concentration in tissues and both ethoxyresorufin *O*-deethylase (EROD) and aryl hydrocarbon hydroxylase (AHH) activities in hepatic microsomes, thus indicating P4501A1 induction. In an earlier study (Ronis *et al.*, 1989) the levels of P4501A1 measured by Western blotting were related to levels of total PCB in body fat in three species of fish-eating seabirds collected from the Irish sea during 1978–88. The species were cormorant, razorbill and puffin. Both studies gave clear evidence for induction of this haem protein in birds in the field, thus raising questions about possible Ah receptor-mediated toxicity.

Another mechanism of action shown by certain coplanar PCBs depends on metabolic activation–conversion of the PCBs to monohydroxy metabolites by cytochrome P4501A1/1A2 (Figure 6.3). 4′-OH-3,3′,4,5′-TCB is a metabolite of 3,3′,4,4′-TCB, which is closely related structurally to thyroxine (T_4). 4′-OH-3,3′,4,5′ competes very strongly with T_4 for binding positions on the protein transthyretin (TTR), (Figure 6.3; Brouwer *et al.*, 1990, 1998; Brouwer, 1991). Some other hydroxy metabolites of PCBs do the same, although less strongly (Lans *et al.*, 1993). When the metabolite binds to TTR, thereby excluding T_4, the associated retinol (vitamin A) binding protein breaks away (Figure 6.5). Thus, the TCB metabolite can reduce the levels of bound thyroxine and retinol in blood with consequent physiological effects (e.g. vitamin A deficiency). These changes have provided the basis for the development of biomarker assays for the toxic action of PCBs (Brouwer, 1991). 3,3′,4,4′-TCB is also an inducer of P4501A1/1A2, and can enhance the rate of its own activation if the levels of exposure are high enough (for further discussion of this question see Walker and Johnston, 1989). In a laboratory study with common tern (*Sterna hirundo*) chicks, dosing with coplanar 3,3′4,4′,5-PCB either alone or in combination with the non-planar 2,2′4,4′,5,5′-HCB caused a reduction in plasma total thyroxine, which was negatively

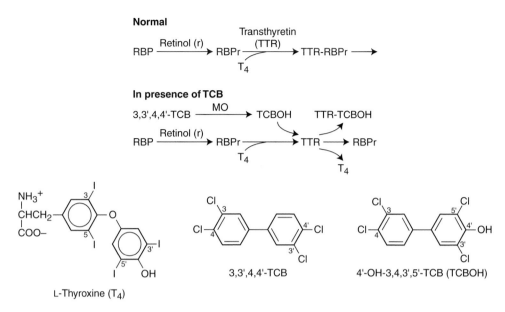

FIGURE 6.5 *Thyroxine antagonism. Mechanism of toxicity of a polychlorinated biphenyl. Retinol (r) binds to retinol-binding protein (RBP), which is then attached to transthyretin (TTR). Thyroxine (T₄) binds to TTR and is transported via the blood in this form. The coplanar PCB, 3,3′,4,4′-tetrachlorobiphenyl (3,3′,4,4′-PCB) is converted into hydroxymetabolites by the inducible cytochrome P450 called P4501A1. The metabolite 4′-OH-3,3′,4,5′-tetrachlorobiphenyl (TCBOH) is structurally similar to thyroxine and strongly competes for thyroxine binding sites. The consequences are loss of thyroxine from TTR, the fragmentation of the TTR–RBP complex, and loss of both thyroxine and retinol from blood. After Brouwer (1991).*

correlated with hepatic TEQ (dioxin equivalents) concentrations (Bosveld *et al.*, 2000). Cytochrome P4501A1 was also induced in these birds.

In mammals, there is evidence that hydroxy PCBs are transferred across the placenta into the fetus, where they accumulate (Morse *et al.*, 1995, 1996). In an experiment with pregnant rats exposed to 3,3′,4,4′-TCB, substantial levels of 4′-OH-3,3′,4,5′-TCB accumulated in the fetus with concomitant reduction in levels of T_4. This was due to competitive binding of the PCB metabolite to TTR, thereby excluding T_4. Similar effects were found when pregnant rats were exposed to Aroclor 1254, where relatively high levels of 4-OH-2,3,3′,4′,5-pentachlorobiphenyls were detected in fetal plasma and brain (Morse *et al.*, 1996). Similar effects have also been demonstrated in pregnant mice that had been fed 3,3′,4,4′-TCB (Danerud *et al.*, 1996). Changes in fetal and neonatal levels of thyroid hormones caused by PCBs may have a number of harmful consequences including effects on brain development and behaviour.

The functional form of thyroxine (T_3) is generated by the deiodination of T_4, and PCBs can influence the tissue levels of this form by disturbing metabolism, as well as by reducing the binding of T_4. PCBs have been shown to inhibit the sulphation of thyroid hormones and the deiodination of T_4 to T_3. They can also induce the glucuronyl transferase that conjugates T_4 (Brouwer *et al.*, 1998).

There is considerable evidence that PCB mixtures and higher chlorinated congeners act as immunosuppressants in mammals (EHC 140). Apart from a number of studies showing the effects upon gross measures of immunological function (e.g. spleen and thymus weights, lymphocyte counts and histology of lymphoid tissue), functional alterations in humoral and cell-mediated immunity have also been demonstrated. Immunosuppression has been reported in harbour seals (*Phoca vitulina*) that had been fed fish diets rich in higher chlorinated PCBs (see Ross *et al.*, 1995; de Voogt, 1996).

Before leaving the question of mechanisms of toxicity, two other issues will be briefly mentioned. The first, which is more relevant to human toxicology than to ecotoxicology, is the question of carcinogenicity. There is no clear evidence that PCB mixtures are primary carcinogens. They do, however, induce P450s that activate carcinogens and mutagens, and there is evidence that treatment with PCBs after exposure to certain carcinogens leads to an enhancement of carcinogenic action (EHC 140). However, there is still uncertainty about the importance of carcinogenicity in wild species. It has been questioned whether it can have a significant effect on population numbers in the wild. The other point deserving mention is the possibility that refractory PCB congeners may promote oxyradical toxicity. When they are bound as substrates to P450s, they are not susceptible to the normal process of oxidation (see Chapter 2), and highly reactive oxygen species, such as the superoxide anion ($O_2^{\cdot -}$), may escape from the vicinity of the haem nucleus and cause damage to neighbouring cellular macromolecules instead of interacting with the PCB. More work needs to be done to test the validity of this theory.

There has been controversy over the question of toxic effects of PCBs on vertebrates in the field. PCB residues found in samples from the field can provide valuable evidence of toxicity, as in the case of other refractory pollutants such as the persistent organochlorine insecticides. Unfortunately, it is difficult to relate residues to toxic effects in the case of PCBs because of the complex mixtures of congeners that are found. The calculation of TEQs has allowed some simplification, and this approach will be discussed in the next section in connection with population effects. Before this option became available, however, there were some poisoning incidents in which PCBs were implicated, but it was difficult to prove causality.

One widely reported incident was a large 'seabird wreck' that occurred in the Irish Sea in the autumn of 1969 (NERC Publication, 1971). An estimated 17 000 dead sea birds were washed ashore along the coastline of the Irish sea. Most of them were guillemots, but there were also razorbills, gannets and herring gulls. The dead birds contained, on average, some 56 ppm of the total PCB in the liver. The significance of such a high concentration seems to have been overlooked in the report, which attempts to relate total body burdens, rather than critical tissue concentrations, to possible effects (Walker, 1990a,b). Unfortunately, in the absence of analytical data on individual congeners and estimations of TEQ values, it is not possible to bring the matter to a definite conclusion. However, certain other points deserve consideration. Some of the birds showed pathological changes in the liver and the kidney that would be consistent with PCB poisoning. Also, some birds contained liver concentrations of total PCB that were high enough to suggest lethal poisoning, on the basis of the limited evidence available. Cormorants experimentally poisoned with Clophen 60 had the following tissue levels: total PCB/liver, 210–285 ppm; total PCB/brain, 76–180 ppm (Koeman *et al.*, 1972). The problem is that in the absence of analytical data for individual congeners the TEQ values for coplanar PCBs in these samples cannot be estimated, and total PCB concentrations provide only a rough guide to likely toxic effects. Differences in susceptibility to PCBs between cormorants and other birds and differences in susceptibility between birds in the field and laboratory are unknown (there are likely to be situations where birds are more susceptible to particular pollutants in the field than in the laboratory because of the impact of other factors, e.g. disease, shortage of food and bad weather). With the benefit of hindsight, it is clear that PCDDs and PCDF residues might have contributed to Ah receptor-mediated toxicity. However, these were not determined because suitable methods of analysis were not available at the time of the seabird wreck. At the end of the day the balance of evidence suggests that PCBs were an important contributory factor to this environmental disaster.

Some deaths of cormorants in the field in Holland were attributed to PCB poisoning on more substantial evidence than was available in the foregoing incident (Koeman, 1972; Walker, 1990a,b). Birds found dead had mean liver and brain levels of 319 ppm and 190 ppm, respectively, which are not very different from those found in experimentally poisoned individuals.

6.2.5 ECOLOGICAL EFFECTS OF PCBs

PCBs have been implicated in a number of population declines reported from the field. Some of the best documented examples are of certain predatory birds inhabiting the Great Lakes. These observations were made between the early 1970s and the mid-1990s (Gilbertson *et al.*, 1998). The herring gull, the Caspian tern (*Hydroprogne caspia*) and the double-crested cormorant (*Phalocrocorax auritus*), which are largely or entirely fish-eating species, showed population declines when PCB levels were high, a situation that was complicated by the presence of relatively high levels of other pollutants (e.g. *p,p′*-DDT and its metabolites). Recoveries have occurred with falling levels of PCBs, *p,p′*-DDE and other pollutants. It is a situation where interpretation of data is made more difficult because of the correlation between the temporal trends of different pollutants. Here, biomarkers of toxic effect are extremely valuable because they can be used to establish causal links between the concentrations of particular chemicals and adverse effects, including population declines (see also sections 7.2.4 and 7.2.5 for joint effects with PCDDs and PCDFs).

With the fall of *p,p′*-DDE levels, and associated eggshell thinning, populations of these affected species have recovered in many areas. However, reproductive failure, and physical deformities (such as crossed bills in double-crested cormorants) remained into the mid-1990s in certain areas where PCB levels are still relatively high. In the Caspian tern there was a high degree of inverse correlation between TEQs (dioxin equivalents) in eggs and embryonic mortality. With the double-crested cormorant there was a negative correlation between TEQ values and reproductive success, and a positive correlation between TEQ and the incidence of crossed bills (Gilbertson *et al.*, 1998). Like the Caspian tern, this species showed a correlation between TEQ values and embryonic mortality; however, the slopes of the regression lines were very different in the two species (Ludwig *et al.*, 1996). There is, therefore, evidence linking the depressed state of certain populations of piscivorous birds in the Great Lakes with Ah receptor-mediated toxicity caused by coplanar PCBs and other planar polyhalogenated aromatic compounds.

It has also been suggested that seal populations in and near the North Sea (Wadden and Baltic Seas) have been adversely affected by PCBs (Brouwer, 1991; Boon *et al.*, 1992; Ross *et al.*, 1995). Thyroid hormone antagonism, skeletal deformities, impaired reproduction and immunosuppression have been reported either in free-living animals or in animals dosed with fish caught in the North Sea that contained high levels of PCBs. It was also suggested that the spread of a disease that caused high mortality in seal populations was promoted by immunosuppression due to relatively high environmental levels of PCBs. Population declines of Californian sea-lions have also been linked to high tissue levels of PCBs (EHC 140).

Despite wide-ranging restrictions, and limitations on their release, levels of PCBs are slow to come down in certain locations. It appears that redistribution of PCBs from sinks is still going on. It has frequently been suggested that they have had – and in some cases are still having – adverse effects on predators at the top of food chains, e.g. fish-eating birds in some parts of the Great Lakes, and in marine mammals. The complexity of PCB pollution, with the possibility of interactive effects between different PCB congeners and/or between PCBs and other persistent pollutants, has made this a difficult point to prove or disprove. There is a need for the development and application of biomarker assays that can provide evidence of causality and link levels of pollutants, taken singly or in combination, with consequent harmful effects, and then to relate the harmful effects to the state of populations using population dynamic models (see Chapters 12 and 15 in Walker *et al.*, 2000).

6.3 *Polybrominated biphenyls*

The principal source of pollution by PBBs has been the commercial mixture Firemaster, which was produced in the USA between 1970 and 1974. Production was discontinued in 1974 after a severe pollution incident in Michigan. Firemaster was accidentally mixed with cattle feed on a Michigan farm. In due course, PBBs entered the human food chain via contaminated animal products. Substantial residues were found in humans from the area, and these were subsequently found to be highly persistent.

Firemaster is a stable solid, resembling a PCB mixture in its lipophilicity, chemical and thermal stability and low vapour pressure. Firemaster contains some 80 out of a possible 209 PBB congeners, but just two of them – 2,2′,4,4′,5,5′-hexabromobiphenyl (HBB) and 2,2′,3,4,4′,5,5′-heptabromobiphenyl – account for around 85% of the commercial product (EHC 152). These two compounds were found to be very slowly eliminated by humans who had been exposed to them during the Michigan incident. A half-life of approximately 69 weeks was estimated for 2,2',4,4',5,5'-HBB.

PBB residues are now widespread in terrestrial and aquatic ecosystems of the USA (Sleight, 1979) and Europe. On account of their lipophilicity and slow elimination by mammals, hexa- and hepta-bromobiphenyls may be expected to undergo biomagnification with movement along food chains in an analogous manner to higher chlorinated PCBs. Although they have been shown to have severe toxic effects on farm animals receiving high doses, nothing is known of any toxic effects that they may have had on wild vertebrates.

6.4 *Summary*

PCB mixtures were once used for a variety of purposes and came to cause widespread environmental pollution. Over 100 different congeners are present in commercial products such as Aroclor 1248 and Aroclor 1254. PCBs are lipophilic, stable and of low vapour pressure. Many of the more highly chlorinated PCBs are refractory, showing very strong biomagnification with movement along food chains.

The toxicology of PCBs is complex and not fully understood. Coplanar PCBs interact with the Ah receptor, with consequent induction of cytochrome P4501A1/2 and Ah receptor-mediated toxicity. Induction of P4501A1 provides the basis of valuable biomarker assays, including bioassays such as CALUX. Certain PCBs, e.g. 3,3',4,4'-TCB are converted to monohydroxymetabolites, which act as thyroxine antagonists. PCBs can also cause immunotoxicity (e.g. in seals).

PCBs have been implicated in the decline of certain populations of fish-eating birds, e.g. in the Great Lakes of North America. Although their use is now banned in most countries, and very little is released into the environment as a consequence of human activity, considerable quantities remain in sinks (e.g. contaminated sediments and landfill sites) from which they are slowly redistributed to other compartments of the environment.

PBB mixtures have been used as fire retardants. Many of their constituent congeners are highly persistent, and there was a major environmental accident in the USA when farm animals and humans became heavily contaminated by them.

6.5 *Further reading*

EHC 140 (1993) *Polychlorinated Biphenyls and Terphenyls*. A detailed reference work giving much information on the environmental toxicology of PCBs.

Safe, S. (1990) An authoritative account of the toxicology of PCBs and the development of toxic equivalency factors (1985–7).

Waid, J.-S. (1985–87) A useful source of information on PCBs.

CHAPTER **7**

Polychlorinated dibenzodioxins and polychlorinated dibenzofurans

7.1 Background

The polychlorinated lipophilic compounds polychlorinated dibenzodioxins (PCDDs) and polychlorinated dibenzofurans (PCDFs) do not occur naturally, and they are not synthesised intentionally. They occur as by-products of chemical synthesis, industrial processes and occasionally of interaction between other organic contaminants in the environment. Because of human exposure to them during the Vietnam war, industrial accidents, and the high mammalian toxicity of some of them, they have received much attention as hazards to human health. There has also been concern about some PCDDs and PCDFs from an ecotoxicological point of view because they combine marked biological persistence with high toxicity. The following account will deal mainly with PCDDs, which have been studied in some detail, and have been important in developing the concept of Ah receptor-mediated toxicity. Less is known about PCDFs, which will be described only briefly.

7.2 *Polychlorinated dibenzodioxins*

7.2.1 ORIGINS AND CHEMICAL PROPERTIES

PCDDs are polychlorinated planar molecules with an underlying structure of two benzene rings linked together by two bridging oxygens, thus creating a third 'dioxin' ring (Figure 7.1). They are sometimes simply termed 'dioxins' (EHC 88). In theory, there are 75 different congeners, but only a few of them are regarded as being important from an ecotoxicological point of view. 2,3,7,8-Tetrachlorodibenzodioxin (TCDD) has received far more attention than the others because of its high toxicity and persistence, and the detection of significant levels in the environment (structure given in Figure 7.1). They are formed when *o*-chlorophenols or their alkali metal salts are heated to a high temperature (see Crosby, 1998). The formation of 2,3,7,8-TCDD from an *ortho* trichlorophenol is shown in Figure 7.1.

PCDDs have been released into the environment in a number of different ways. Sometimes this has been due to the use of a pesticide that is contaminated with them. 2,4,5-Trichlorophenoxyacetic acid (2,4,5-T) and related phenoxyalkanoic herbicides have been contaminated with them as a consequence of the interaction of chlorophenols used in the manufacturing process. Relatively high levels of 2,3,7,8-TCDD occurred in a herbicide formulation containing 2,4,5-T, which was known as Agent Orange. Agent Orange was sprayed as a defoliant on extensive areas of jungle during the Vietnam War. Consequently, many humans, as well as wild animals and plants, were exposed to dioxin. PCDDs are also present in pentachlorophenols, which are used as pesticides.

Industrial accidents have also led to environmental pollution by PCDDs. In 1976 there was an explosion at a factory in Seveso, Italy, that was concerned with the production of trichlorophenol antiseptic. A cloud containing chlorinated phenols and

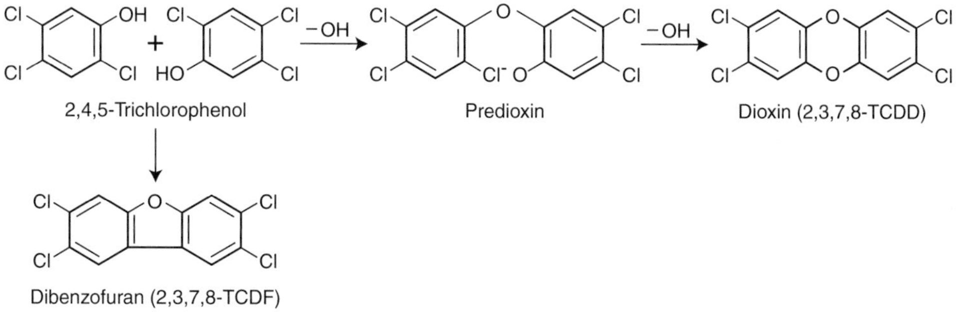

FIGURE 7.1 *Formation of dioxin and dibenzofuran. After Crosby (1998).*

dioxins was released and caused severe pollution of neighbouring areas. People who had been exposed showed typical symptoms of early PCDD intoxication (chloracne). Another source of PCDDs is the effluent from paper mills, where wood pulp is treated with chlorine (Sodergren, 1991). This has been a problem in northern Europe (including Russia) and in North America. Evidently, chlorine interacts with phenols derived from lignin to generate chlorophenols, which then interact to form dioxins. Finally, PCDDs and PCDFs can be generated during the disposal of PCB residues by combustion in specially designed furnaces. If combustion is incomplete in furnaces, the residues can be released into the air to pollute surrounding areas. Presumably chlorinated phenols are first produced, which then interact to form PCDDs and PCDFs. Investigation of such cases of pollution have sometimes led to the closure of the commercial operations responsible.

2,3,7,8-TCDD has been more widely studied than other PCDDs, and it will be taken as an example for the whole group of compounds. It is a stable solid with a melting point of 306°C. Its water solubility is very low and has been estimated to be 0.01–0.2 µg/L; its log K_{ow} is 6.6. More highly chlorinated PCDDs are even less soluble in water.

7.2.2 METABOLISM OF PCDDs

2,3,7,8-TCDD has been much more widely studied than other PCDDs and will be taken as representative of the group. Metabolism is slow in mammals. Because of the high toxicity of the compound, only low doses can be given in *in vivo* experiments, making the quantification of metabolites difficult. However, there appear to be two distinct types of metabolite: (1) monohydroxylated PCDD derivatives, and (2) products of the cleavage of one of the ether bonds of the dioxin ring. Both types can be generated after epoxidation of the aromatic ring. Epoxidation is thought to precede the formation of both (1) and (2) (see EHC 88; Poiger and Buser, 1983). Hydroxymetabolites excreted in rat bile are present largely as glucuronide or sulphate conjugates.

7.2.3 ENVIRONMENTAL FATE OF PCDDs

Higher chlorinated PCDDs, including 2,3,7,8-TCDD, are lipophilic and biologically stable, and are distributed in the environment in a similar fashion to higher chlorinated PCBs, reaching relatively high levels at the top of food chains. Within vertebrates, however, they show a greater tendency to be stored in the liver and a lesser tendency to be stored in fat depots than do most PCBs.

Some biological half-lives for 2,3,7,8-TCDD are given in Table 7.1. Because of its high toxicity there is concern about very low levels of 2,3,7,8-TCDD in biota. This raises analytical problems, and high-resolution capillary gas chromatography is needed to obtain reliable isomer-specific analyses at low concentrations. In the analysis of

TABLE 7.1 *Half-lives of 2,3,7,8-TCDD*

Species	Dose (route)	Half-life (days)
Rat (different strains)	1–50 µg/kg (oral)	17–31
Mouse (different strains)	0.5–10 µg/kg (i.p.)	10–24
Hamster	650 µg/kg (oral)	15
Guinea-pig	0.56 µg/kg (i.p.)	94

herring gull eggs collected from the Great Lakes, Herbert *et al.* (1994) reported the following residues of 2,3,7,8-TCDD:

Lake Ontario (early 1970s)	2–5 µg/kg (i.e. 0.002–0.005 ppm by weight)
Lake Ontario (1984/85)	0.08–0.1 µg/kg
Lake Michigan (1971)	0.25 µg/kg
Lake Michigan (1972)	0.07 µg/kg
Lake Michigan (1984/85)	0.001–0.002 µg/kg

In another study conducted during 1983–5, fish from the Baltic Sea were found to contain 0.003–0.029 µg/kg of 2,3,7,8-TCDD (Rappe *et al.*, 1987).

7.2.4 TOXICITY OF TCDDs

2,3,7,8-TCDD is a compound of very high toxicity to certain mammals, and there has been great interest in elucidating its mode of action. The situation is complicated, at least on the surface, by the variety of symptoms associated with dioxin toxicity. Symptoms include dermal toxicity, immunotoxicity, reproductive defects, teratogenicity and endocrine toxicity (Ahlborg *et al.*, 1994). Many of these effects have been attributed to the very strong binding affinity of this molecule for the Ah receptor (Ah receptor-mediated toxicity). Apart from 2,3,7,8-TCDD, other PCDDs, PCDFs and coplanar PCBs also interact with the Ah receptor, causing induction of P4501A1 as well as toxic effects. There are, however, large differences between individual compounds, both in their affinity for the receptor and in their toxic potency. To overcome this problem, and develop a practical approach for risk assessment, the concept of 'toxic equivalency factors' was introduced (Safe, 1990; Ahlborg *et al.*, 1994).

Toxic equivalency factors (TEFs) are estimated relative to 2,3,7,8-TCDD. They are measures of the toxicity of individual compounds relative to that of 2,3,7,8-TCDD. A variety of toxic indices, measured *in vivo* or *in vitro*, have been used to estimate TEFs, including reproductive effects (e.g. embryo toxicity in birds), immunotoxicity and effects on organ weights. The induction of P4501A1 has also been used to estimate TEFs. The usual approach is to compare a dose–response curve for a test compound with that of the reference compound, 2,3,7,8-TCDD, to establish the concentrations (or doses) that are required to elicit a standard response. The ratio, concentration test compound/concentration of 2,3,7,8-TCDD, when both compounds produce the same response is the TEF. Once determined, TEFs can be used to convert concentrations of chemicals found in environmental samples to TEQs. Thus, $[C] \times TEF = TEQ_{dioxin}$, where $[C]$ = environmental concentration of planar polychlorinated compound.

The criteria for including a compound in the above scheme, and assigning it a TEF value, were set out in a WHO–European Centre for Environmental Health consultation in 1993 (Ahlborg *et al.*, 1994). They are as follows:

1   The compound should show a structural relationship to PCDDs and PCDFs.
2   The compound should bind to the Ah receptor.
3   The compound should elicit biochemical and toxic responses that are characteristic of 2,3,7,8-TCDD.
4   The compound should be persistent and accumulate in the food chain.

Some examples of TEF values used in an environmental study on polyhalogenated aromatic hydrocarbons (PHAHs) in fish are given in Table 7.2 (data from Giesy, 1997). The first point to notice is that values for PCDDs and PCDFs are generally much higher than those for PCBs. Even the most potent of the PCBs, 3,3',4,4',5-PCB, only has a TEF value of 2.2×10^{-2}, which is lower than nearly all the values for PCDDs and PCDFs. That said, PCBs tend to be at much higher concentrations in environmental samples than the other two groups, with the consequence that they have been found to contribute higher overall TEQ values than PCDDs or PCDFs in many environmental samples despite the low TEF values.

Up to this point, discussion of TEQs has been restricted to the estimation of them from concentrations of individual compounds determined chemically, using TEFs as conversion factors. It is also possible to measure the total TEQ value directly, by means of a bioassay. Rat and mouse hepatoma lines, which contain the Ah receptor, show P4501A1 induction when exposed to planar PHAHs. The degree of induction can be measured in terms of the increase in EROD activity (see Chapter 2). One example of a cellular line of this type is the rat H4IIE line, which has come to be widely used for environmental bioassays (see, for example, Giesy, 1997; Koistinen, 1997; Whyte *et al.*, 1998). A development of this approach is the 'chemically activated luciferase gene expression' (CALUX) assay (Aarts *et al.*, 1993; Garrison *et al.*, 1996). Here, a reporter gene for the enzyme luciferase is linked to the operation of the Ah receptor, so that the degree of induction is indicated by the quantity of light that is emitted by the cells. This system has the advantage of not requiring an EROD assay to determine TEQ values. Values obtained by the direct measurement of TEQ (i.e. TCDD equivalents) using cellular systems can be directly compared with values estimated from chemical data using TEFs. When using TEFs to estimate TEQs, it is assumed that all of the compounds are acting by a common mechanism, through a common receptor, and that effects of individual components in a mixture are simply additive, without potentiation or antagonism. There are, however, reservations about adopting such a simple approach. In the first place the mechanism of toxicity has not

TABLE 7.2 *TEF values for polyhalogenated aromatic hydrocarbons (PHAHs) used in a study of residues in fish (Giesy et al., 1997)*

Compound	TEF
PCDDs	
1,2,3,7,8-PCDD	4.2×10^{-1}
1,2,3,4,7,8-HCDD	8.3×10^{-2}
1,2,3,4,6,7-HpCDD	2.3×10^{-2}
PCDFs	
2,3,7,8-TCDF	2.0×10^{-1}
2,3,4,7,8-PCDF	2.8×10^{-1}
1,2,3,6,7,8-HCDF	6.0×10^{-1}
1,2,3,4,7,8,9-HpCDF	2.0×10^{-2}
Non-*ortho* PCBs	
3,4,4′,5,-TCB	1.9×10^{-3}
3,3′,4,4′-TCB	1.8×10^{-5}
3,3′,4,4′,5-PCB	2.2×10^{-2}
Mono-*ortho* PCBs	
2,3,3′,4,4′,5-PCB	3.5×10^{-7}
2,3,3′,4,4′-PCB	8.0×10^{-6}
2,3,3′,4,4′,5-HCB	5.5×10^{-5}
2,2′,3,4,4′,5′-HCB	1.5×10^{-5}

been elucidated, and there may be toxic mechanisms operating that do not involve the Ah receptor. It is true that early application of the approach in human toxicology produced relatively consistent results. In studies with experimental animals (mainly rodents) the summation of TEQ values for mixtures of compounds have often been found to relate reasonably well to toxic effects (see also pp. 131–132). However, there is evidence suggesting that PHAHs can express toxicity by other mechanisms, for example in certain cases of developmental neurotoxicity and carcinogenicity (Brouwer, 1996; Verhallen *et al.*, 1997). Also, PCBs in mixtures have sometimes shown antagonistic effects, so additivity cannot be automatically assumed (Davis and Safe, 1990; Giesy, 1997).

The use of TEQs in environmental risk assessment has great attractions. It offers a way of tackling the problem of determining the biological significance of levels of diverse PAHs found in mixtures. There are, however, practical and theoretical problems that need to be resolved before it can be widely used with confidence. First, to what extent can the variety of toxic effects caused by PAHs in vertebrates generally be explained in terms of Ah-mediated toxicity? Only limited work has been carried out on birds or fish, and practically nothing on amphibians or reptiles. Second, even restricting the argument to effects that are directly connected with binding of PHAH to the Ah receptor, how comparable are the Ah receptors of unrelated species? The limited data available so far suggest that the Ah receptor is highly conserved, and may not differ very much between different groups of vertebrates. However, in field studies there have sometimes been large differences between species in the relationship between TEQ values and toxic effects (see Ludwig *et al.*, 1996; McCarty and Secord, 1999; and further discussion in sections 6.2.4 and 7.2.5). More work needs to be carried out to clarify this issue. At the moment, the TEF values in use have been obtained for a very limited number of species. There needs to be caution in using them for calculating TEQs in untested species.

TEQ values based on 2,3,7,8-TCDD have been estimated or measured in a number of field studies on fish, birds and marine mammals. Although some encouraging progress has been made, there have also been a number of problems. Not infrequently, TEQ values determined by bioassay have considerably exceeded values calculated from chemical data using TEFs. These discrepancies have not been explicable in terms of antagonistic effects, and the balance of evidence suggests that environmental compounds, other than the PHAHs determined by analysis, have contributed to measured TEQ values. This has been observed in studies on fish (Giesy, 1997) and white-tailed sea eagles (*Haliaetus albicilla*) (Koistinen, 1997). In the study on fish, as much as 75% of the measured TEQ could not be accounted for using chemical data. Because of the concern over the human health hazards associated with PCDDs, many toxicity tests have been performed on rodents. Some toxicity data are given for 2,3,7,8-TCDD below:

Acute oral LD_{50}/rat	22–297 µg/kg (different strains were tested)
Acute oral LD_{50}/mice	114–2570 µg/kg (different strains were tested)
Acute oral LD_{50}/guinea-pig	0.6–19 µg/kg

A variety of toxic symptoms were shown, the pattern differing between species. There were often long periods between commencement of dosing and death. There were also large species differences in toxicity, the guinea-pig being extremely susceptible, the mouse far less so. However, the critical point is that 2,3,7,8-TCDD is an exceedingly toxic compound – even to the mouse! With such differences in toxicity between closely related species it seems probable that there will be even larger differences across the wide range of vertebrate species found in nature.

7.2.5 ECOLOGICAL EFFECTS RELATED TO TEQs FOR 2,3,7,8-TCDD

Estimates of TCDD toxicity in field studies have been bound up with the estimate of TEQ values in an attempt to relate Ah receptor-mediated toxicity caused by all PHAHs to effects on individuals and populations. The results of a number of studies are summarised in Table 7.3.

The results for the double-crested cormorant and the Caspian tern in the Great Lakes show a relationship between TEQs and reproductive success. They were obtained

TABLE 7.3 *TEQ values found in field studies and ecological effects associated with them*

Area	Species	TEQ (pg/g) (method of determination)	Observations	Reference
Great Lakes	Double-crested cormorant (eggs)	100–300 (bioassay) Only PCBs assayed	Egg mortalities of 8–39% correlated well with TEQ values Relationship to continued poor breeding success	Tillett *et al.* (1992)
	Caspian tern (eggs)	170–400 (chemical analysis) Only PCBs assayed 2700	Related to embryonic mortality Total reproductive failure	Ludwig *et al.* (1996) Ludwig *et al.* (1993)
Baltic Sea	White-tailed sea eagle (egg and muscle)	< 1220 (bioassay) < 1040 (chemical determination)	PCB fraction accounted for 75%+ of TEQ by either assay Reduced productivity of birds in this area	Koistinen (1997)
Upper Hudson River, USA	Tree swallows (*Tachycineta bicolor*) (nestlings)	410–25 400 (chemical determination)	TEQs mainly due to PCBs especially 3,3',4,4'-TCB Reduced reproductive success but less effect than expected from high TEQs	Secord *et al.* (1999) McCarty and Secord (1999)
Saginaw Bay, Great Lakes, USA	Fish (whole body) Sampled 1990	11–348 (bioassay) 14–70 (chemical determination)	PCDDs, PCDFs and PCBs made variable, but on the whole similar contributions to TEQ values. Levels probably not high enough to adversely affect fish populations	Giesy (1997)

during the late 1980s, at a time when DDE-related thinning of eggshells had fallen and TEQ values were based on PCBs alone. However, in certain areas PHAH levels remained high, and populations of these two species were still depressed. These investigations suggested that populations were still being adversely affected by PHAHs as a consequence of Ah receptor-mediated toxicity. The data for white-tailed sea eagles from the Baltic coast also relate to an area where, at the time of the investigation, there was evidence of reduced breeding success in the local population – at a time when the species was increasing elsewhere in Scandinavia. The TEQs in the most highly contaminated individuals were high enough to support the suggestion that PHAHs were contributing to lack of breeding success.

The results for tree swallows are surprising in that the birds were still able to breed with TEQs far above the levels that had severe/fatal effects on other species of birds! However, there was evidence of reduced reproductive success in 1994 and of high rates of abandonment of nests and supernormal clutches in 1995 (McCarty and Secord, 1999). It would appear that tree swallows are particularly insensitive to this type of toxic action. This finding raises questions about the wider applicability of the use of TEQ values and calls to mind the large interspecific differences in TCDD toxicity found in some toxicity tests (section 7.2.4).

A noteworthy finding of the investigations thus far is the considerable variation in the relative contributions of PCBs, PCDDs and PCDFs to TEQ values in wild vertebrates, where a distinction has been made between them. Thus, the TEQ values for white-tailed sea eagles and tree swallows were accounted for, very largely, by coplanar PCBs. Fish from Saginaw Bay, however, showed a different picture. TEQs were lower, and PCDDs, PCDFs and PCBs all made similar contributions to TEQs. There was considerable variation between species and age groups of fish, with contributions to total TEQs in the following ranges:

PCDD	5–38%
PCDF	13–69%
PCB	10–50%

7.3 *Polychlorinated dibenzofurans*

PCDFs are similar in many respects to PCDDs, although less intensively studied, and will be mentioned only briefly here. The chemical structure is shown in Figure 7.1. Like PCDDs, they can be formed by the interaction of chlorophenols, and they are found in commercial preparations of chlorinated phenols and in products derived from phenols (e.g. 2,4,5-T and related phenoxyalkanoic herbicides). They are also present in commercial PCB mixtures and can be formed during the combustion of PCBs. They have similar physical properties to PCDDs, with low water solubility and high K_{ow} values (2,3,7,8-TCDF has a log K_{ow} value of 5.82). Some PCDFs bind very strongly to the Ah receptor. (TEF values are given in Table 7.2.)

2,3,7,8-TCDF is not very toxic to the mouse (acute oral LD_{50} < 6000 µg/kg) but is highly toxic to the guinea-pig (acute oral LD_{50} 5–10 µg/kg). Symptoms of toxicity in the guinea-pig were similar to those found for 2,3,7,8-TCDD. Thus, the selectivity pattern was similar to that for 2,3,7,8-TCDD, but toxicity was considerably less (see EHC 88).

In some studies of PHAH levels in wild vertebrates, PCDFs have been found to make a significant contribution to TEQ values determined chemically (see, for example, Giesy, 1997).

7.4 *Summary*

Both PCDDs and PCDFs are refractory lipophilic pollutants formed by the interaction of chlorophenols. They have entered the environment as a consequence of their presence as impurities in pesticides, after certain industrial accidents, in effluents from pulp mills and because of the incomplete combustion of PCB residues in furnaces. Although present at very low levels in the environment, some of them (e.g. 2,3,7,8-TCDD) are highly toxic and undergo biomagnification in food chains.

PCDDs and PCDFs, together with coplanar PCBs, can express Ah receptor-mediated toxicity. TCDD (dioxin) is used as a reference compound in the determination of TEFs, which can be used to estimate TEQs (toxic equivalents) for residues of PHAHs found in wildlife samples. Biomarker assays for Ah receptor-mediated toxicity have been based on the induction of P4501A1. TEQs measured in field samples have sometimes been related to toxic effects upon individuals and associated ecological effects (e.g. reproductive success).

7.5 *Further reading*

Ahlborg, U. G. *et al.* (1996) Discusses the wider use of TEFs.
EHC 88 (1989) *PCDDs and PCDFs*. Gives information on the environmental toxicology of PCDDs and PCDFs.
Safe, S. (1990) An authoritative account of the development of TEFs.

CHAPTER 8

Organometallic compounds

8.1 *Background*

Metalloids such as arsenic and antimony, and metals such as mercury, lead and tin, which occupy a similar location to metalloids in the periodic system (some authorities regard tin as a metalloid), all tend to form stable covalent bonds with organic groups. By contrast, metals such as sodium, potassium, calcium, strontium, barium, and other metals that belong to groups 1 and 2 of the periodic system, do not form covalent bonds with organic groups. The compounds used as examples here all possess covalent linkages between a metal and an organic group – most commonly an alkyl group. The elements in question are mercury, tin, lead and arsenic, all of which are appreciably toxic in their inorganic forms as well as their organometallic forms. The attachment of the organic group to the metal can bring fundamental changes in chemical properties, and consequently in environmental fate and toxic action. In particular, the attachment of alkyl or other non-polar groups to metals increases lipophilicity and thereby enhances movement into and across biological membranes, storage in fat depots and adsorption by the colloids of soils and sediments. Thus, the question of speciation is critical to understanding the ecotoxicology of these metals.

In the first place organometallic compounds of mercury, tin, lead and arsenic have been produced commercially, mainly for use as pesticides, biocides or bactericides. In addition, methyl mercury and methyl arsenic are generated from their inorganic forms in the environment, so residues of them may be both anthropogenic and natural in origin. Most of the following account will be devoted to organomercury and organotin compounds, which have been extensively studied. Organolead and organoarsenic compounds have received less attention from an ecotoxicological point of view, and will be dealt with only briefly.

8.2 *Organomercury compounds*

8.2.1 ORIGINS AND CHEMICAL PROPERTIES

A range of organomercury compounds has been produced commercially since early in the twentieth century, principally for use as antifungal agents. Most of them have the general formula R–Hg–X, where R is an organic group and X is usually an inorganic group (occasionally a polar organic group such as acetate). The organic group is non-polar (or relatively so), and gives the molecule a lipophilic character. The most common organic groups are alkyl, phenyl and methoxyethyl (see EHC 86). The solubility of organomercury compounds depends primarily on the nature of the X group; nitrates and sulphates tend to be 'salt-like' and relatively water soluble, whereas chlorides are covalent, non-polar compounds of low water solubility. Methyl mercury compounds tend to be more volatile than other organomercury compounds.

The structures of some organomercury compounds are shown in Figure 8.1, and some physical properties are given in Table 8.1. The R–Hg bond is chemically stable, and is not split by water or by weak acids or bases. This is a reflection of the low affinity of Hg for oxygen. It can, however, be broken biochemically. Organomercury, like other organometallic compounds, has a strong affinity for –SH groups of proteins and peptides:

$$R\text{–}Hg\text{–}X + protein\text{–}SH \rightarrow R\text{–}Hg\text{–}S\text{–}protein + X^- + H^+$$

This tendency to interact with –SH groups is the principle behind the mode of toxic action of organomercury compounds; it is also the basis for one mechanism of detoxication. Apart from the release of man-made organomercurial compounds, methyl mercury can also be generated from inorganic mercury in the environment as indicated in the following equation:

$$Hg \rightarrow Hg^{2+} \rightarrow CH_3Hg^+ \rightarrow (CH_3)_2Hg$$

Thus, both elemental mercury and the mineral form cinnabar (HgS) can release Hg^{2+}, the mercuric ion. Bacteria can then methylate it to form sequentially CH_3Hg^+, the methyl mercuric cation, and dimethyl mercury. The latter, like elemental mercury, is volatile and tends to pass into the atmosphere when formed. The methylation of mercury can be accomplished in the environment by bacteria, notably in sediments.

General formula   RHgX

    where R = C_nH_{2n+1}, C_6H_5 or $CH_3OC_2H_5$

CH_3HgCl Methylmercuric chloride

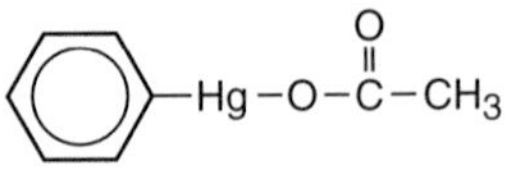

$$\text{(phenyl ring)}-Hg-O-\overset{\displaystyle O}{\overset{\|}{C}}-CH_3$$ Methylmercuric acetate

FIGURE 8.1 *Organomercury compounds.*

TABLE 8.1 *Properties of organomercury compounds*

Compound	Water solubility (mg/L)	Vapour pressure (mmHg)
Methylmercuric chloride	1.4	8.5×10^{-3}
Phenylmercuric acetate	4400	

A form of vitamin B_{12} can produce methyl carbanion, a reactive species that is responsible for methylation of Hg^{2+} (see Figure 8.2; Craig, 1986; IAEA Technical Report 137, 1972). Methyl carbanion acts as a nucleophilic agent towards Hg ions.

It is difficult to establish to what extent methyl mercury residues found in the environment arise from natural as opposed to human sources. There is no doubt, however, that natural generation of methyl mercury makes a significant contribution to these residues. Samples of tuna fish caught in the late nineteenth century, before the synthesis of organomercury compounds by man, contain significant quantities of methyl mercury.

8.2.2 METABOLISM OF ORGANOMERCURY COMPOUNDS

As mentioned above, methyl mercury compounds can undergo further methylation to generate highly volatile dimethyl mercury. Organomercury compounds can also be converted back into inorganic forms of mercury by enzymic action. Oxidative metabolism is important here, and has been reported in both microorganisms and invertebrates. Methyl mercury is slowly degraded by α-oxidation; other alkyl forms are subject to more rapid β-oxidation. This may explain why methyl mercury is degraded more slowly than other forms and is correspondingly more persistent. Phenyl mercury is degraded relatively rapidly to inorganic mercury by vertebrates, and is generally less persistent than alkyl mercury.

$$R\text{--}Hg \overset{O}{\rightarrow} Hg^{2+}$$

Another type of detoxication involves the production of cysteine conjugates, which are readily excreted. (Again, organomercury compounds show their affinity for –SH groups!) Methyl mercury cysteine is an important biliary metabolite in the rat and is

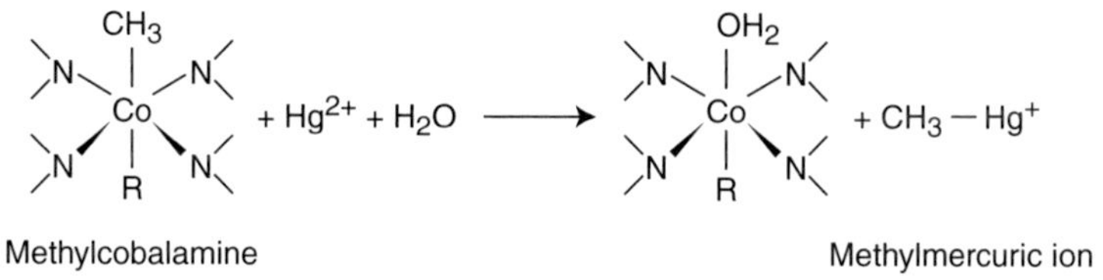

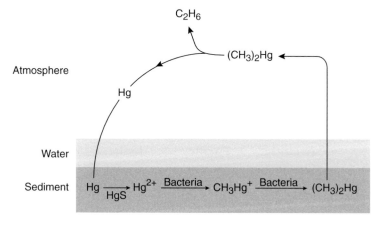

FIGURE 8.2 *Methylation of inorganic mercury by methylcobalamine. From Crosby (1998).*

degraded within the gut (presumably by microorganisms) to release inorganic mercury (see IAEA Technical Report 137, 1972).

The following ranges of half-lives have been reported for vertebrate species, which are, presumably, related to rates of biotransformation since the original lipophilic compounds show little tendency to be excreted unchanged: alkyl mercury, 15–25 days; phenyl mercury, 2–5 days.

8.2.3 ENVIRONMENTAL FATE OF ORGANOMERCURY

As noted above, organomercury is both released into the environment as a consequence of human activity and biosynthesised there in the form of methyl mercury. Furthermore, the inorganic mercury from which methyl mercury is synthesised is released into the environment by both natural processes (e.g. weathering of minerals) and human activity (mining, factory effluents, etc.).

The environmental cycling of methyl mercury is summarised in Figure 8.3. Dimethyl mercury, being highly volatile, tends to move into the atmosphere after its generation in sediments; once there it can be converted back into elemental mercury by the action of UV light. Some dimethyl mercury is taken up by fish and transformed into a methyl cysteine conjugate that is excreted. However, the most important species of

FIGURE 8.3 *Environmental fate of methyl mercury. Adapted from Crosby (1998).*

methyl mercury in aquatic and terrestrial food chains is CH_3Hg^+, which exists in various states of combination: with S– groups of proteins and peptides, and with inorganic ions such as chloride. The total methyl mercury of tissues, sediments, etc. is determined by chemical analysis, but the state of combination is not usually known. Some free forms of methyl mercury, e.g. CH_3HgCl, are highly lipophilic and undergo bioaccumulation and bioconcentration with progression along food chains in similar fashion to lipophilic polychlorinated compounds.

In a report from the USEPA (1980), fish contained between 10 000 and 100 000 times the concentration of methyl mercury present in ambient water. In a study of methyl mercury in fish from different oceans, higher levels were reported in predators than in non-predators (see Table 8.2). Taken overall, these data suggest that predators have some four- to eightfold higher levels of methyl mercury than do non-predators, and it appears that there is marked bioaccumulation with transfer from prey to predator.

In a laboratory study (Borg *et al.*, 1970), bioaccumulation of methyl mercury was studied in the goshawk (*Accipiter gentilis*). The details are shown in Table 8.3. Chickens bioaccumulated methyl mercury to approximately twice the level in their food, and goshawks bioaccumulated methyl mercury about fourfold over a similar period of exposure. This provides further evidence for the slow elimination of methyl mercury by vertebrates, and the relatively poor detoxication capacity of predatory birds towards lipophilic xenobiotics compared with non-predatory birds (see Chapters 2 and 5). In a related study with ferrets that had been fed chicken contaminated with methyl mercury, a somewhat higher bioaccumulation factor was indicated (approximately sixfold), albeit over the somewhat longer exposure period of 35–58 days. This provided further evidence for strong bioaccumulation by predators.

Apart from CH_3Hg^+, other forms of $R–Hg^+$, which originate from anthropogenic sources and are not known to be generated from inorganic mercury in the natural environment, occur in terrestrial and aquatic food chains. Usually, the R group is phenyl, alkoxyalkyl or higher alkyl (ethyl, propyl, etc.). These forms behave in a similar manner to CH_3HgX and do tend to undergo biomagnification, but they are generally more easily biodegradable to inorganic mercury and tend to bioaccumulate less strongly.

A major source of organomercury pollution in Western countries was once their use as fungicidal seed dressings for cereals and other agricultural crops (see IAEA Technical Report 137, 1972). Another important source was organomercury antifungal

TABLE 8.2 *Methyl mercury residues in fish (mg Hg/kg wet weight)*

Type of fish	Atlantic Ocean	Pacific Ocean	Indian Ocean	Mediterranean Sea
Non-predators	0.03–0.27	0.03–0.25	0.005–0.16	0.1–0.24
Predators	0.3–1.3	0.3–1.6	0.004–1.5	1.2–1.8

Data from EHC 101.

TABLE 8.3 *Bioaccumulation of methyl mercury in the goshawk*

Material/species	CH_3Hg (as ppm Hg)	Duration of feeding (days)	Approximate bioaccumulation factor
Dressed grain	8		
Muscle of chickens fed dressed grain	10–40	40–44	2
Chicken tissue fed to goshawks	10–13		
Muscle from goshawks fed chicken tissue	40–50	30–47	4

agents used in the wood pulp and paper industry. Most of these uses were discontinued by the 1970s, but certain practices have continued into the 1990s, including the use of phenylmercury fungicides as seed dressings in Britain and some other countries. In the 1950s and early 1960s, Sweden and other Scandinavian countries had serious pollution problems due to the use of methyl mercury compounds as seed dressings. The deaths of seed-eating birds and the raptors preying upon them were attributed to methyl mercury poisoning (Borg *et al.*, 1969). Thus, as with dieldrin, bioaccumulation led to secondary poisoning. Interestingly, seed-eating rodents contained lower mercury levels than did seed-eating birds. Some data for total mercury levels found in Swedish birds during the mid-1960s are shown in Figure 8.4. Most of the mercury was in the methyl form. Findings such as these led to the banning of methyl mercury seed dressings in Scandinavia. Other forms of organomercury seed dressing (e.g. phenyl mercury) were not implicated in poisoning incidents and continued to be marketed in many Western countries after methyl mercury compounds were banned.

8.2.4 TOXICITY OF ORGANOMERCURY COMPOUNDS

The toxicity of organomercury, like that of certain other types of organometals, is due to a tendency to combine with functional –SH groups of proteins. A prime site of action is the nervous system, especially the central nervous system, and there is evidence that Na^+,K^+-ATPase is a critical target (Clarkson, 1987). Here lies an important distinction between the toxicity of organic and inorganic mercury salts. Inorganic mercury forms can also bind to –SH groups; however, inorganic mercury, when in ionic form, does not readily cross the blood–brain barrier, so shows less tendency than lipophilic organic mercury to reach the central nervous system (CNS) and cause toxic effects there. Rather, it expresses its toxicity elsewhere (e.g. on the kidney and on cardiac function). The neurotoxicity of organomercury was graphically illustrated in

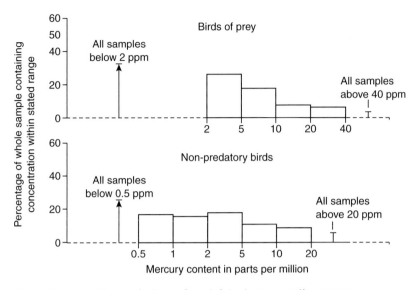

FIGURE 8.4 *Mercury residues in the livers of Swedish birds. From Walker (1975).*

an environmental disaster at Minamata Bay in Japan in the late 1950s and early 1960s. Release of both organic and inorganic mercury from a factory led to the appearance of high levels of methyl mercury in the neighbouring marine ecosystem. Levels were high enough in fish to cause lethal intoxication of local people for whom fish was the main protein source. People died as a consequence of brain damage caused by methyl mercury. The victims had brain levels of Hg in excess of 50 ppm. In mammals, methyl mercury toxicity is mainly manifest as damage to the CNS with associated behavioural effects (Wolfe *et al.*, 1998). Initially, animals become anorexic and lethargic, and, with the progression of toxicity, muscle ataxia and visual impairment are seen. Finally, convulsions occur that lead to death. In dosing experiments with mink (*Mustela vison*), dietary levels of methyl mercury of 1.1 ppm fed over a period of 93 days produced subclinical neurological lesions (Wobeser *et al.*, 1976), which has been proposed as a *lowest observed adverse effect level* (LOAEL). In another study, otters (*Lutra canadensis*) were dosed with 2, 4 or 8 ppm methyl mercury in the diet (O'Connor and Nielsen, 1981). Anorexia and ataxia were reported at 2 ppm in two out of three individuals, anorexia, ataxia and neurological lesions at 4 ppm, and all above symptoms leading to death at 8 ppm. The brain Hg concentrations (ppm per unit wet weight) at dose levels of 2, 3 and 8 ppm were 13.3, 21.0 and 23.7 ppm respectively. Thus, symptoms of neurotoxicity were observed in individuals containing about half the brain concentration of Hg associated with lethal toxicity. Interestingly, the proportion of the total mercury accounted for as organomercury declined with time, indicating that demethylation slowly occurred in the brain. Captive goshawks dying from methyl mercury poisoning contained 30–40 ppm of Hg in the brain and 40–50 ppm of Hg in muscle (Borg *et al.*, 1970; see also section 7.2.4). In this study and others (Wolfe *et al.*, 1998), it became apparent that birds, like mammals, experience a range of sublethal effects before tissue levels became high enough to cause death. The first symptoms of methyl mercury poisoning in birds are reduced food consumption and weakness of the extremities. Muscular coordination is poor, there is ataxia, and birds can neither walk nor fly (see Rissanen and Mietinnen in IAEA Technical report 137, 1972). In view of the sublethal neurotoxic effects produced by methyl mercury, it seems unlikely that birds in general, and predatory birds in particular, would often bioaccumulate such lethal brain concentrations in the field. More probably, they would die from starvation as a result of sublethal effects before building up lethal concentrations such as these. Predators would lose their ability to catch prey once muscular coordination was affected, and the capacity of all birds to feed would be adversely affected by difficulties with flight. These feeding skills are not tested in laboratory trials where birds are presented with food, and where they may be expected to tolerate relatively high levels of methyl mercury in tissues before losing their ability to feed.

The acute toxicity of different types of organomercury compounds to mammals, expressed as mg/kg, falls into the following ranges:

Methyl mercury compounds	16–32
Ethyl mercury compounds	16–28
Phenyl mercury compounds	5–70

Thus, there is not a great deal of difference between the three classes in acute toxicity – all are highly toxic. However, methyl mercury is more persistent than the other two types, so has the greater potential to cause chronic toxicity. The latter point is important when considering the possibility of sublethal effects.

8.2.5 ECOLOGICAL EFFECTS OF ORGANOMERCURY COMPOUNDS

Of the different forms of organomercury, methyl mercury is the one most clearly implicated in toxic effects in the field. When methyl mercury seed dressings were used in Sweden and other northern European countries during the 1950s and 1960, many deaths of seed-eating birds, and of predatory birds feeding upon them, were attributed to methyl mercury poisoning. There was evidence of birds experiencing sublethal effects, e.g. inability to fly. There may well have been local declines in the bird populations consequent upon these effects, but these were not clearly established. Methyl mercury seed dressings were subsequently banned in Western countries, so the question is now rather an academic one.

The other major problem concerning methyl mercury was the severe pollution of Minamata Bay in Japan. Here fish, fish-eating and scavenging birds, and humans feeding upon fish all died from organomercury poisoning. Again there may have been localised declines of marine species as a result of methyl mercury, but there is no clear evidence of this.

Methyl mercury continues to be synthesised in the environment from inorganic mercury, which originates from both natural and human sources. There continues to be concern about the relatively high levels of organomercury in certain marine areas (e.g. the Mediterranean Sea; see Renzoni in Walker and Livingstone, 1992) and in the Great Lakes of North America (Meyer, 1998). Apart from the question of possible direct toxic effects caused by methyl mercury, there is the possibility of adverse interactive effects (potentiation) with other pollutants that reach significant levels in marine organisms, e.g. PCBs, PCDDs, PCDFs, *p,p'*-DDE and heavy metals (see Walker and Livingstone, 1992), or selenium (Heinz and Hoffman, 1998).

8.3 *Organotin compounds*

8.3.1 CHEMICAL PROPERTIES

Like mercury, tin is a metal that has a tendency to form covalent bonds with organic groups. The compounds to be discussed here are tributyl derivatives of tetravalent tin. The general formula for them is $(nC_4H_9)_3Sn–X$, where X is an anion.

The most important of them from an ecotoxicological point of view, and the one that will be used here as an example, is tributyltin oxide (TBTO). Its structure is shown in Figure 8.5.

TBTO is a colourless liquid of low water solubility and low polarity. Its water solubility varies between < 1.0 and > 100 mg/L, depending on the pH, temperature and presence of other anions. Anions determine the speciation of TBT in water. Thus, in seawater, TBT exists largely as hydroxide, chloride and carbonate, the structures of which are given in Figure 8.5. At pH values below 7.0 the predominant forms are the chloride and the protonated hydroxide, at pH 8 they are the chloride, hydroxide and carbonate, and at pH values above 10 they are the hydroxide and the carbonate (EHC 116). The K_{ow} for TBTO expressed as log P_{ow} lies between 3.19 and 3.84 for distilled water, and is approximately 3.54 for seawater. TBTO is adsorbed strongly to particulate matter. Another type of organotin compound, triphenyl tin (TPT), has been used as a fungicide.

8.3.2 METABOLISM OF TRIBUTYLTIN

In rats and mice TBT compounds are hydroxylated by microsomal monooxygenase attack (see EHC 116). Hydroxylation can occur on either the α- or β-carbon of the butyl group, i.e. α or β in relation to the Sn atom (see Figure 8.5). After hydroxylation, the hydroxylated moiety breaks away to leave behind dibutyl tin.

General formula

$$(C_4H_9)_3Sn - X \quad \text{where X is usually an anion}$$

Tributyltin oxide

$$(C_4H_9)_3Sn - O - Sn(C_4H_9)_3$$

Forms existing in water

$$(C_4H_9)_3 - SnOH \qquad \text{Hydroxide}$$

$$(C_4H_9)_3 - SnOH_2^+ \qquad \text{Protonated hydroxide}$$

$$(C_4H_9)_3 - SnCl \qquad \text{Chloride}$$

$$\left[(C_4H_9)_3 - Sn\right]_2 CO_3^- \qquad \text{Carbonate}$$

Metabolism

$$(C_4H_9)_3Sn^+ \xrightarrow[\text{(P450)}]{O} (C_4H_9)_2Sn^{2+}$$

FIGURE 8.5 *Tributyltin.*

TBTs also cause inhibition of the P450 of monooxygenases. In fish, and in the common whelk, TBT causes a conversion of P450s to the inactive P420 form (Mensink, 1997; Fent *et al.*, 1998). In fish, inactivation was also found with TPT and was related to the inhibition of EROD activity. In these studies organotin compounds act as both substrates and deactivators of the haem protein (cf. the interaction of organophosphates with 'B' type esterases).

TBT, like most other organic pollutants, is metabolised more rapidly by vertebrates than by aquatic invertebrates such as gastropods.

8.3.3 ENVIRONMENTAL FATE OF TRIBUTYLTIN

The main uses of TBT compounds arise from their toxic properties. They have been used as antifoulants on boats, ships, buoys and crab pots, as biocides for cooling systems, pulp and paper mills, breweries, leather-processing plants and textile mills and as molluscicides. Of particular interest and importance is their incorporation into antifouling paints used on boats of many kinds, ranging from small leisure craft to large ocean-going vessels. Release of TBT from antifouling paints has provided a small, yet highly significant, source of pollution to surface waters.

When TBTO is released into ambient water, a considerable proportion becomes adsorbed to sediments, as might be expected from its lipophilicity. Studies have shown that between 10% and 95% of TBTO added to surface waters becomes bound to sediment. In the water column it exists in several different forms, principally the hydroxide, the chloride and the carbonate (Figure 8.5). Once TBT has been adsorbed, loss is almost entirely due to slow degradation, leading to desorption of diphenyl tin

(DPT). The distribution and state of speciation of TBT can vary considerably between aquatic systems, depending on pH, temperature, salinity and other factors.

TBT levels have been monitored in coastal areas of Western Europe and North America. These have ranged upwards to 5.34 µg/L in Western Europe and 1.71 µg/L in North America (EHC 116). The highest levels were recorded in the shallow waters of estuaries and harbours where there were large numbers of small boats.

TBT is taken up by aquatic organisms directly from water and from food. Comparison of concentrations in molluscs with concentrations in ambient water indicate very strong bioconcentration/bioaccumulation. When molluscs such as the edible mussel (*M. edulis*) and the Pacific oyster (*Crassostrea gigas*) were exposed experimentally to TBTO in ambient water, bioconcentration factors (BCFs) ranging between 1000-fold and 7000-fold were found (EHC 116). With mussels exposed to relatively low levels of TBT, tissue levels were still increasing after 7 weeks of exposure, no plateau level having been reached. Exposure of *M. edulis* under natural conditions indicated higher BCFs than this, ranging from 5000 to 60 000 (Cheng and Jensen, 1989). Further investigation has shown that uptake of TBT from food can be greater than uptake directly from water. Thus, BCFs are a reflection more of bioaccumulation than of bioconcentration in this case!

8.3.4 TOXICITY OF TRIBUTYLTIN

Mechanistic studies have shown that TBT and certain other forms of trialkyl tin have two distinct modes of toxic action. On the one hand, they act as inhibitors of oxidative phosphorylation in mitochondria (Aldridge and Street, 1964). Inhibition is associated with reduction of ATP synthesis, disturbance of ion transport across the mitochondrial membrane, and swelling of the membrane. Oxidative phosphorylation is a vital process in animals and plants, and trialkyl tin compounds act as wide-ranging biocides.

As already noted, TBT and other organotin compounds inhibit P450 activity. Apart from the examples discussed earlier in section 8.3.2, TBTO has been shown to inhibit P450 activity in cells from various tissues of mammals, including liver, kidney and small intestine mucosa, both *in vivo* and *in vitro* (Rosenberg and Drummond, 1983; EHC 116). This effect may be the underlying reason for the action of TBT compounds as endocrine disruptors in molluscs. The conversion of testosterone to oestradiol is catalysed by a P450-based monooxygenase termed aromatase, which is inhibited by TBT. Inhibition of aromatase can, in principle, lead to an increase in cellular levels of testosterone (Matthiessen and Gibbs, 1998).

It is now known that over 100 species of gastropods worldwide suffer from a condition described as 'imposex', the development of male characteristics by females, the basis of a valuable biomarker assay. It has already been established for some species (e.g. dog whelk) that TBT is the cause of this, and it is suspected that organotin compounds account for most cases of imposex world wide (Matthiessen and Gibbs, 1998).

Some examples of masculinisation of female gastropods caused by TBT are given in Table 8.4.

It can be seen that there are differences between species in the physiological changes that are caused. However, the common factor is the development of male characteristics, usually with an adverse effect on breeding (Matthiessen and Gibbs, 1998). The underlying cause appears to be elevated levels of testosterone following exposure to TBT, an effect that has now been observed in several species of gastropods. The most widely held explanation of this is inhibition of the P450 isozyme known as aromatase by TBT, which fits in with the well-documented ability of TBT to deactivate P450s. In some species, inhibition of aromatase can lead to an increase in the cellular level of testosterone, although there is still some debate over the importance of this mechanism in molluscs. Amongst other things, there are a number of P450 forms involved in the metabolism of steroid hormones apart from aromatase, and TBT appears to act upon several different forms (Fent, 1996; Mensink, 1997; Fent *et al.*, 1998). Thus, inhibition of P450s other than aromatase might cause elevation of testosterone levels.

Another mechanism for testosterone elevation by TBT has been proposed for the periwinkle by Ronis and Mason (1996). They suggest that TBT raises testosterone levels by inhibiting its conjugation to sulphate.

Apart from gastropods, harmful effects of TBT have also been demonstrated in oysters (EHC 116; Thain and Waldock, 1986). Early work showed shell thickening in adult Pacific oysters (*Crassostrea gigas*) caused by the development of gel centres when exposed to 0.2 μg/L of TBT fluoride (Alzieu *et al.*, 1982). Subsequent work established the NOEL (no observed effect level) for shell thickening in this, the most

TABLE 8.4 *Masculinisation of gastropods caused by TBT*

Species	Effects	Comment
Dog whelk (*Nucella lapillus*)	TBT causes imposex Development of penis blocks oviduct	Cause of population declines along south coast of England
Sting whelk (*Ocenebra erinacea*)	Similar effect to that observed in dog whelk	
Common whelk (*Buccinum undatum*)	Also causes imposex	Effect only seen in juvenile Effects reported from North Sea in the vicinity of shipping lanes
Periwinkle (*Littorina littorea*)	Gross malformation of oviduct usually without development of penis Termed 'intersex'	Has caused infertility and population decline in England, but some recovery after 1987 ban on TBT use on small vessels

sensitive of the tested species, at about 20 ng/L. It has been suggested that shell thickening is a consequence of the effect of TBT on mitochondrial oxidative phosphorylation (Alzieu *et al.*, 1982). Reduced ATP production may retard the function of Ca^{2+}-ATPase, which is responsible for the Ca^{2+} transport that leads to $CaCO_3$ deposition, during the course of shell formation. Abnormal calcification causes distortion of the shell layers.

Other disturbances have been shown to be caused by TBT in oysters. These include the effects on gonad development and gender in adult oysters, and upon settlement, growth and mortality of larval forms. In one experiment, European flat oysters (*Ostrea edulis*) were exposed to 0.24 or 2.6 µg/L TBT over a period of 75 days (Thain and Waldock, 1986). A great number of larvae were produced in a related control group, but none in either of the treated groups. On subsequent examination of the treated oysters, no females were found in either of the two treated groups (20% of the control group were females). In the group receiving the highest exposure to TBT, 72% were undifferentiated.

TBTO has appreciable toxicity to fish, LC_{50} values ranging from 1 to 30 µg/L for most of the species that have been tested. Acute oral LD_{50} values for the rat and the mouse are in the range 85–240 mg/kg.

8.3.5 ECOLOGICAL EFFECTS OF TRIBUTYLTIN

A striking example of the harmful effects of low levels of TBT was the decline of dog whelk populations around the shores of southern England (Gibbs and Bryan, 1986; Matthiessen and Gibbs, 1998; see also Walker *et al.*, 2000). Dog whelks disappeared completely from certain shallow waters (e.g. marinas, estuaries and harbours) where there were large numbers of small boats. It was found that females showed the first signs of imposex with levels of TBT as low as 1 ng/L in water, and that development of male characteristics increased progressively with dose. TBT concentrations exceeding 5 ng/L caused blockage of the oviduct because of the proliferation of the vas deferens and led to breeding failure. Before 1987 levels high enough to have this effect were common in shallow waters with large numbers of small craft. In 1987 a ban on the use of TBT as an antifoulant on small boats was introduced in Britain, after which the population began to recover.

An interesting local phenomenon connected with TBT pollution in southern England is the so-called 'Dumpton syndrome'. In the eponymous coastal area in Kent, a local population survived levels of TBT that caused extinction of the population elsewhere. This was found to be related to a local genetic deficiency of male dog whelks, which had an underdeveloped genital tract (small or absent penis and incomplete gonoducts). Some females in the area did not develop imposex and were able to breed successfully. The tentative interpretation was that there was a genetic deficiency in certain males and females from the Dumpton area that caused low levels of testosterone; thus, neither males nor females could properly develop male characteristics. It is thought that Dumpton females were able to breed with males that were unaffected by Dumpton syndrome (Gibbs, 1993).

In the late 1970s there were severe problems with oyster populations along certain stretches of the French coast, notably in the Bay of Arcachon (Alzieu *et al.*, 1982). Poor shell growth, shell malformations and very poor spatfalls were all observed. Subsequent investigations attributed these harmful effects largely or entirely to TBT. As with imposex in the dog whelk, affected areas had relatively large numbers of small boats and associated (relatively) high levels of TBT in ambient water. As with the dog whelk, populations recovered in badly affected areas after a ban on the use of TBT on small craft (< 25m), and there was a consequent reduction in environmental levels of TBT. In the Bay of Arcachon, the percentages of the population showing deformities were 95–100%, 70–80% and 45–50% in 1980/1, 1982 and 1983 respectively. The spatfall was excellent by 1983, having failed completely in 1980 and 1981. Harmful effects of TBTs on oysters have been reported from many other locations, including other coastal areas of Western Europe, the USA and Japan.

8.4 *Organolead compounds*

Lead tetraalkyl compounds, of which lead tetraethyl is the best known, were once widely used as 'antiknock' compounds, i.e. they were added to petrol to control semiexplosive burning. In many countries this use has been greatly curtailed with the large-scale introduction of lead-free petrol. The reduction in use of tetraalkyl lead has been part of a wider aim of reducing human exposure to lead. In fact, lead tetraalkyls in petrol are broken down to a considerable degree during the operation of the internal combustion engine, so that most of the lead in car exhausts is in the inorganic form.

The main concern about tetraalkyl lead has been about human health hazards, which has resulted in the progressive replacement of leaded petrol by unleaded petrol in most countries (EHC 85). There has been particular concern about possible brain damage to children in polluted urban areas. Little work has been done on the effects of organolead compounds on wildlife or ecosystems, so the following account will be brief.

Lead tetramethyl and lead tetraethyl are covalent lipophilic liquids of low water solubility. Certain inorganic forms of lead, e.g. lead tetrachloride, have similar properties but other forms, such as lead nitrate and lead dichloride, are ionic and water soluble. Covalent and lipophilic forms of lead, like lipophilic forms of organomercury and organotin, can readily cross membranous barriers such as the blood–brain barrier. Consequently, they readily enter the CNS of animals, where they cause damage by combining with sulphydryl groups.

Lead tetraethyl is oxidised by the P450-based monooxygenase system to form the lead triethyl cation:

$$Pb(C_2H_5)_4 \rightarrow Pb(C_2H_5)_3{}^+$$

There is only limited information on the ecotoxicity of organolead compounds. The toxicity of tetraethyl lead to fish ranged from 0.02 to 2.0 mg/L in tests on three different species (see EHC 85). These and other data suggest that alkyl lead compounds are 10–100 times more toxic to fish than is inorganic lead. Turning to birds, toxicity tests upon starlings (*Sturnus vulgaris*) were carried out with trimethyl and triethyl lead (Osborn *et al.*, 1983). Birds were dosed with each compound at two different rates, 0.2 and 2.0 mg/day for 11 days. The high dose corresponded to 28 mg/kg/day, and for both compounds all birds died within 6 days of commencement of dosing. Symptoms of neurotoxicity and behavioural effects were seen before death, most noticeably with trimethyl lead. In 1979, a major poisoning incident involving over 2000 birds of different species, the majority of them dunlin (*Calidris alpina*), was attributed to alkyl lead poisoning (Bull *et al.*, 1983; Osborn *et al.*, 1983; EHC 85). It occurred on the Mersey estuary, UK, where there were periodically high levels of alkyl lead originating from the effluent of a petrochemical works. Casualties included various species of gulls, waders and duck. Smaller numbers of dead birds were found in incidents that occurred over the following 2 years. The total lead content of dead birds averaged about 11 ppm in liver, most of it in the form of alkyl lead. Surviving birds that showed symptoms of poisoning contained about 8.9 ppm of lead. One important food source for the waders was the mollusc *Macoma baltica*, which was found to contain about 1 ppm of lead at the time of the first incident, suggesting that there may have been strong bioaccumulation in the birds.

8.5 *Organoarsenic compounds*

Organoarsenic compounds have been of importance in human toxicology, but have not as yet received much attention in regard to environmental effects. Like methyl mercury compounds, they are synthesised in the environment from inorganic forms and are released into the environment as a consequence of human activity (EHC 18). They can cause neurotoxicity.

Concerning anthropogenic sources, methyl arsenic compounds such as methyl arsenic acid and dimethyl arsenic acid have been used as herbicides, and were once a significant source of environmental residues. Dimethyl arsenic acid (Agent Blue) was used as a defoliant during the Vietnam War.

Methyl arsenic, like methyl mercury, is generated from inorganic forms by methylation reactions in soils and sediments. However, the mechanism is evidently different from that for mercury, depending on the attack by a methyl carbonium ion rather than a methyl carbanion (Craig, 1986; Crosby, 1998). Methylation of arsenic occurs with the trivalent rather than the pentavalent form, and up to three methyl groups can be bound to one As atom (Figure 8.6). The final product, trimethyl arsine, is both volatile and highly toxic to mammals. It has been implicated in cases of human

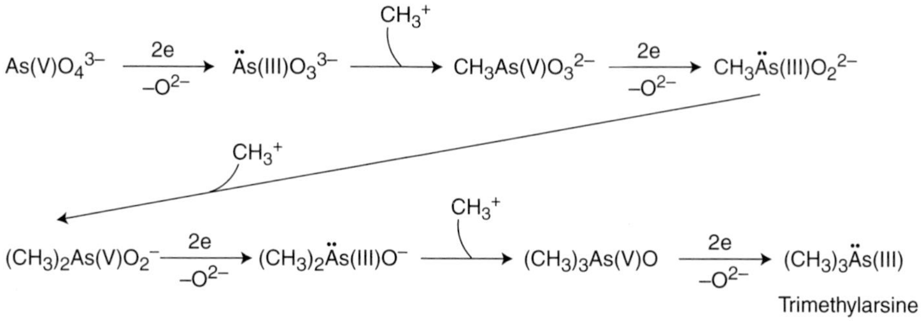

$$As(V)O_4^{3-} \xrightarrow[-O^{2-}]{2e} \ddot{A}s(III)O_3^{3-} \xrightarrow{CH_3^+} CH_3As(V)O_3^{2-} \xrightarrow[-O^{2-}]{2e} CH_3\ddot{A}s(III)O_2^{2-}$$

$$(CH_3)_2As(V)O_2^- \xrightarrow[-O^{2-}]{2e} (CH_3)_2\ddot{A}s(III)O^- \xrightarrow{CH_3^+} (CH_3)_3As(V)O \xrightarrow[-O^{2-}]{2e} (CH_3)_3\ddot{A}s(III)$$

Trimethylarsine

FIGURE 8.6 *Methylation of arsenate. After EHC 18.*

poisoning after its generation by microorganisms from inorganic arsenic (Paris Green) in old wallpaper.

Significant quantities of organoarsenic compounds have been found in marine organisms, and questions have been asked about the possible health risk to humans consuming sea food.

8.6 *Summary*

Mercury, tin, lead, arsenic and antimony form toxic lipophilic organometallic compounds, which have a potential for bioaccumulation/bioconcentration in food chains. Apart from anthropogenic organometallic compounds, methyl derivatives of mercury and arsenic are biosynthesised from inorganic precursors in the natural environment.

Methyl mercury compounds are neurotoxic and can cause behavioural effects. They act by binding to the sulphydryl groups of proteins. Cases of human poisoning (Minamata incident) and poisoning of wildlife (Sweden in the 1950s and 1960s) have been caused by methyl mercury.

Tributyltin compounds used as antifouling agents on boats have had serious toxic effects upon many molluscs, including populations of oysters and dog whelks. Females of the latter species developed a condition known as imposex, which rendered them infertile, and caused local extinction of the population in shallow coastal waters. Imposex provides the basis for a valuable biomarker assay.

Alkyl lead compounds are also highly neurotoxic and have been implicated in large-scale kills of wading birds.

8.7 *Further reading*

The following issues of Environmental Health Criteria (EHC) are valuable sources of information on the environmental toxicology of organometallic compounds: 18, *Arsenic*; 85, *Lead – Environmental Aspects*; 86, *Mercury – Environmental Aspects*; 101, *Methylmercury*; 116, *Tributyltin*.

Craig, P. J. (1986) *Organometallic Compounds in the Environment*. A collection of detailed chapters on the environmental chemistry and biochemistry of organometallic compounds.

International Atomic Energy Agency (1972) Technical Report Series no. 137 contains some useful accounts of work carried out in Sweden on ecotoxicology of organomercury compounds that is difficult to find in the general literature.

Matthiessen, P. and Gibbs, P. E. (1998) A concise review of effects of TBT on molluscs.

CHAPTER 9

Polycyclic aromatic hydrocarbons

9.1 Background

Hydrocarbons are compounds composed of carbon and hydrogen alone. They may be classified into two main groups.

1 Aromatic hydrocarbons, which contain ring systems with delocalised electrons, e.g. benzene.
2 Non-aromatic hydrocarbons, which do not contain such a ring system. Included here are alkanes, which are fully saturated hydrocarbons, alkenes, which contain one or more double bonds, and alkynes, which contain one or more triple bonds.

Some examples of different types of hydrocarbons are given in Figure 9.1. Non-aromatic compounds without a ring structure are termed 'aliphatic', and those with a ring structure (e.g. cyclohexane) are termed 'alicyclic'. Aromatic hydrocarbons often consist of several fused rings, as in the case of benzo(*a*)pyrene.

Both classes of hydrocarbon occur naturally, notably in oil and coal deposits. Aromatic compounds are also products of incomplete combustion of organic compounds, and they are released into the environment both by the activities of man and by certain natural events, e.g. forest fires and volcanic activity.

Aromatic hydrocarbons, the subject of this chapter, are of particular concern because of the carcinogenic and/or mutagenic properties of some of them. Inevitably, there

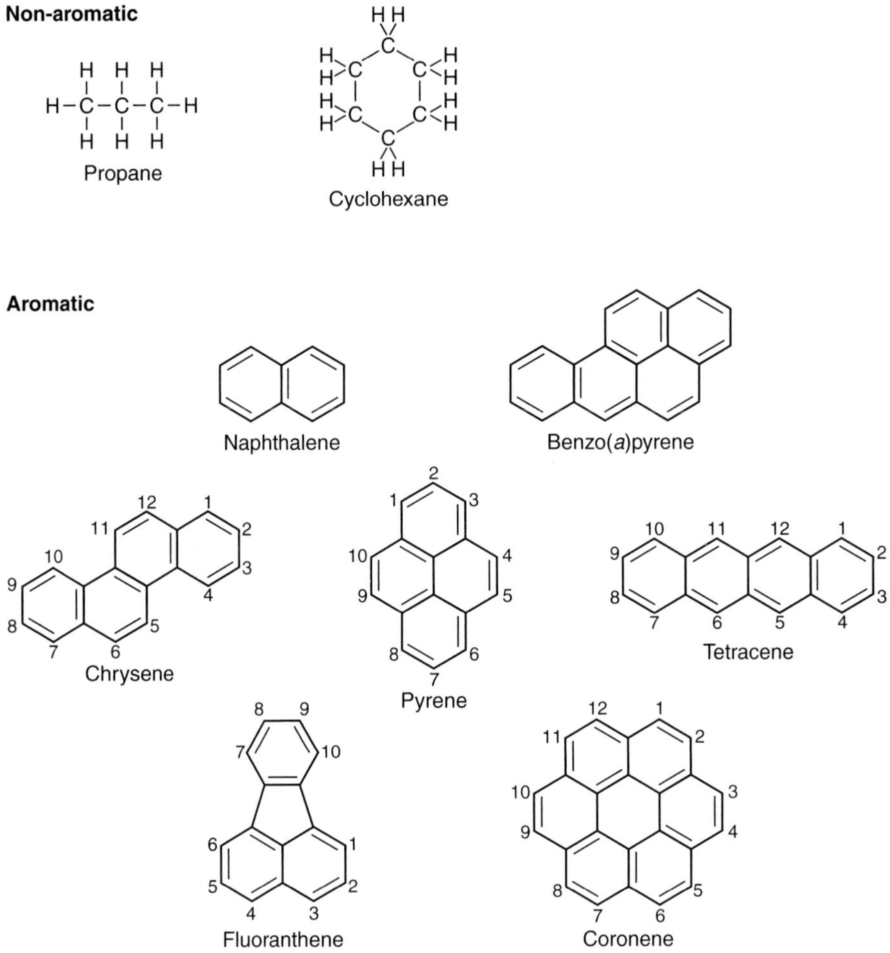

Non-aromatic

Propane

Cyclohexane

Aromatic

Naphthalene

Benzo(*a*)pyrene

Chrysene

Pyrene

Tetracene

Fluoranthene

Coronene

FIGURE 9.1 *Structures of some hydrocarbons.*

are questions about possible harmful effects on ecosystems that are exposed to high levels of them. Non-aromatic hydrocarbons are usually of low toxicity, have not received much attention in ecotoxicology and will not be discussed further in the present text. It should, however, be remembered that crude oil consists mainly of alkanes, and large releases into the sea because of the wreckage of oil tankers have caused the death of many sea birds and other marine organisms as a result of their physical effects of 'oiling', or 'smothering' (see Clark, 1992). Also, released crude oil may act as a vehicle for other lipophilic pollutants, which it dissolves (e.g. organotin compounds and PCBs, both of which may be present in or on wrecked vessels).

9.2 *Origins and chemical properties of PAHs*

The largest releases of PAHs are due to the incomplete combustion of organic compounds during the course of industrial processes and other human activities. Important sources include the combustion of coal, crude oil and natural gas for both industrial and domestic purposes, the use of such materials in industrial processes (e.g. the smelting of iron ore), the operation of the internal combustion engine and the combustion of refuse (see EHC 202). The offshore oil industry and the wreckage of oil tankers are important sources of PAHs in certain areas. Forest fires, which may or may not be the consequence of human activity, are a significant and usually unpredictable source of PAHs. In general, environmental contamination is by complex mixtures of PAHs, not by single compounds.

The structures of some PAHs of environmental interest are given in Figure 9.1. Naphthalene is a widely distributed compound consisting of only two fused benzene rings. It is produced commercially for incorporation into mothballs. Many of the compounds with marked genotoxicity contain three to seven fused aromatic rings. Benzo(*a*)pyrene is the most closely studied of them, and it will be used as an example in the following account.

Because of the fusion of adjacent rings, PAHs tend to have a rigid planar structure. As a class, they are of low water solubility and marked lipophilicity, and have high K_{ow} values (Table 9.1); the higher the molecular weight, the higher the lipophilicity and the higher the log K_{ow}. Vapour pressure is also related to molecular weight; the higher the molecular weight, the lower the vapour pressure. PAHs have no functional groups and are chemically rather unreactive. They can, however, be oxidised both in the natural environment and biochemically (see Figure 9.2 and p. 173). Photo-decomposition can occur in air and sunlight to yield oxidative products such as quinones and endoperoxides. Nitrogen oxides and nitric acid can convert PAHs to nitro derivatives, and sulphur oxides and sulphuric acids can produce sulphanilic and sulphonic acids.

9.3 *Metabolism of PAHs*

Because of the absence of functional groups, primary metabolic attack on PAHs is limited to oxidation, usually catalysed by cytochrome P450. As with coplanar PCBs, oxidative attack involves P450 forms of more than one gene family, including members of gene families 1, 2 and 3. The position of oxidative attack on the ring system (regioselectivity) depends upon the P450 form to which a PAH is bound. P4501A1 is particularly implicated in the metabolic activation of carcinogens such as benzo(*a*)pyrene, where oxygen atoms can be inserted into the critical bay region positions. The metabolism of benzo(*a*)pyrene has been studied in some depth and detail, and it will be used as an example of the metabolism of PAHs more generally (Figure 9.2).

TABLE 9.1 *Properties of some PAHs*

Compound	Vapour pressure Pa at 25°C	Log K_{ow}	Water solubility at 25°C (μg/L)
Naphthalene	10.4	3.4	3.17×10^{-4}
Anthracene	8.4×10^{-4}	4.5	73
Pyrene	6.0×10^{-4}	5.18	135
Chrysene	8.4×10^{-5}	5.91	2.0
Benzo(*a*)pyrene	7.3×10^{-7}	6.50	3.8
Dibenz(*a,h*)anthracene	1.3×10^{-8}	6.50	0.5

Data from EHC 202.

Initial metabolic attack can be upon one of a number of positions on the benzo(*a*)pyrene molecule to yield various epoxides. Epoxides tend to be unstable and can quickly rearrange to form phenols. They may also be converted to *trans*-dihydrodiols by epoxide hydrolase and/or to glutathione conjugates by the action of glutathione-S-transferases. Hydroxymetabolites are more polar than the parent molecules and can be converted into conjugates such as glucuronides and sulphates by the action of conjugases. Two important oxidations of benzo(*a*)pyrene are shown in Figure 9.2: formation of the 7,8 oxide and the 4,5 oxide. The 4,5 oxide is unstable under cellular conditions, undergoing rearrangement to form a phenol and biotransformation to a *trans*-dihydrodiol or a glutathione conjugate. *In vitro*, it shows mutagenic properties, e.g. in the Ames test. However in vertebrates, *in vivo*, it appears to be detoxified very effectively, thus preventing the formation of DNA adducts to any significant degree. The 7,8 oxide is converted to the 7,8-*trans*-dihydrodiol by the action of epoxide hydrolase. The 7,8-*trans*-dihydrodiol is a substrate for P4501A1 and consequent oxidation yields the highly mutagenic 7,8-diol-9,10-oxide, a metabolite that, under cellular conditions, interacts with guanine residues of DNA. Thus, a mutagenic diol epoxide generated in the endoplasmic reticulum is able to escape detoxication *in situ*, or elsewhere in the cell as it migrates to the nucleus to interact with DNA. The isomer of the 7,8-diol-9,10-oxide responsible for adduction (Figure 9.2) is a poor substrate for epoxide hydrolase. It should be added that benzo(*a*)pyrene is an inducer of P4501A1, so it can increase the rate of it own activation! The toxicological significance of this type of interaction will be discussed in section 9.5.

In terrestrial animals, the excreted products of PAHs are mainly conjugates formed from oxidative metabolites. These include glutathione conjugates of epoxides and sulphates, and glucuronide conjugates of phenols and diols.

PAHs such as benzo(*a*)pyrene and 3-methylcholanthrene induce cytochrome P4501A1/2 (Chapter 2). Induction of P4501A1/2 is used as the basis for biomarker assays for PAHs and other planar organic pollutants such as coplanar PCBs, PCDDs and PCDFs.

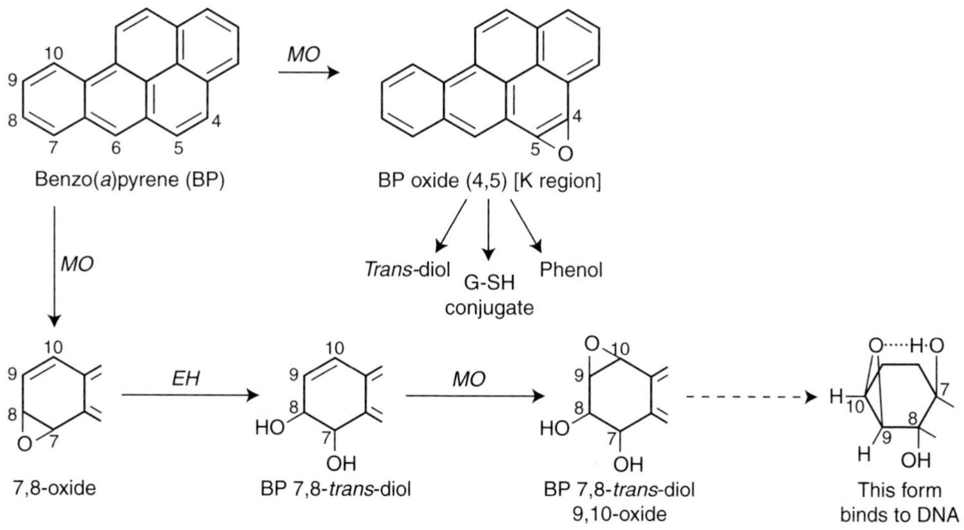

FIGURE 9.2 *Metabolism of benzo(a)pyrene.*

9.4 *Environmental fate of PAHs*

Viewed globally, the largest emissions of PAHs are into the atmosphere, and the main source is incomplete combustion of organic compounds. As mentioned above, emissions are mainly the consequence of human activity, although certain natural events, e.g. forest fires, are sometimes also important. Emissions into the air are of complex mixtures of different PAHs, including particulate matter as in smoke. PAHs in the vapour phase can be adsorbed on to airborne particles. Airborne PAHs eventually enter surface waters due to precipitation of particles or to diffusion. Once there, because of their high K_{ow} values, they tend to become adsorbed to the organic material of sediments and taken up by aquatic organisms. Similarly, airborne PAHs can eventually reach soil to become adsorbed by soil colloids and absorbed by soil organisms.

Apart from release into air, which is important globally, the direct transfer of PAHs to water or land surfaces can be very important locally. The wreckages of oil tankers and discharges from oil terminals cause marine pollution by crude oil, which contains appreciable quantities of PAHs. Disposal of waste containing PAHs around industrial premises has caused serious pollution of the land in some localities.

When crude oil is released into the sea, oil films ('slicks') can spread over a large area, the extent and direction of movement being determined by wind and tide (see Clark, 1992). The hydrocarbons of lowest molecular weight have the highest vapour pressures and tend to volatilise, leaving behind the least volatile components of crude oil. Eventually, the residue of relatively involatile hydrocarbons will sink to become associated with sediment. Thus, long after the surface film of oil has disappeared, residues of PAHs will exist in sediment, where they are available to bottom-dwelling organisms. To illustrate the range of PAHs found in sediments, some values for PAH residues detected in sediment from the highly polluted Duwamish waterway in the USA follow (source Varanasi *et al.*, 1992). All concentrations are given as mean values expressed as ng/g dry weight.

Naphthalene	400
Fluorene	390
Phenanthrene	2400
Anthracene	610
Fluoranthene	3900
Benz(*a*)anthracene	2000
Chrysene	2900
Benzo(*a*)pyrene	2300
Pyrene	4800
Dibenz(*a*,*h*)anthracene	470
Perylene	900
Benzofluoranthenes	4900

PAHs can be bioconcentrated/bioaccumulated by certain aquatic invertebrates low in the food chain that lack the capacity for effective biotransformation (Walker and Livingstone, 1992). Molluscs and *Daphnia* spp. are examples of organisms that readily bioconcentrate PAHs. On the other hand, fish and other aquatic vertebrates readily biotransform PAHs, so biomagnification does not extend up the food chain as it does in the case of persistent polychlorinated compounds. As noted earlier (pp. 8 and 69), P450-based monooxygenases are not well represented in molluscs and in many other aquatic invertebrates, so this observation is unsurprising. P450 attack is the principal (probably the only) effective mechanism of primary metabolism of PAHs.

An example of PAH residues in polluted marine ecosystems is given in Table 9.2. They are from studies carried out at different coastal sites in the USA (Varanasi *et al.*, 1992) The residues are categorised into lower aromatic hydrocarbons (LAHs), composed of one to three rings, and higher aromatic hydrocarbons (HAHs), composed of four to seven rings. HAHs predominate in sediment, showing levels in the range 1800–12 600 ng/g, i.e. 1.8–12.6 ppm by weight. As mentioned above, some invertebrates can bioaccumulate PAHs, which is the main reason for significant levels of PAHs in the food remains found within fish stomachs. Analysis of invertebrates from some of these areas showed levels of 300–3500 ng/g total aromatic hydrocarbon in their tissues, of which over 90% was accounted for as HAHs. The concentrations of aromatic hydrocarbons are, however, substantially below those found in samples of sediment. Thus, HAH levels range from 135 to 2700 ng/g in stomach contents. Interestingly, no residues of aromatic hydrocarbons were detected in the livers of the fish analysed in this study, illustrating their ability to rapidly metabolise both LAHs and HAHs. It appears that organisms at the top of aquatic food chains are not exposed to substantial levels of PAHs in food because of the detoxifying capacity of organisms beneath them in the food chain. On the other hand, fish, birds and aquatic mammals

TABLE 9.2 *Concentration of aromatic hydrocarbons (AHs) in samples from US coastal sites*

Sample	Total AH (ng/g)	I–3 rings (LAH) (ng/g)	4–7 rings (HAH) (ng/g)
Sediments	700–14 000	200–1400	1800–12 600
Fish stomach contents	150–3000	15–300	135–2700
Fish liver	Nil	Nil	Nil

Extracted from data presented by Varanasi *et al.* (1992).

feeding on molluscs and other invertebrates are in a different position. Their food may contain substantial levels of PAHs. Although they can achieve rapid metabolism of dietary PAHs, it should be remembered that oxidative metabolism causes activation as well as detoxication. The types of P450 involved determine the position of metabolic attack (see section 9.3). Cytochrome P4501A1, for example, activates benzo(*a*)pyrene by oxidising the bay region to form a mutagenic diol epoxide. The tendency for PAHs to be activated as opposed to detoxified depends on the balance of P450 forms, and this balance is dependent upon the state of induction. Many PAHs, PCDDs, PCDFs, coplanar PCBs and other planar organic pollutants are inducers of P450s belonging to family 1A. Thus, activation of PAH may be enhanced because of the presence of pollutants that induce P4501A1/1A2, forms that are particularly implicated in the process of activation (Walker and Johnston, 1989).

9.5 *Toxicity of PAHs*

In themselves, PAHs are rather unreactive and appear to express little toxicity. Toxicity is the consequence of their transformation to more reactive products, by chemical or biochemical processes. In particular, the incorporation of oxygen into the PAH ring structure has a polarising effect; the electron-withdrawing properties of oxygen leads to the production of reactive species such as carbonium ions. This is evidently the reason why PAHs become more toxic to fish and *Daphnia* after exposure to UV radiation (Oris and Giesy, 1986, 1987); photo-oxidation of PAHs to reactive products increases toxicity.

Much research on the toxicity of PAHs has been concerned with human health hazards and has focused on their mutagenic and carcinogenic action. These two properties are to some extent related because there is growing evidence that certain DNA adducts formed by metabolites of carcinogenic PAHs become fixed as mutations of oncogenes or tumour-suppressor genes, which are found in chemically produced cancers (Purchase, 1994). Typical mutations occur at specific codons in the *ras*, *neu* or *myc* oncogenes or in *p53*, retinoblastoma or APC tumour-suppressor genes. These genes code for proteins involved in growth regulation, with the consequence that mutated cells have altered growth control. One example of such a carcinogenic metabolite is the 7,8-diol-9,10-oxide of benzo(*a*)pyrene (Figure 9.2). More generally, many compounds found to be mutagenic in bacterial mutation assays (e.g. the Ames test) are also carcinogenic in long-term dosing tests with rodents. This said, many carcinogens act by non-genotoxic mechanisms (Purchase, 1994).

Benzo(*a*)pyrene is converted to its 7,8-diol-9,10-oxide by the action of cytochrome P4501A1 and epoxide hydrolase, as shown in Figure 9.2. In one of its enantiomeric forms, this metabolite can then form DNA adducts by alkylating certain guanine residues (Figure 9.3). The metabolite acts as an electrophile, owing to strong carbonium ion formation on the 10 position of the epoxide ring, which is located in the bay region. The epoxide ring cleaves, and a bond is formed between C-10 of the PAH ring and the free amino group of guanine. The oxygen atom of the cleaved epoxide ring acquires a proton, thus leaving a hydroxyl group attached to C-9. This adduct, like others formed between reactive metabolites of PAHs and DNA, is bulky and can be detected by ^{32}P post-labelling, and by immunochemical techniques (e.g. Western blotting). It has been proposed that there is a particular tendency for strong carbonium ion formation to occur on the bay region of PAHs, and that such bay region epoxides have a strong tendency to form DNA adducts (see Hodgson and Levi, 1994).

There is strong evidence that DNA adduction by these bulky reactive metabolites of PAHs is far from random, and that there are certain hot spots that are preferentially attacked. Differential steric hindrance and the differential operation of DNA repair mechanisms ensure that particular sites on DNA are subject to stable adduct formation (Purchase, 1994). DNA repair mechanisms clearly remove many PAH/guanine adducts very quickly, but studies with ^{32}P post-labelling have shown that certain adducts can be very persistent – certainly over many weeks. Evidence for this has been produced in studies on fish and *Xenopus* (an amphibian) (Reichert *et al.*, 1991; Waters *et al.*, 1994).

Although genotoxicity is of central importance in human toxicology, its significance in ecotoxicology is controversial. However, PAH has been shown to cause tumour development in fish in response, for example, to oral, dermal or intraperitoneal administration of benzo(*a*)pyrene and 3-methylcholanthrene. Hepatic tumours have

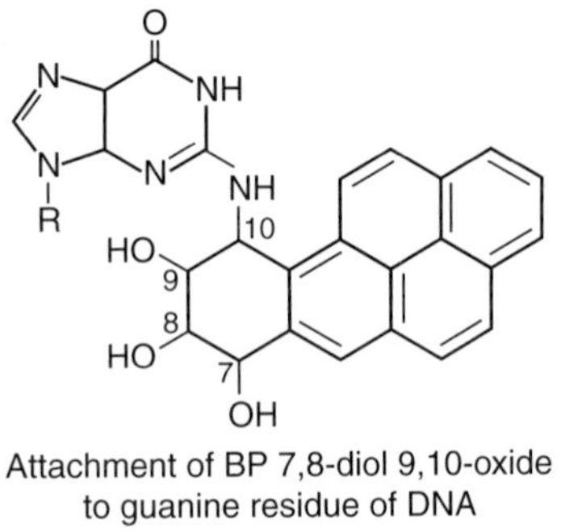

Attachment of BP 7,8-diol 9,10-oxide
to guanine residue of DNA

FIGURE 9.3 *DNA adduct formation.*

been reported in wild fish exposed to sediment containing approximately 250 mg/kg of PAH (EHC 202). K-*ras* mutations occurred in pink salmon embryos (*Onchorhynchus gorbuscha*) after exposure to crude oil from the tanker *Exxon Valdez*, which caused extensive pollution of coastal regions of Alaska (Roy *et al.*, 1999). However, it is not clear whether cancer is a significant factor in determining the survivorship or reproductive success of free-living vertebrates or invertebrates. Cancers usually take a long time to develop, and the lifespan of free-living animals is limited by factors such as food supply, disease, predation, etc. Do they live long enough for cancers to be a significant cause of population decline?

Apart from carcinogenicity, there is the wider question of other possible genotoxic effects in free-living animals, effects that may be heritable if the mutations are in germ cells. Studying aquatic invertebrates exposed to PAHs, Kurelec (personal communication, 1991) noticed a number of longer-term physiological effects, which he termed collectively 'genotoxic disease syndrome'. Although the basis for these effects has not yet been elucidated, the observations raise important questions that should be addressed. PAHs have been shown to form a variety of adducts in fish and amphibians, so there is a strong suspicion that some of these may lead to the production of mutations. If mutations occur in germ cells, there are inevitably questions about their effects on progeny. Most mutations are not beneficial!

There is evidence for immunosuppressive effects of PAHs in rodents (Davila *et al.*, 1997). For example, strong immunosuppressive effects were reported in mice that had been dosed with benzo(*a*)pyrene and 3-methylcholanthrene, effects which persisted for up to 18 months (EHC 202). Multiple immunotoxic effects have been reported in rodents, and there is evidence that these result from the disturbance of calcium homeostasis (Davila *et al.*, 1997). PAHs can activate protein tyrosine kinases in T cells, which initiates the activation of a form of phospholipase C. Consequently, release of inositol triphosphate is enhanced, a molecule that immobilises Ca^{2+} from storage pools in the endoplasmic reticulum.

Turning to the acute toxicity of PAHs, terrestrial organisms will be dealt with before considering aquatic organisms, to which somewhat different considerations apply. The acute toxicity of PAHs to mammals is relatively low. Naphthalene, for example, has a mean oral LD_{50} of 2 700 mg/kg to the rat. Similar values have been found with other PAHs. LC_{50} values of 150 mg/kg and 170–210 mg/kg have been reported for phenanthrene and fluorene, respectively, in the earthworm. The NOEL level for survival and reproduction in the earthworm was estimated to be 180 mg/kg dry soil for benzo(*a*)pyrene, chrysene and benzo(*k*)fluoranthene (EHC 202).

Toxicity of PAHs to aquatic organisms depends on the level of UV radiation to which the test system is exposed. PAHs can become considerably more toxic in the presence of radiation, apparently because photooxidation transforms them into toxic oxidative products. PAHs such as benzo(*a*)pyrene can have LC_{50} values as low as a few micrograms per litre in fish that have been exposed to UV radiation (Oris and Giesy, 1986, 1987). PAHs can also show appreciable toxicity to sediment-dwelling invertebrates. LC_{50} values of 0.5–10 mg/kg (reference concentration in sediment) have been reported for marine amphipods for benzo(*a*)pyrene, fluoranthene and phenanthrene, used singly or in mixtures. These values are much lower than the LC_{50} or NOEL concentrations for earthworms, quoted above, which were exposed to contaminated soil. It has also been shown that exposure of adult fish to anthracene and artificial UV radiation can impair egg production (Hall and Uris, 1991).

9.6 *Ecological effects of PAHs*

Serious ecological damage has been caused locally by severe oil pollution The wreckages of the oil tankers the *Torrey Canyon* (Cornwall, UK, 1967), the *Amoco Cadiz* (Brittany, France, 1978), the Ex*xon Valdez* (Alaska, USA, 1989) and the *Sea Empress* (South Wales, UK, 1996) all caused serious pollution locally. Less dramatically, leakages from offshore oil operations have also caused localised pollution problems. Most of the reported harmful effects have been due to the physical action of the oil rather than the toxicity of PAHs. Fish, however, may have been poisoned when there was strong UV radiation (see section 9.5). The oiling of sea birds and other marine organisms has been the cause of some local population declines. Sometimes the reduction in invertebrate herbivores of polluted beaches and rock pools has led to an upsurge of the plants upon which they feed. The flourishing of seaweeds and of algal blooms has sometimes followed such environmental disasters, to be reversed when pollution is reduced and the herbivores recover. In the neighbourhood of oil terminals, the diversity of benthic fauna has been shown to decrease (see Clark, 1992).

Although it has been relatively easy to demonstrate local short-term effects of oil pollution on sea bird populations, establishing longer-term effects on marine ecosystems has proved more difficult, notwithstanding the persistence of PAH residues in sediments. In one study, the edible mussel (*M. edulis*) was used as an indicator organism to investigate PAH effects along a pollution gradient in the neighbourhood of an oil terminal at Sullom Voe, Shetlands, UK (Moore *et al.*, 1987; Livingstone *et al.*, 1988). The impact of PAHs was assessed using a suite of biomarker assays. One of the assays was 'scope for growth', an assay that seeks to measure the extra available energy of the organism that can be used for growth and reproduction; extra, that is, in relation to the basic requirement for normal metabolic processes. A strong negative relationship was shown between scope for growth and the tissue concentration of two- and three-ring PAHs (Figure 9.4). Although this observation might be criticised on the grounds that other pollutants could have followed the same pollution gradient and had similar effects upon scope for growth, there was some supporting evidence from a controlled mesocosm study, which showed a similar dose–response relationship over part of the range and gave a similar regression line. Also, controlled laboratory studies with *M. edulis* showed that scope for growth could be reduced by dosing with diesel oil to give tissue levels of two- and three-ring PAHs similar to those found in the field study. In addition to the reduction in scope for growth, some biomarker responses were observed

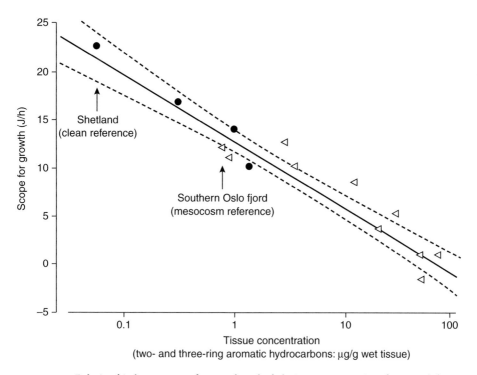

FIGURE 9.4 *Relationship between scope for growth and whole tissue concentration of two- and three-ring aromatic hydrocarbons in* M. edulis *(mean ± 95% confidence limits).* △, *Data from Solbergstrand mesocosm experiment, Oslo Fjord, Norway.* ●, *Data from Sullom Voe, Shetland Islands (Moore* et al.*, 1987).*

(Livingstone, 1985). The lowest level of dosing with diesel oil (29 ppb for 4 months) caused a doubling of P450 levels, a reduction in lysosomal stability and an increase in cytochrome P450 reductase. Thus, strong evidence was produced for the harmful effects of PAH on *M. edulis* near an oil terminal, but it was not established whether they were sufficient to cause a population decline.

As discussed earlier (section 6.2.5), several studies have linked the presence of high PAH levels in the marine environment with a high incidence of tumours in fish. The ecological significance of these observations, however, is not known.

9.7 *Summary*

On the global scale, most PAH release is the consequence of incomplete combustion of organic compounds. In the marine environment, however, there can be significant levels of PAH pollution locally, as a result of the large-scale release of crude oil, especially due to the wreckage of oil tankers, but also to leakage of crude oil during offshore oil operations. PAHs can be biomagnified by some aquatic invertebrates, but not in organisms higher in the food chain, where they undergo relatively rapid metabolism.

PAHs do not have high acute toxicity to terrestrial animals. To fish, however, they can show considerable toxicity in the presence of UV light as a consequence of their photooxidation. In human toxicology, the main concern has been about the mutagenic and carcinogenic properties of some PAHs. The metabolic activation of such compounds leads to the formation of DNA adducts that can be stabilised as mutations of oncogenes or tumour-suppressor genes, genes that have a role in growth regulation. There is growing evidence that PAHs can also form DNA adducts in wild vertebrates; however, there is controversy about the significance of this from an ecological point of view. PAHs cause the induction of P4501A1/2, a response which has been utilised in the development of biomarker assays for these and other planar lipophilic organic pollutants.

9.8 *Further reading*

Clark, R. B. (1992) *Marine Pollution*. A readable account of marine pollution caused by crude oil, but it does not deal with biochemical aspects.

EHC 202 (1998) *Non Heterocyclic Polycyclic Aromatic Hydrocarbons.* A very detailed account of the environmental toxicology of PAHs seen from the global point of view. However, it largely ignores marine pollution caused by oil spills.

CHAPTER 10

Organophosphorous and carbamate insecticides

10.1 *Background*

Organophosphorous insecticides (OPs) and carbamate insecticides are dealt with here in a single chapter because they share a common mode of action: cholinesterase inhibition. Unlike DDT and most of the cyclodiene insecticides, they do not have long biological half-lives or present problems of biomagnification along food chains. When organochlorine insecticides such as DDT and dieldrin began to be phased out during the 1960s, they were often replaced by OPs or carbamates, which were seen to be more readily biodegradable and less persistent, although not necessarily as effective for controlling pests, parasites or vectors of disease. They replaced organochlorine insecticides as the active ingredients in crop sprays, sheep dips, seed dressings, sprays used for vector control and various other insecticidal preparations.

When organochlorine insecticides were phased out, the less persistent insecticides that replaced them were thought to be more 'environmentally friendly'. However, some of the insecticides that were used as replacements also presented problems, e.g. because of very high acute toxicity. The insecticides to be discussed in the present chapter illustrate well the ecotoxicological problems that can be associated with compounds that have low persistence but high neurotoxicity.

Organophosphorous insecticides were first developed during the Second World War, both as insecticides and as chemical warfare agents. During this time, several new insecticides were synthesised by G. Schrader working in Germany, prominent among which was parathion, an insecticide that came to be widely used in agriculture after the war. In the post-war years, many new OPs were introduced and used for a wide range of applications. Early insecticides had only 'contact' action when applied to crops in the field, but later ones such as dimethoate, metasystox, disyston and phorate had systemic properties. Systemic compounds can enter the plant to be circulated in the vascular system. Sap-feeding insects such as aphids and whitefly are then poisoned by insecticides (or their toxic metabolites), which circulate within the plant. Some OPs were developed that were more selective and had low mammalian toxicity (e.g. malathion and pirimiphos methyl), making them suitable for certain veterinary uses and for protecting stored grain against insect pests.

The rapidly expanding use of OPs and the proliferation of new active ingredients and formulations was not without its problems. Some OPs proved to be too hazardous to operators because of very high acute toxicity. A few were found to cause delayed neurotoxicity, a condition not caused by cholinesterase inhibition (e.g. mipafox, leptophos). There was also the problem of the development of resistance, e.g. by cereal aphids. In due course, other insecticides such as carbamates were developed and replaced OPs for certain uses when there were problems. New carbamate insecticides were introduced and took a significant share of the market. Some had the advantages of being nematicides and/or molluscicides as well as being insecticides. Some had systemic action (e.g. aldicarb and carbofuran). Sometimes they overcame problems of resistance that arose because of the intensive use of OPs, for example in cereal aphids such as *M. persicae*. Unfortunately, some carbamates also caused environmental problems because of high vertebrate toxicity.

In the following account OPs will be discussed before considering carbamates.

10.2 *The OPs*

10.2.1 CHEMICAL PROPERTIES

The OPs to be discussed here correspond to one or other of the two following structural formulae:

$$
\begin{array}{c}
R1 \\
 \diagdown \\
  \overset{\overset{\displaystyle O}{\parallel}}{P} \text{--X} \\
 \diagup \\
R2
\end{array}
\qquad\qquad
\begin{array}{c}
R1 \\
 \diagdown \\
  \overset{\overset{\displaystyle S}{\parallel}}{P} \text{--X} \\
 \diagup \\
R2
\end{array}
$$

Compounds corresponding to structure (1) are referred to as 'oxons'; when R1, R2 and X are all linked to P through oxygen, the compound is a triester of orthophosphoric acid and may be termed a phosphate. If only one or two of these links are through oxygen, then the compounds are termed 'phosphinate' or 'phosphonate' respectively. Compounds corresponding to structure (2) are termed 'thions', and when R1, R2 and X are all linked to P through oxygen the compound is a triester of phosphorothioic acid and may be termed a 'phosphorothioate'. If one of the links is through S, then the molecule is a phosphorodithioate. R1 and R2 are usually alkoxy groups, whereas X is usually a more complex group linked through oxygen or sulphur. X is sometimes termed the 'leaving group', because it can be removed by hydrolytic attack, either chemically or biochemically.

Some properties of OPs are given in Table 10.1, and some structures in Figure 10.1. Useful texts on the properties and toxicology of OPs are Eto (1974), Fest and Schmidt (1982) and Ballantyne and Marrs (1992). There is some variation in the values quoted for the above properties in the literature reflecting purity of sample, accuracy of method, etc. The above are representative values and are not necessarily the most accurate ones for the purest samples.

Of the compounds listed in Table 10.1, all except dimethoate and azinphos-methyl exist as liquids at normal temperature and pressure. Looking through Table 10.1, it can be seen that there is considerable variation in both water solubility and vapour pressure. Thus, dimethoate and demeton-S-methyl have appreciable water solubility and show marked systemic properties, whereas parathion, chlorfenvinphos and azinphos-methyl have low water solubility and are not systemic. Disulfoton, although of low water solubility in itself, undergoes biotransformation in plants to yield more polar metabolites, including sulphoxides and sulphones, which are systemic. In general, OPs are considerably more polar and water soluble than organochlorine insecticides.

The relatively high vapour pressure of most OPs limits their persistence when sprayed on exposed surfaces (e.g. on crops, seeds or farm animals). Chlorfenvinphos, however, has relatively low vapour pressure and is consequently reasonably persistent and has been used as a replacement for organochlorine compounds, e.g. as an insecticidal seed dressing and as a sheep dip.

The environmental fate and behaviour of compounds is a function of their physical, chemical and biochemical properties. Being a diverse group, individual OPs differ considerably from one another in their properties and consequently in their environmental behaviour and the way that they are used as pesticides. Pesticide chemists and formulators have been able to exploit the properties of individual OPs to serve

TABLE 10.1 *Properties of some OP insecticides*

Compound	Water solubility (µg/mL at 25°C)	Log K_{ow}	Vapour pressure (mmHg at 25°C)
Parathion	11	3.83	6.7×10^{-6}
Diazinon	40	3.40	1.4×10^{-4}
Dimethoate	35 000		8.5×10^{-6}
Azinphos-methyl	33		3.8×10^{-4}
Malathion	145	2.36	3.98×10^{-5}
Disyston	25		1.8×10^{-4}
Demeton-S-methyl	3300	1.32	3.6×10^{-4}
Chlorfenvinphos	145		3.0×10^{-6}

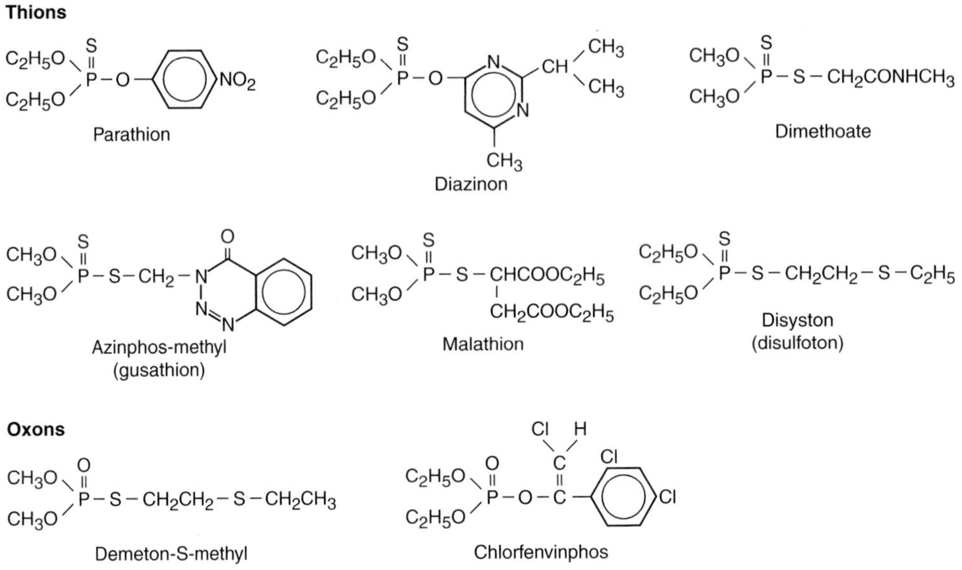

Thions

Parathion

Diazinon

Dimethoate

Azinphos-methyl
(gusathion)

Malathion

Disyston
(disulfoton)

Oxons

Demeton-S-methyl

Chlorfenvinphos

FIGURE 10.1 *Some OPs.*

the purposes of more effective, and more environmentally friendly, pest control, e.g. in the development of compounds such as chlorfenviphos, which has enough stability and a sufficiently low vapour pressure to be effective as an insecticidal seed dressing, but, like other OPs, it is readily biodegradable; thus it was introduced as an 'environmentally friendly' alternative to persistent organochlorine insecticides for these purposes.

Of the compounds shown in Figure 10.1, six are thions and only two (demeton-S-methyl and chlorfenvinphos) are oxons. Four of the thions possess two sulphur linkages to P and are phosphorodithionates. The oxons tend to be more unstable and reactive than the thions, and they are much better substrates for esterases, including AChE! Oxygen has stronger electron-withdrawing power than sulphur, so oxons tend to be more polarised than thions. In fact the thions are not effective anticholinesterases in themselves and need to be converted to oxons by monooxygenases before toxicity is expressed (see section 10.2.4). As technical products, thions have an advantage over most oxons in being more stable.

Organophosphorous insecticides as a class are chemically reactive, and they are not very stable either chemically or biochemically. The leaving group (X in structural formula) can be removed hydrolytically, and OPs generally are readily hydrolysed by strong alkali. (See Figure 10.3 for examples of enzymic hydrolysis.) After OPs have been released into the environment they undergo chemical hydrolysis in soils, sediments and surface waters. The rate of hydrolysis depends on the pH; in most cases, the higher the pH, the faster the hydrolysis of the OP. Demeton-S-methyl, for example, has half-lives in aqueous solution of 63, 56 and 8 days at pH values of 4, 7 and 9 respectively (EHC 197). Thus, most OPs are not very persistent in alkaline soils or waters.

Thions are prone to oxidation, and can be converted to oxons under environmental conditions. Also, some OPs can undergo isomerisation under the influence of sunlight and/or high temperatures, a well-documented example being the conversion of malathion to iso-malathion. Although malathion is a thion of low mammalian toxicity, iso-malathion is an oxon of high mammalian toxicity. Cases of human poisoning have been the consequence of malathion undergoing this conversion in badly stored grain.

Another group of organophosphorous anticholinesterases deserving brief mention are certain chemical warfare agents, often termed 'nerve gases'. Examples are soman, sarin and tabun. These compounds have, as befits their intended purpose, very high mammalian toxicity and high vapour pressure. All the examples given are oxons, which tend to have greater mammalian toxicity than thions. In addition, they are phosphinates rather than phosphates, having only one P linkage through oxygen or sulphur.

10.2.2 METABOLISM OF OPs

As examples of OP metabolism, the major metabolic pathways of malathion, diazinon and disyston are shown in Figure 10.2, identifying the enzyme systems involved. Organophosphorous insecticides are highly susceptible to metabolic attack, and metabolism is relatively complex, involving a variety of enzyme systems. The interplay between activating transformations on the one hand and detoxifying transformations on the other determines toxicity in particular species and strains (see Walker, 1991). Because of this complexity, knowledge of the metabolism of most OPs is limited. Further information on OP metabolism may be found in Eto (1974), Fest and Schmidt (1982) and Hutson and Roberts (1999).

All three insecticides shown in Figure 10.2 are thions, and all are activated by conversion to their respective oxons. Oxidation is carried out by the P450-based microsomal monooxygenase system, which is well represented in most land vertebrates and insects, but less well represented in plants where activities are very low. Oxidative desulphuration of thions to oxons does occur slowly in plants, and may be due to either or both monooxygenase attack and peroxidase attack (Drabek and Neumann, 1985; Riviere and Cabanne, 1987). Compounds such as disyston, which have thioether bridges in their structure, can undergo sequential oxidation to sulphoxides and sulphones. Other examples are demeton-S-methyl (Figure 10.1) and phorate. The oxon forms of OP sulphoxides and sulphones can be potent anticholinesterases, and sometimes make an important contribution to the systemic toxicity of insecticides such as demeton-S-methyl, disyston and phorate.

The oxidation of OPs can bring detoxication as well as activation. Oxidative attack can lead to the removal of R groups (oxidative dealkylation), leaving behind P–OH, which ionises to PO⁻. Such a conversion looks superficially like a hydrolysis and was

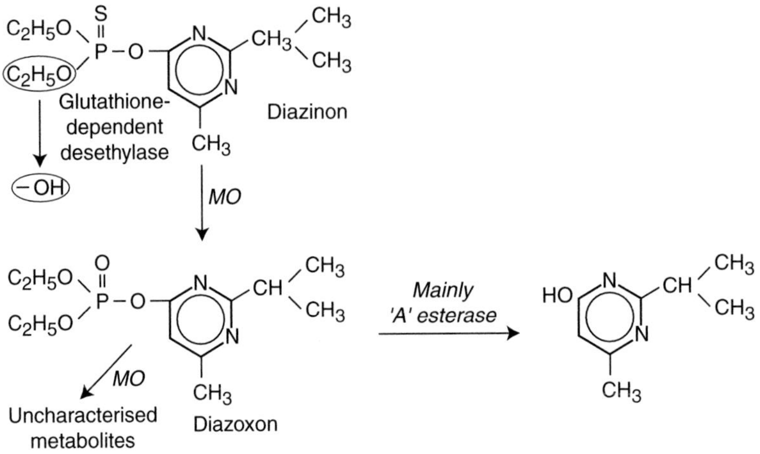

C_2H_5O ... $\overset{S}{\underset{\|}{P}}$—O— ... CH$_3$
C_2H_5O ... CH$_3$—CH$_3$
Glutathione-
dependent
desethylase ... Diazinon

—OH

CH$_3$

MO

C_2H_5O ... $\overset{O}{\underset{\|}{P}}$—O— ... CH$_3$
C_2H_5O ... CH—CH$_3$

*Mainly
'A' esterase*

HO ... CH$_3$
CH—CH$_3$

MO ... CH$_3$

Uncharacterised ... Diazoxon
metabolites

CH$_3$

C_2H_5O ... $\overset{S}{\underset{\|}{P}}$—S—CH$_2CH_2$—S—C$_2H_5$ ... Disyston
C_2H_5O

MO ... MO

C_2H_5O ... $\overset{S}{\underset{\|}{P}}$—S—CH$_2CH_2$—$\overset{O}{\underset{\|}{S}}$—C$_2H_5$
C_2H_5O

C_2H_5O ... $\overset{O}{\underset{\|}{P}}$—S—CH$_2CH_2$—S—C$_2H_5$
C_2H_5O ... Oxon form

MO ... MO ... MO

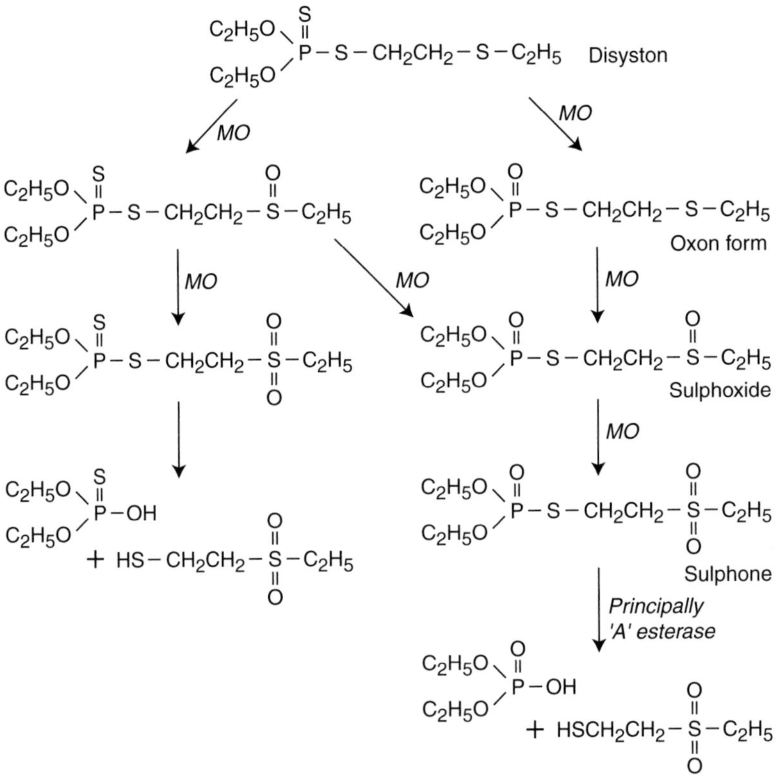

C_2H_5O ... $\overset{S}{\underset{\|}{P}}$—S—CH$_2CH_2$—$\overset{O}{\underset{\overset{\|}{\underset{O}{\|}}}{S}}$—C$_2H_5$
C_2H_5O

C_2H_5O ... $\overset{O}{\underset{\|}{P}}$—S—CH$_2CH_2$—$\overset{O}{\underset{\|}{S}}$—C$_2H_5$
C_2H_5O ... Sulphoxide

MO

C_2H_5O ... $\overset{S}{\underset{\|}{P}}$—OH
C_2H_5O

+ HS—CH$_2$CH$_2$—$\overset{O}{\underset{\overset{\|}{\underset{O}{\|}}}{S}}$—C$_2H_5$

C_2H_5O ... $\overset{O}{\underset{\|}{P}}$—S—CH$_2CH_2$—$\overset{O}{\underset{\overset{\|}{\underset{O}{\|}}}{S}}$—C$_2H_5$
C_2H_5O ... Sulphone

*Principally
'A' esterase*

C_2H_5O ... $\overset{O}{\underset{\|}{P}}$—OH
C_2H_5O

+ HSCH$_2$CH$_2$—$\overset{O}{\underset{\overset{\|}{\underset{O}{\|}}}{S}}$—C$_2H_5$

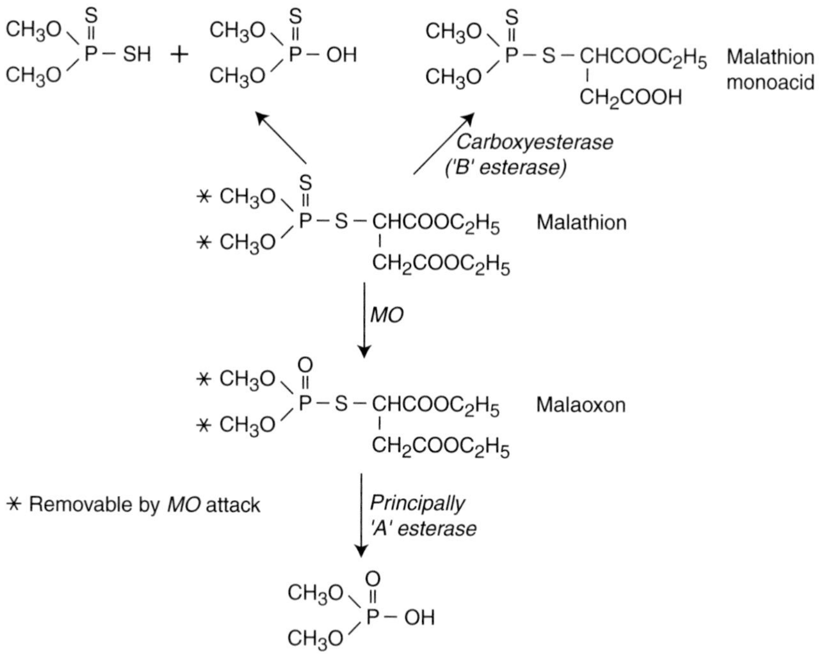

CH_3O — P(=S) — SH + CH_3O — P(=S) — OH → CH_3O — P(=S) — S — CHCOOC$_2$H$_5$ / CH$_2$COOH Malathion monoacid

↖ ↗ Carboxyesterase ('B' esterase)

* CH_3O, * CH_3O — P(=S) — S — CHCOOC$_2$H$_5$ / CH$_2$COOC$_2$H$_5$ Malathion

↓ MO

* CH_3O, * CH_3O — P(=O) — S — CHCOOC$_2$H$_5$ / CH$_2$COOC$_2$H$_5$ Malaoxon

* Removable by *MO* attack

↓ Principally 'A' esterase

CH_3O, CH_3O — P(=O) — OH

FIGURE 10.2 *Metabolism of OPs.*

sometimes confused with it before the great diversity of P450-catalysed biotransformations became known. Oxidative deethylation yields polar ionisable metabolites and generally causes detoxication (Eto, 1974; Batten and Hutson, 1995). Oxidative demethylation (O-demethylation) has been demonstrated during the metabolism of malathion.

The bond between P and the leaving group (X) of oxons is susceptible to esterase attack, the cleavage of which represents a very important detoxication mechanism. Examples include the hydrolysis of malaoxon and diazoxon (see Figure 10.2). Such hydrolytic attack depends upon the development of d+ on P as a consequence of the electron-withdrawing effect of oxygen. By contrast, thions are less polarised and are not substrates for most esterases. Two types of esterase interact with oxons (see Figure 2.9 and 'Esterases and hydrolases'). 'A' esterases continuously hydrolyse them, yielding a substituted phosphoric acid and a base derived from the leaving group as metabolites. 'B' esterases, on the other hand, are inhibited by them, the oxons acting as 'suicide substrates'. With cleavage of the ester bond and release of the leaving group, the enzyme becomes phosphorylated and is reactivated only very slowly. If 'ageing' occurs it is not reactivated at all. Thus, continuing hydrolytic breakdown of oxons by B esterases is, at best, slow and inefficient. Nevertheless, B esterases produced in large quantities by resistant aphids can degrade/sequester OPs to a sufficient degree to substantially lower their toxicity and thereby provide a resistance mechanism

(Devonshire and Sawicki, 1979; Devonshire, 1991). Acetylcholinesterase, the site of action of OPs, is a B esterase that is highly sensitive to inhibition by oxons.

In addition to ester bonds with P (section 10.2.1, Figures 10.1 and 10.2), some OPs have other ester bonds not involving P, which are readily broken by esteratic hydrolysis to bring a loss of toxicity. Examples include the two carboxylester bonds of malathion, and the amido bond of dimethoate (Figure 10.1). The two carboxylester bonds of malathion can be cleaved by B esterase attack, a conversion that provides the basis for the marked selectivity of this compound. Most insects lack an effective carboxylesterase, and for them malathion is highly toxic. Mammals and certain resistant insects, however, possess forms of carboxylesterase that rapidly hydrolyse these bonds and are accordingly insensitive to malathion toxicity.

Organophosphorous compounds are also susceptible to glutathione-S-transferase attack. Both R groups and X groups can be removed by transferring them to reduced glutathione to form a glutathione conjugate. As with oxidative dealkylation, an ionisable P–OH group remains after removal of the substituted group, and the result is detoxication. Diazinon, for example, can be detoxified by glutathione-dependent desethylase in mammals and resistant insects.

Looking at the overall pattern of OPs metabolism, it can be seen that there is often competition between activating and detoxifying metabolic processes. Moreover, many of these processes occur relatively rapidly. There are often marked differences in the balance of these processes between species and strains, differences that may be reflected in marked selectivity. As mentioned above, malathion is highly selective between insects and mammals because most insects lack a carboxylesterase that can detoxify the molecule. Some strains of insects (e.g. *Tribolium castaneum*) owe their resistance to the presence of such an esterase. Inhibition of B esterase activity with another OP [e.g. ethyl-*p*-nitrophenylthionobenzene phosphate (EPN)] can remove this resistance mechanism and make the resistant strain susceptible again to malathion. Likewise, malathion becomes highly toxic to mammals if administered together with a B esterase inhibitor. The inhibitor acts as a synergist. When rapid detoxication by carboxylesterase is blocked, considerable quantities of malathion are activated by monooxygenase to form malaoxon, and toxicity is enhanced.

Diazinon, and the related insecticides pirimiphos-methyl and pirimiphos-ethyl, are selectively toxic between birds and mammals (EHC 198). All possess leaving groups derived from pyrimidine, and their oxon forms are excellent substrates for mammalian A esterases. Selectivity is largely explained by the absence of significant A esterase activity from the plasma of birds, an activity well represented in mammals (Machin *et al.*, 1975; Brealey, 1980; Brealey *et al.*, 1980; Walker, 1991). A esterase activity is also low in avian liver relative to that in mammalian liver. Diazinon is activated to diazoxon in the liver, and toxicity then depends on the efficiency with which the latter can be transported by the blood to its site of action (primarily AChE in the brain). In mammals rapid detoxication of oxons in the liver and blood gives effective protection against low doses of these OPs. Birds are not so well protected; many species lack detectable plasma A esterase activity against oxon substrates

(Mackness *et al.*, 1987) and, on available evidence, activity in liver is relatively low (Brealey, 1980; Walker, 1991). Other OPs whose oxons are not good substrates for A esterase (e.g. parathion) do not show such selectivity between birds and mammals, providing further evidence for the importance of A esterase activity in determining the relatively low toxicity of diazinon and related insecticides to mammals. A number of cases of diazinon resistance have been reported in insects (Brooks, 1972). Resistance mechanisms include detoxication by deethylation of diazinon mediated by glutathione-S-transferase, and oxidative detoxication of diazoxon mediated by monooxygenase.

10.2.3 ENVIRONMENTAL FATE OF OPs

In general, the OPs differ from the persistent organochlorine insecticides in their environmental fate and distribution. Because they are degraded relatively rapidly by most animals, they tend not to undergo biomagnification in terrestrial or aquatic food chains. They are not very persistent in soils; hydrolysis, volatilisation and metabolism by soil microorganisms and soil animals ensure relatively rapid removal. Persistence in surface waters and sediments is also limited because of relatively rapid degradation and metabolism. Although most OPs do tend to volatilise as a consequence of their appreciable vapour pressures, they are susceptible to photodecomposition and to hydrolysis when in the atmosphere. Thus, they are not stable enough to undergo extensive long-range transport (cf. many polyhalogenated compounds). For these reasons, any harmful effects produced by OPs are likely to be limited both in time and in space; limited, that is, to the general area in which they are applied, and limited to a relatively short period of time after their release.

However, a few complicating factors need to be considered. Organophosphorous insecticides are sometimes formulated in a way that increases their persistence. Thus, the highly toxic compounds disyston and phorate are formulated as granules for application to soil or directly to certain crops. The insecticides are incorporated within a granular matrix from which they are only slowly released to become exposed to the usual processes of chemical and biochemical degradation. Insecticidal action may thereby be prolonged for a period of 2–3 months – much longer than would occur if they were formulated in other ways (e.g. as emulsifiable concentrates), where release into the environment is more rapid. Also, notwithstanding the limited persistence of OPs generally, they have occasionally been implicated in the poisoning of predatory birds. This may be due to direct contact of predators with spray residues and/or to prey carrying sufficiently high pesticide burdens to affect predators. The latter may be the consequence of prey (e.g. large insects immediately after OP spraying) carrying quantities of insecticide externally that are far in excess of the levels needed to poison them. If predation occurs very soon after exposure of prey to OPs, the levels in tissues may sometimes be high enough to cause poisoning because there has been insufficient time for effective detoxication. Even though insects generally are poor vectors of insecticides because of their sensitivity to them, some strains have acquired resistance to OPs because they have insensitive forms of cholinesterase (see next section) and so are able to tolerate relatively high tissue levels of them. Consequently the development of this type of resistance may increase the risk of secondary poisoning of insectivores by OPs.

The release of OPs into the environment is very largely intentional, with the objective of controlling pests. Invertebrate pests of crops, forests and stored products, and invertebrate vectors of disease, are the principal targets. The organisms in question are mainly insects, but other types of invertebrates (e.g. *Acarina*) are sometimes controlled with OPs. Some OPs (e.g. chlorfenvinphos) have been used to control ectoparasites of sheep and other livestock and there is concern about the illegal disposal of unused dips into water courses.

Organophosphorous insecticides are mainly used for pest or vector control on land, but they have been used to a small extent in the aquatic environment, for example to control parasites of salmon farmed in the marine environment. Organophosphorous insecticides of relatively low mammalian toxicity (e.g. malathion) have sometimes been released into surface waters to control insect pests, e.g. in watercress beds. Apart from the very small direct application of OPs to surface waters, there is continuing concern about unintentional contamination. Overspraying of surface waters, run-off from land and movement of insecticides through fissures in agricultural soil and so into water courses are all potential sources of contamination with OPs, as indeed they are for agricultural pesticides more generally.

Apart from their principal use to control invertebrate pests, OPs have also been used to a limited extent in the control of vertebrate pests. Birds regarded as pests (e.g. *Quelea* spp. in Africa) have been controlled by aerial spraying of roosts with parathion and fenthion (Bruggers and Elliott, 1989). The use of poisoned bait containing, for example, phosdrin to control predators of game birds has become a contentious issue in Western countries. In Britain, the poisoning of protected species such as the red kite (*Milvus milvus*) and the golden eagle (*Aquila chrysaetos*) is illegal, and there have been examples of gamekeepers found doing this being prosecuted and fined.

There continues to be public concern about the possible use of chemical weapons. Although the major powers have refrained from using them, and are in the process of destroying their stockpiles of them, there continues to be uncertainty about the activities of certain extreme regimes which possess chemical weapons, such as that of Saddam Hussein in Iraq. According to reports in the world press, people living in a Kurdish village in Iraq were lethally poisoned by an OP. There has also been controversy about the reported accidental exposure of soldiers to such compounds during the Gulf War, and the related issue of 'Gulf War syndrome'. In the post-Cold War era, the environmental risks associated with the disposal of the large stockpiles of chemical weapons held by the major powers have gained attention.

OPs are often applied as sprays. Commonly the formulations used for spraying are emulsifiable concentrates, where the OP is dissolved in an organic liquid which acts as a carrier. OPs are also used as seed dressings and as components of dips to protect livestock against ectoparasites. As mentioned above, some OPs are incorporated into granular formulations for application to soil or to certain crops.

10.2.4 TOXICITY OF OPS

The primary site of action of OPs is AChE, with which they interact as suicide substrates (see also section 10.2.2 and Figure 2.11). Like other B-type esterases, AChE has a reactive serine residue located at its active site, and the serine hydroxyl is phosphorylated by organophosphates. Phosphorylation causes loss of AChE activity and, at best, the phosphorylated enzyme reactivates only slowly. The rate of reactivation of the phosphorylated enzyme depends on the nature of the X groups, being relatively rapid with methoxy groups (t_{50} 1–2 h), but slower with larger alkoxy groups. Alkyl groups of phosphoryl moieties bound to AChE tend to be lost with time, leaving behind the charged group P–O⁻. The process is termed 'ageing', and once it has occurred reactivation virtually ceases.

In AChE isolated from *Torpedo californica*, reactive serine is one of three amino acids constituting a catalytic triad (Sussman *et al.*, 1991, 1993) (Figure 10.3). The catalytic triad is located at the bottom of a deep and narrow hydrophobic gorge lined with the rings of 14 aromatic amino acids. The catalytic triad is composed of residues of serine, histidine and glutamic acid. Histidine is in close proximity to serine (Figure 10.3), and may therefore draw protons away from serine hydroxyl groups, thereby facilitating ionisation and electrophilic attack of acetylcholine upon CO⁻. During normal hydrolysis of acetylcholine, which occurs very rapidly, the ester bond is broken, the serine residue is acetylated and choline is released. Finally, acetate is released from the enzyme, a proton is returned to serine, and activity is quickly restored. Organophosphates are also treated as substrates by AChE, but the essential difference here is that the phosphorylated enzyme is only reactivated very slowly, if at all.

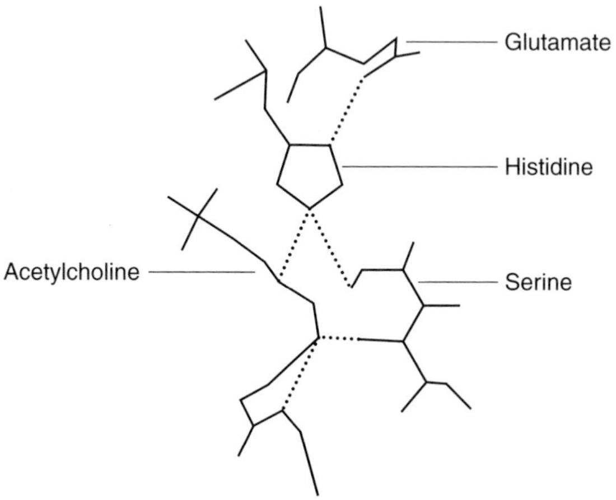

FIGURE 10.3 *Acetylcholinesterase: structure of catalytic triad. The structure of the catalytic triad of the active centre of the enzyme is shown. After Sussman* et al. *(1991).*

The inhibition of AChE can cause disturbances of transmission across cholinergic synapses. AChE is bound to the postsynaptic membrane (Figure 10.4), where it has an essential role in hydrolysing acetylcholine released into the synaptic cleft from the presynaptic membrane. The rapid destruction of such acetylcholine is necessary to ensure that synaptic transmission is quickly terminated. Acetylcholine interacts with nicotinic and muscarinic receptors of the postsynaptic membrane to generate action potentials that pass along postsynaptic nerves. If stimulation of these cholinergic receptors is not quickly terminated, synaptic control is lost. If synaptic transmission continues for too long, depolarisation of the postsynaptic membrane and synaptic block will follow. Synaptic block on the neuromuscular junction results in tetanus; death due to asphyxiation follows if the diaphragm muscles of vertebrates are affected. Organophosphorous insecticides can disturb synaptic transmission in both central and peripheral nerves.

Vertebrates can tolerate a certain degree of inhibition of brain AChE before toxic effects are seen. A typical dose–response curve for the inhibition of AChE by an OP is shown in Figure 10.5. The relationship between the degree of inhibition and the nature and severity of toxic effects is indicated in Figure 10.5. In general, the effects increase in severity with increasing dose, but the exact quantitative relationship between the per cent inhibition and the effects varies between compounds and between species. A typical situation in birds is as follows. At around 40–50% inhibition, mild physiological and behavioural disturbances are seen. Above this, more serious disturbances occur; above 70% inhibition deaths from anticholinesterase poisoning begin (Grue *et al.*, 1991).

There is much evidence from studies with laboratory animals that mild neurophysiological effects and associated behavioural disturbances are caused by levels

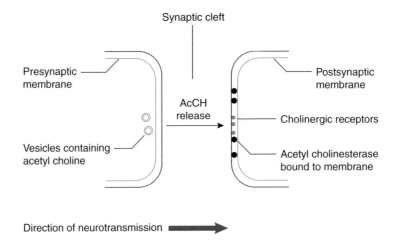

FIGURE 1O.4 *Diagram of cholinergic synapse.*

BOX 10.1 *Antidotes to cholinesterase poisoning by OPs.*

Because of the high human risks associated with both OPs and the related nerve gases, antidotes have been developed to counteract poisoning by them. Basically, these are of two different kinds.

1 Reactivators of phosphorylated cholinesterase. Pyridine aldoxime methiodide (PAM) and related compounds are the best known. They reactivate the phosphorylated enzyme so long as ageing has not occurred. They do not, however, reactivate the aged enzyme. Cholinesterase phosphorylated with certain nerve gases ages rapidly!

2 Atropine acts as an antagonist of acetylcholine at muscarinic receptors, but not at nicotinic receptors. By acting as an antagonist, it can prevent overstimulation of muscarinic receptors by the excessive quantities of acetylcholine remaining in the synaptic cleft when AChE is inhibited. The dose of atropine needs to be carefully controlled because it is toxic.

Antidotes are administrated to patients after there has been exposure to OPs. They are also sometimes given as a protective measure when there is a risk of exposure, e.g. they were given to troops fighting in the Gulf War.

Of the two types of antidote mentioned above, only atropine is effective against carbamate poisoning.

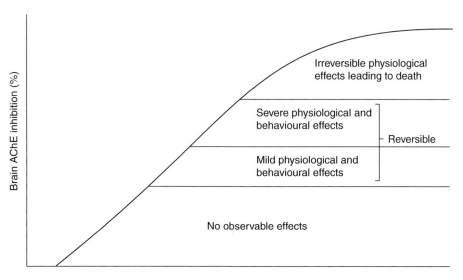

FIGURE 10.5 *Stages in the progression of OP intoxication.*

of OPs well below lethal doses (see EHC 63 for examples). These include effects on EEG patterns, changes in conditioned motor reflexes and changes in performance in behavioural tests (e.g. maze running by rats). Many of these observations were made after exposures that were too low to cause overt symptoms of intoxication. In a recent study with larval rainbow trout (*Oncorhynchus mykiss*), diazinon and malathion caused behavioural disturbances at quite low levels of brain AChE inhibition (Beauvais *et al.*, 2000). With diazinon, the maximum level of inhibition (mean value) of brain cholinesterase was less than 50%. There was a strong negative correlation between speed and distance of swimming and brain AChE inhibition, even down to values of approximately 20%. Similar results were obtained with malathion.

The measurement of inhibition of AChE is a valuable biomarker assay for OPs and carbamates and is not just an index of exposure. Being an assay based upon the principal molecular mechanism of toxicity, it has the advantage of providing an index of the different stages of the manifestation of toxicity, including early neurophysiological and behavioural effects; it can also provide an index of the potentiation of the toxicity of anticholinesterases by other environmental chemicals (section 14.4; Walker, 1996). The main problem is that the assay depends on destructive sampling and cannot, therefore, be used serially on individual animals. The determination of plasma butyryl cholinesterase inhibition is another biomarker assay that is non-destructive and can be used serially. However, this is only a biomarker of exposure, and there are no general rules linking inhibition of cholinesterases in the blood with inhibition of those in the brain because of the complexity of the toxicokinetics, which differ markedly between species.

A few OPs have been shown to cause toxicity by interacting with a receptor other than AChE. Mipafox, leptophos and methamidophos can all cause delayed neuropathy (Johnson, 1992). The target in this case is an esterase located in the nervous system termed 'neuropathy target esterase' (NTE). It is membrane bound and yields a subunit of approximately 155 kDa when solubilised, which is high for an esterase. No symptoms are seen immediately after phosphorylation of the enzyme. Some 2–3 weeks after exposure, long after residues of the chemicals have disappeared from the body, paralysis of muscles of distal extremities of limbs are seen together with a selective degeneration of neurones that supply them. Distal degeneration of long axons of the spinal cord and the peripheral nervous system occurs. In most, but not all, examples of OP-induced delayed neuropathy (OPIDN), symptoms only start to appear with ageing of the phosphorylated enzyme.

With recent growing concern about possible long-term neurological effects on sheep farmers exposed to OPs, there has been increased interest in alternative sites of action of OPs. It has been shown that there are sites in the rat brain that are more sensitive than AChE to phosphorylation by certain OPs (Richards *et al.*, 1999). This and the knowledge of OPIDN has led to speculation about longer-term sublethal effects of OPs in the natural environment due to mechanisms other than cholinesterase inhibition. The chicken is particularly sensitive to OPIDN, which is why it is the species of choice when testing pesticides for their ability to cause this condition. Are birds more sensitive to OPIDN than other groups? Concerns such as these strengthen the case for reducing the use of OPs in agriculture.

Some acute toxicity data for OPs are given in Table 10.2, and compounds of particularly high OP toxicity are highlighted in Table 10.3. Some OPs of exceptionally high toxicity such as parathion are no longer used in Western countries because they are too hazardous. The relatively high toxicity of diazinon, pirimiphos-methyl, pirimiphos-ethyl and dimethoate to birds (see EHC 198) and the mechanistic reasons for this were discussed in the previous section. Such compounds sometimes constitute serious hazards to birds when used in the field. Disyston (disulfoton) and the related compound phorate (thimet) are highly toxic to vertebrates and are normally formulated as granules, which only slowly release the insecticide. Granules limit the availability of the insecticide, and are therefore safer to use and present less risk to the environment than more readily available formulations, e.g. emulsifiable concentrates. Organophosphorous insecticides are often very toxic to fish (see, for example, the data for demeton-S-methyl and diazinon).

OPs can be highly toxic to aquatic invertebrates, with toxic effects reported for concentrations down to 0.01 mg/L in ambient water (see EHC 63). LD_{50} values of diazinon and dimethoate for four species of birds have been expressed as a mean to facilitate comparison with data for the rat. Estimation of the toxicity expressed by OPs in the field is complicated by the fact that the presence of other compounds (e.g. in tank mixes or on seed dressings) may cause potentiation. As explained earlier, the fact that OP toxicity is regulated by relatively rapid metabolic transformation brings the risk that other compounds may inhibit detoxication and/or enhance activation. For example, it has been shown in laboratory studies that exposure of the red-legged partridge (*Alopecurus rufus*) to prochloraz and other EBI fungicides can enhance the toxicity of malathion and dimethoate (section 2.6; Johnston *et al.*, 1989, 1994a,b).

TABLE 10.2 *Toxicity of OPs*

Compound	Species	Type of measurement (units)		Value
Parathion	Rat	Acute oral LD_{50}	(mg/kg)	3–6
Malathion	Rat	Acute oral LD_{50}	(mg/kg)	480–5600
Dimethoate	Rat	Acute oral LD_{50}	(mg/kg)	150–300
Dimethoate	Birds (4)	Acute oral LD_{50}	(mg/kg)	(26)
Diazinon	Rat	Acute oral LD_{50}	(mg/kg)	235–1250
Diazinon	Birds (4)	Acute oral LD_{50}	(mg/kg)	(4.5)
Diazinon	Fish	96 h LC_{50}	(mg/L)	0.09–2.76
Demeton-S-methyl	Rat	Acute oral LD_{50}	(mg/kg)	35–129
Demeton-S-methyl	Birds	Acute oral LD_{50}	(mg/kg)	10–50
Demeton-S-methyl	Fish	96 h LC_{50}	(mg/L)	0.6–60
Disyston	Rat	Acute oral LD_{50}	(mg/kg)	12.5

TABLE 10.3 *Some non-persistent anticholesterinase insecticides of high toxicity to vertebrates*

| | Acute oral LD$_{50}$ (mg/kg) | | | |
| | Rat | | Birds | |
Insecticide	< 10	10–20	< 10	10–20
Organophosphorous				
Disulfoton (Disyston)		✓	✓	✓
Phorate	✓		✓	
Phosdrin (Mevinphos)	✓		✓	✓
Parathion	✓		✓	
Carbophenothion			✓	
Diazinon			✓	✓
Fenthion			✓	
Carbamates				
Aldicarb (Temik)	✓		✓	
Carbofuran	✓		✓	✓
Methiocarb (molluscicide)			✓	✓

This effect was attributed to induction of several forms of P450 by the fungicides and consequent increased activation of the OPs. The extent to which such interactions may occur under field conditions remains a matter of speculation. It is difficult to establish the occurrence of potentiation in the field.

There have been many examples of birds and other vertebrates dying as a consequence of exposure to OPs in the field. Worldwide, hundreds of incidents have been reported involving the poisoning of birds on agricultural land by OPs or carbamates (Hill, 1992). In a survey of the literature, Grue *et al.* (1991) concluded that birds dying as a result of acute OP poisoning had at least 50% inhibition of brain AChE activity in the great majority of cases, most of them showing 70% inhibition or more. Seventy per cent inhibition of brain AChE or more has sometimes been regarded as diagnostic of OP poisoning in birds. A similar picture emerges from laboratory studies. Identifying OPs as the cause of acute poisoning in the field depends on linking exposure to an OP (e.g. by analysis of crop or gut contents) to a level of inhibition of brain AChE high enough to suggest lethal toxicity (Grue *et al.*, 1991; Greig-Smith *et al.*, 1992a; Thompson and Walker, 1994). Many poisoning incidents reported in the field in Western Europe and North America have been traced to the consumption by birds of material containing high levels of OP, such as seed dressed with carbophenothion, chlorfenvinphos and fonophos, baits containing phosdrin and occasionally granules containing OPs. Victims range from grain-eating species poisoned by the dressed seed to eagles, kites and buzzards killed by poisoned baits. Spraying of OPs in the field has also been linked to lethal anticholinesterase poisoning. Thus, red-tailed hawks (*Buteo jamaicensis*) were found to be poisoned by OP sprays in almond orchards in California (Hooper *et al.*, 1989), and canopy-living birds were poisoned by fenitrothion applied to forests in New Brunswick, Canada, to control spruce bud worm (see Chapter 15; Walker *et al.*, 2000). Although inhibition of brain AChE is a valuable biomarker assay for identifying lethal and other toxic effects of OPs in vertebrates, it needs to be used with discretion. The degree of inhibition of brain AChE associated with lethality varies between species and compounds; sometimes cholinergic effects appear to be more important at extracerebral sites, including the peripheral nervous system, than they are in the brain itself. Also, post-mortem changes can occur in the field to confound analysis: loss of enzyme activity, and reactivation of the inhibited can both occur after death. Identifying OPs as the cause of lethal intoxication in the field is made easier if typical symptoms of cholinesterase poisoning are observed before death.

Inevitably, terrestrial invertebrates are susceptible to the toxicity of OPs used in the field. The honeybee is one species of particular importance, and the use of OPs, and other insecticides, on agricultural land has been restricted to minimise toxicity to this species. One practice has been to avoid application of hazardous chemicals to crops when there are foraging bees. The use of some compounds, e.g. triazophos, has been restricted because of very high toxicity to honeybees.

OPs are known to be highly toxic to aquatic invertebrates and to fish. This has been demonstrated in field studies. For example, malathion applied to watercress beds caused lethal intoxication of *Gammarus pulex* located downstream (Crane *et al.*, 1995). Marine invertebrates have been found dead after the application of OPs. Accidental release of OPs into rivers, lakes and bays has sometimes caused large-scale fish kills (see EHC 63).

10.2.5 ECOLOGICAL EFFECTS OF OPs

Population dynamics

Some OPs are prime examples of pollutants that are highly toxic but of low persistence, and they serve as useful models for compounds of similar ilk which have been less well investigated. Because of their limited persistence, toxic effects are expected to be localised and of limited duration. Since the compounds degrade quickly in tissues, residues in carcasses of animals or birds found in the field do not provide reliable evidence of the cause of death (cf. the persistent organochlorine insecticides). Supporting evidence such as inhibition of brain AChE activity is usually needed to establish causality. From an ecological point of view such compounds appear less hazardous than compounds such as dieldrin or heptachlor epoxide, which are both highly toxic and persistent. There are, however, situations in which they may still cause ecological problems. If they are applied to an area of farmland or to an orchard several times a year, over several years effects may be seen on species of limited mobility, which are slow to recolonise treated areas after OP residues have declined. Such a problem may be compounded if other non-persistent insecticides (e.g. carbamates and pyrethroids) are also used. Effects of this kind have been reported from the Boxworth experiment – a long-term field experiment conducted by the Ministry of Agriculture, Fisheries and Food (MAFF) in eastern England during the period 1982–90 (Greig-Smith *et al.*, 1992b). In areas where OPs, pyrethroids and carbamates were extensively used ('insurance areas') the numbers of some non-dispersive species, such as the ground beetles *Bembidium obtusum* and *Notiophilus biguttatus*, fell drastically during the first 3 years, and they remained low or totally absent until the end of the experiment. In general, there was a decline in predatory invertebrates in the area receiving the highest input of pesticide (cf. the control area), and this was associated with a reduced level of predation upon aphids.

There is concern from an ecological point of view if a high proportion of the population of a protected species is present in a particular area when a highly toxic chemical is being used. An example of this problem was the heavy mortality of wintering greylag geese (*Anser anser*) and pink-footed geese (*Anser brachyrhynchus*) in east-central Scotland during 1971–2 (Hamilton *et al.*, 1976). Deaths were due to consumption by the geese of the OP carbophenothion, used as a seed dressing for winter wheat and barley. The geese consumed uncovered seed, and also seedlings with the contaminated seed coat still attached. It transpired that carbophenothion was particularly toxic to geese belonging to the genus *Anser*, more toxic than had been realised in the original risk assessment of the OP. Geese belonging to the genus *Branta* such as the Canada goose (*Branta canadensis*) were found to be less susceptible. It was estimated that 60 000–65 000 wintering greylag geese, representing about two-thirds of the entire British population, came to this area of Scotland during autumn in the early 1970s. Hundreds of birds died, and it was concluded that carbophenothion represented an unacceptable hazard to wintering *Anser* geese in east-central Scotland. Subsequently, the use of carbophenothion as a seed dressing for winter wheat or barley was banned in the affected area.

Another example of OP spraying evidently causing ecological problems was the large-scale application of fenitrothion to forests in New Brunswick, Canada (Ernst *et al.*, 1989; Chapter 15 in Walker *et al.*, 2000). As described earlier (section 10.2.4) the deaths of individual birds were attributed to acute poisoning by the OP. The mortality rate due to poisoning, however, was not known, although the levels of cholinesterase inhibition measured in surveys suggested that it may have been high. There was evidence for severe reproductive impairment in the white-throated sparrow (*Zonotrichia albicollis*) associated with a mean brain AChE inhibition of 42%. In general, many birds sampled in the area had 50%, or more, inhibition of brain AChE, and it was suspected that sublethal effects on birds were widespread. Apart from birds, there was clear evidence for declines in populations of honeybees and wild bees as a result of the application of fenitrothion.

Population genetics

There have been many reports of insect pest species developing resistance to OPs, to the extent that control of the pest has been lost. A detailed account of resistance lies outside the scope of the present book, and readers are referred to specialised texts by Georghiou and Saito (1983), Brown (1971) and Otto and Weber (1992). A few examples will now be considered which illustrate the mechanisms by which insects become resistant to OPs.

In Europe, one of the most widely studied cases of resistance was that developed to OPs in general by cereal aphids (Devonshire, 1991). Existing OPs became ineffective for aphid control in some areas, and there was a need to find effective alternatives, e.g. carbamates or pyrethroids, that were not susceptible to the same resistance mechanism. It was found that there were a number of different clones of the peach potato aphid (*M. persicae*) with differing levels of resistance to OPs. The level of resistance was related to the number of copies of a gene for a carboxylesterase (see earlier discussion in sections 2.5 and 10.2.2). In general, the larger the number of copies of the gene, the greater the activity of the carboxylesterase and the greater the level of OP resistance (in certain instances, transcriptional control was also found to be important). Resistance had evidently been acquired through gene replication, not through the appearance of a novel esterase gene absent from susceptible aphids. Some strains of mosquitoes have also been shown to develop OP resistance by this mechanism.

A number of other examples are known where genetically based resistance has been due to enhanced detoxication of OPs. These include malathion resistance in some stored product pests due to carboxylesterase activity, and the resistance of strains of the housefly to diazinon due to detoxication by specific forms of a glutathione-S-transferase and monooxygenase (Brooks, 1972).

In some strains of insects, resistance has been related to the presence of genes that code for insensitive 'aberrant' forms of the AChE. Interestingly, it has been shown that resistance may be the consequence of the change of a single amino acid residue in AChE. Sequence analysis of the AChE gene from resistant strains of *D. melanogaster* and the house fly has identified six point mutations that are associated with resistance (Devonshire *et al.*, 1998; Salgado, 1999). All of these mutations bring changes in amino acid residues located near to the active site of AChE according to the torpedo enzyme model described above (p. 187). According to the model, all the changes would cause steric hindrance of the relatively bulky insecticides, but not (to any important extent) of acetyl choline itself. Thus, the insensitive enzyme could continue to function as AChE. The existence of more than one of these point mutations brings a higher level of resistance than does a single point mutation.

Apart from the importance of OP resistance in pest control, ecotoxicologists have become interested in the development of resistance as an indication of the environmental impact of insecticides. Thus, the development of esteratic resistance mechanisms by aquatic invertebrates may provide a measure of the environmental impact of OPs in fresh water (Parker and Callaghan, 1997).

10.3 *Carbamate insecticides*

The chemical and biological properties of carbamate (CB) insecticides are described in some detail in the texts of Kuhr and Dorough (1976) and Ballantyne and Marrs (1992). An early model for their development was physostigmine (Figure 10.6), a natural product found in Calabar beans. Many CB insecticides came into use during the 1960s, sometimes as substitutes for banned organochlorine insecticides.

10.3.1 CHEMICAL PROPERTIES

The general structure of CB insecticides, with some examples, is given in Figure 10.6. As can be seen, they are derivatives of carbamic acid, the unstable monoamide of carbonic acid. CB insecticides have one, occasionally two, methyl groups attached to the nitrogen atom. A range of differing organic groups is linked to the oxygen atom. The nature of the R group is an important determinant of the properties of CB insecticides. Distinct from these compounds are carbamate herbicides, many of which have relatively complex R groups attached to nitrogen (see Figure 13.1 and Hassall, 1990, for further details).

The properties of some CB insecticides are given in Table 10.4. Of the examples given, carbaryl is less polar, and accordingly less water soluble and more volatile than the other three compounds. The reason for this is the non-polar character of the naphthyl R group, which contrasts with the more polar groups of the other compounds. In general, CB insecticides are more polar and water soluble than organochlorine insecticides, although there are marked contrasts between different members of the group. In most cases, the R group links to oxygen through carbon, as with carbaryl, propoxur and carbofuran in Figure 10.6. A few compounds, e.g. aldicarb, are linked through nitrogen, and the R group is an oxime residue. Apart from the examples given, pirimicarb, methiocarb and methomyl are also CB insecticides that have been widely used in pest control. Methomyl is a further example of an oxime carbamate.

Carbamates are subject to chemical hydrolysis, which takes place relatively slowly under neutral or acid conditions, but more rapidly under alkaline conditions.

10.3.2 METABOLISM OF CB INSECTICIDES

Carbamates are metabolised relatively rapidly, and metabolism tends to be complex. The present account will be restricted to major routes of biotransformation that are important in determining toxicity. The metabolism of carbaryl, aldicarb and carbofuran is outlined in Figure 10.7. With carbaryl, primary attack in vertebrates is principally oxidation by the monooxygenase system (Hutson and Paulson, 1995). This can occur both on the naphthyl ring and on the methyl group attached to nitrogen. Attack on ring positions leads to the formation of unstable epoxides, which may either rearrange

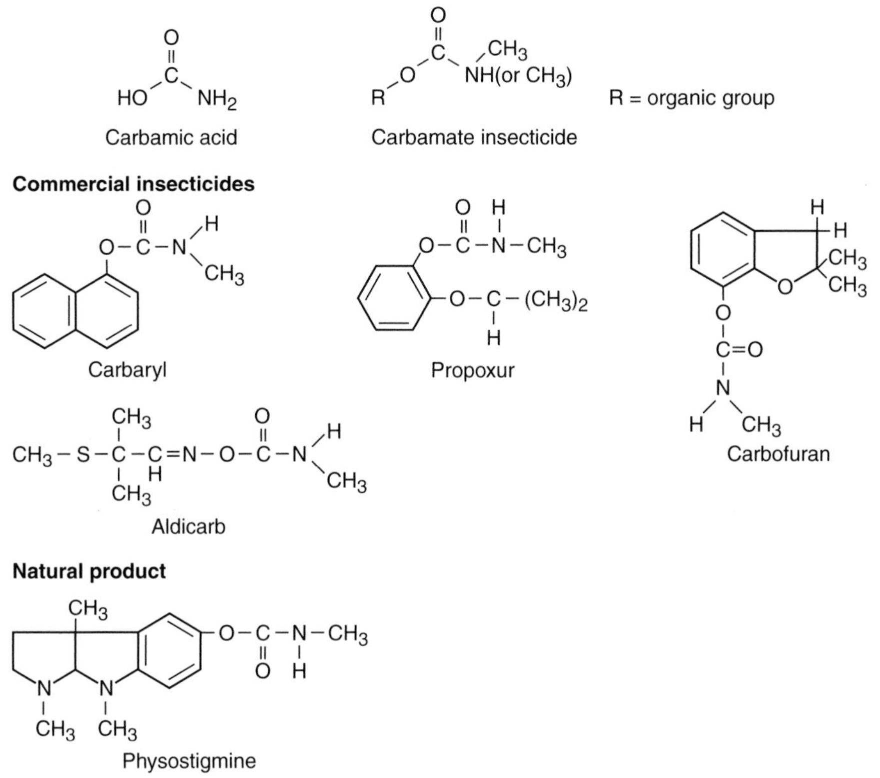

FIGURE 10.6 *Some insecticidal carbamates.*

TABLE 10.4 *Properties of some carbamate insecticides*

Compound	Water solubility (μg/mL at 25°C)	Log K_{ow}	Vapour pressure (mmHg at 25°C)
Carbaryl	40	2.36	3×10^{-3}
Propoxur	1000		3×10^{-5} (30°C)
Aldicarb	6000	1.36	1×10^{-4}
Carbofuran	700		1×10^{-5}

to form phenols, undergo hydration to diols by the action of epoxide hydrolase or be converted to glutathione conjugates. The importance of primary oxidative attack in certain insects is illustrated by the fact that inhibitors of P450-based monooxygenases, e.g. piperonyl butoxide, are powerful synergists of carbaryl (synergistic ratios of more than several hundred have been reported, Kuhr and Dorough, 1976). There is uncertainty about the importance of hydrolysis as a primary mechanism of metabolic attack in vertebrates *in vivo*. It may be that oxidative attack tends to precede hydrolytic

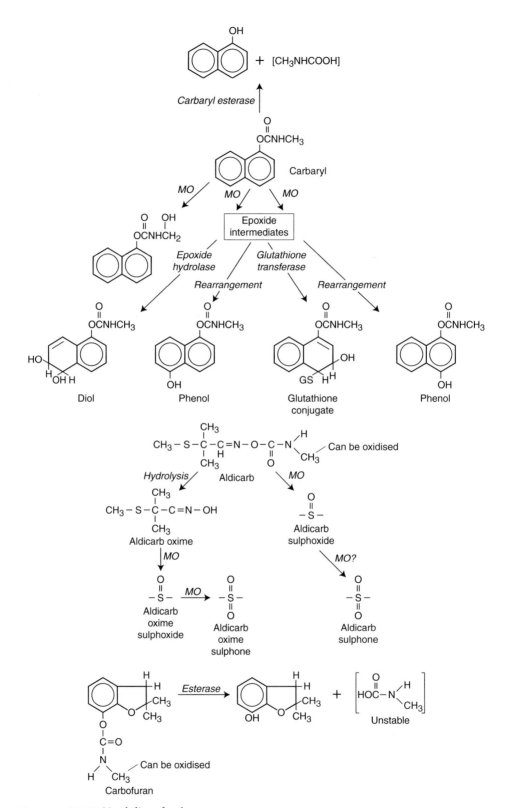

FIGURE 10.7 *Metabolism of carbamates.*

cleavage. For example, the hydrolysis of carbaryl by rat liver microsomes has a requirement for NADPH as a cofactor (see EHC 153). Oxidative metabolites are more polar than the original carbaryl molecule and may be better substrates for esterases. All of the biotransformations of carbaryl featured here cause an increase in polarity and have a detoxifying function.

With aldicarb, primary metabolic attack is again by oxidation and hydrolysis. Hydrolytic cleavage yields an oxime and represents a detoxication. Oxidation to aldicarb sulphoxide and sulphone, however, yields products that are active anticholinesterases. Carbofuran is detoxified by both hydrolytic and oxidative attack.

10.3.3 ENVIRONMENTAL FATE OF CB INSECTICIDES

CB insecticides have been widely used in agriculture as insecticides, molluscicides and acaricides. They have been applied as sprays and as granules or pellets. Highly toxic compounds such as aldicarb and carbofuran are usually only available as granules because other formulations are regarded as being too hazardous. Granules ensure slow release of the active ingredient under field conditions, which can have the benefits of longer-term control of pests and of reduced environmental hazards. However, there have occasionally been environmental problems with granules because (1) birds sometimes consume them and (2) when fields are flooded, the active ingredient may be dissolved in the water of large puddles on the land surface making it available to animals and birds (Hardy, 1990).

Like OPs, CBs tend not to be very persistent in food chains, and do not undergo biomagnification with passage along them. Some of them (e.g. aldicarb and carbofuran) are systemic so may be taken up by insects feeding on plant sap. This may occur with non-target species feeding on weeds, as well as pest species feeding upon crops.

10.3.4 TOXICITY OF CB INSECTICIDES

Carbamates, like OPs, act as inhibitors of cholinesterase. They are treated as substrates by the enzyme and carbamylate the serine of the active site (Figure 10.8). Generally speaking, carbamylated AChE reactivates more rapidly than phosphorylated AChE. (After ageing has occurred, phosphorylation of the enzyme is effectively irreversible; see section 10.2.4.) Carbamylated AChE reactivates when preparations are diluted with water, a process that is accelerated in the presence of acetyl choline, which competes as a substrate. Thus, the measurement of AChE inhibition is complicated by the fact that reactivation occurs during the course of the assay. Carbamylated AChE is not reactivated by pyridine aldoxime methiodide (PAM) and related compounds, which are used as antidotes to OP poisoning (see Box 10.1).

Carbamates vary greatly in their toxicity to vertebrates. Some examples are given in Table 10.5. The most striking feature of the data is the very high acute toxicity of the two systemic carbamates aldicarb and carbofuran; carbaryl is far less toxic to

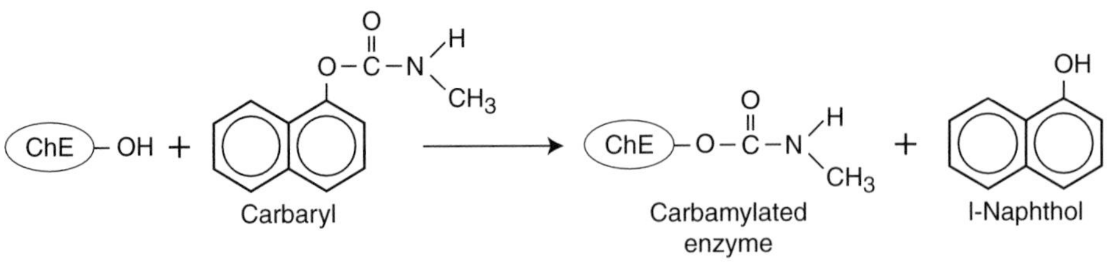

FIGURE 10.8 *Carbamylation of cholinesterase (ChE).*

mammals, birds or fish. Propoxur is substantially less toxic to mammals and birds. The toxicity to mammals and birds looks broadly similar for the four compounds represented here. However, a comparison of the toxicity of 20 CB insecticides to the red-winged blackbird (*Agelaius phoeniceus*) and starling (*S. vulgaris*) with that towards the rat showed that in over 85% of cases the carbamate was more toxic to the bird than to the rat (Walker, 1983). A possible factor here is the relatively low monooxygenase activity found in these and many other species of birds compared with that of the rat (see Chapter 2). Detoxication by monooxygenase appears to be an important factor in determining CB toxicity (see section 2.6).

As noted earlier, the toxicity of carbamates can be potentiated by other compounds, a consequence of their relatively rapid metabolic detoxication. Synergists such as piperonyl butoxide and other methylene dioxyphenyl compounds can greatly increase the toxicity of carbaryl and other CB insecticides by inhibiting detoxication by monooxygenases, up to the extent of several hundred fold (Kuhr and Dorough, 1976). Although synergised CB insecticides have not been marketed because of the environmental risks associated with the release of piperonyl butoxide and related compounds, the possibility remains that carbamates may be potentiated by other pesticides under field conditions. Multiple exposures often occur in the field because of the use of pesticide mixtures (in formulations, tank mixes and seed dressings), or because of sequential exposure as birds and other mobile species move from field to field. In one study of the red-legged partridge (*A. rufa*), pre-exposure to the OP malathion markedly increased the toxicity of carbaryl (Johnston *et al.*, 1994c). Dosing with malathion alone produced no inhibition of brain cholinesterase and no symptoms of cholinesterase poisoning. Dosing with carbaryl alone caused 56% inhibition of brain cholinesterase but again no visible symptoms of toxicity. The combination of malathion with carbaryl caused 88% inhibition of brain cholinesterase and extensive toxicity. Thirty-three per cent of birds died, and a further 50% showed clear symptoms of cholinesterase poisoning. Birds dosed with both compounds contained greater than sevenfold higher carbaryl residues in brain than did those dosed with carbaryl alone. The potentiation of toxicity was attributed to a failure of oxidative detoxication in the liver. Malathion, like other thions, can deactivate P450 forms during the course of oxidative desulphuration (de Matteis, 1974). This finding raises wider questions about potentiation of the toxicity of pesticides in the field. Many OPs are thions (see section

TABLE 10.5 *Toxicity of some carbamates*

Compound	Species	Toxicity test	Value	Units
Carbaryl	Rodents	Acute oral LD_{50}	206–963	mg/kg
	Other mammals	Acute oral LD_{50}	700–2000	mg/kg
	Birds	Acute oral LD_{50}	56 to >5000	mg/kg
	Fish	96 h LC_{50}	0.7–108	mg/L
Propoxur	Rat	Acute oral LD_{50}	95–175	mg/kg
	Birds	Acute oral LD_{50}	12–60	mg/kg
Aldicarb	Rat	Acute oral LD_{50}	0.1–7.7	mg/kg
	Birds	Acute oral LD_{50}	0.8–5.3	mg/kg
	Fish	96 h LC_{50}	0.05–2.4	mg/L
Carbofuran	Rat	Acute oral LD_{50}	6–14	mg/kg
	Birds	Acute oral LD_{50}	0.4–4.2	mg/kg

Data from EHC 64, 121 and 153. Walker (1983), Kuhr and Dorough (1976).

10.2.1), and inhibition of vertebrate monooxygenases in the field may occur where animals or birds experience sufficiently high sublethal exposures to them. It should be emphasised that lethal exposures to OPs have been widely reported (section 10.2.4), and sublethal exposures must have been much more common. Also, the inhibition of oxidative detoxication can bring potentiation of other readily degradable insecticides in addition to that of CBs, e.g. pyrethroids and certain OPs. The problem is that such potentiation is difficult to establish in the field.

There have been a number of examples of birds and mammals being poisoned in the field by the more toxic CBs used in the approved manner. One example was the poisoning by aldicarb of approximately 100 black-headed gulls (*Larus ridibundus*) on agricultural land (Hardy, 1990). The compound had been applied as a granular formulation to wet soil to control nematodes and insects in a sugar beet crop. Birds apparently died from consuming granules directly and from feeding upon contaminated earthworms. In a field study conducted at Boxworth farm, Cambridgeshire, UK, the highly toxic CB methiocarb was shown to cause lethal poisoning of wood mice (*Apodemus sylvaticus*) when used as a molluscicide (Greig-Smith *et al.*, 1992b). The compound had been broadcast on the soil surface as a 4% pelleted formulation. This example will be considered again in the next section (section 10.3.5). In a further example, the movement of pesticides was studied from the land surface through a soil that had developed fissures, and then into neighbouring water courses (Matthiessen *et al.*, 1995). After heavy rains, the elution of carbofuran from a granular formulation into water courses was sufficiently high to kill freshwater shrimps (*Gammarus* spp.) that had been deployed there.

10.3.5 ECOLOGICAL EFFECTS OF CB INSECTICIDES

Although the ability of highly toxic CBs to cause lethal poisoning in the field has been clearly demonstrated, the ecological significance of such effects remains unclear. In the case of the methiocarb poisoning of wood mice mentioned above (p. 201), population numbers before and after the application of the molluscicide were estimated by trapping (Greig-Smith *et al.*, 1992b). There was a rapid decline in numbers immediately after application, but there was also a rapid recovery within a week or so. In the longer term, the use of the molluscicide did not affect the numbers of mice in the treated area. In a more detailed study, it was found that the broadcasting of molluscicide pellets altered the structure of the population; there was a higher proportion of juveniles in the wood mouse population after broadcasting than in a control population. This change was not seen if pellets were drilled instead of broadcast.

The repeated use of carbofuran and other carbamates has been associated with changes in the metabolic capacity of soil microorganisms (Suett, 1986). Carbofuran granules were found to lose their effectiveness in some 'problem' soils where the insecticide was regularly used. The soils showed enhanced capacity to degrade carbofuran, an effect attributed to either or both of the following. (1) The increase in numbers of pre-existing species or strains capable of metabolising the carbamate, and, presumably, using it as a nutrient/energy source. (2) The induction of enzyme systems capable of metabolising the carbamate (adaptive enzymes). Such effects have been found also with herbicides such as MCPA and 2,4-D, where the enhancement of metabolic capacity of the soil is lost after a period of time, if there is no further exposure to the insecticide.

10.4 *Summary*

The organophosphorous and carbamate insecticides are anticholinesterases, are readily biodegradable and do not tend to bioaccumulate in food chains. They are generally seen as being more environmentally friendly than the persistent organochlorine insecticides, which they came to replace for many purposes during the 1960s, and they have been widely used in agriculture and for certain other purposes worldwide. However, some of them are highly toxic to vertebrates and beneficial invertebrates and have caused mortality of animals and birds in the field – albeit in a limited area and over a limited time span. There can be adverse effects upon immobile invertebrates when these, and other non-persistent insecticides, are repeatedly used during the course of intensive agriculture. A further cause of concern is that they are neurotoxic and can have behavioural effects in the field – effects that are difficult to quantify, but may be ecologically important. Recently, there has been growing concern about effects upon sites other than AChE in the nervous system (e.g. on neuropathy target esterase)

– effects that may have longer-term consequences for affected individuals than sublethal cholinesterase inhibition.

A few OP compounds are chemical warfare agents.

10.5 *Further reading*

Ballantyne, B. and Marrs, T. C. (1992) *Clinical and Experimental Toxicology of Organophosphates and Carbamates*. An in-depth account of the toxicology of both groups of compounds including some information about the so-called nerve gases.

EHC 121 (1991) *Aldicarb*. A useful reference on one of the most toxic insecticides.

EHC 63 (1986) *Organophosphorous Insecticides* and 64 (1986) *Carbamate Pesticides: A General Introduction*. Valuable sources of information on organophosphorous and carbamate insecticides respectively.

Eto, M. (1974) *Organophosphorous Insecticides* and Fest, C. and Schmidt, K.-J. (1982) *Chemistry of Organophosphorous Compounds*. Valuable texts on the basic properties of the OPs.

Kuhr, R. J. and Dorough, H. W. (1976) *Carbamate Insecticides*. A good text of similar ilk on carbamate insecticides.

CHAPTER 11

The anticoagulant rodenticides

11.1 *Background*

Warfarin, which was introduced to the market in the late 1940s, was the first of a series of anticoagulant rodenticides (ARs) related in structure to dicoumarol (Figure 11.1) (Meehan, 1986; Buckle and Smith, 1994). All of them are toxic because they act as anticoagulants – extending the clotting time of blood and so causing haemorrhaging. The anticoagulant properties of naturally occurring dicoumarol were discovered in the USA early in the twentieth century, when it was found to be the causal agent in cases of fatal haemorrhaging of cattle that had been fed spoiled clover. Subsequently, it was discovered that dicoumarol and rodenticides related to it have anticoagulant action because they act as vitamin K antagonists.

For some years warfarin was by far the most widely used rodenticide of this type. In time, however, strains of rats resistant to warfarin began to appear, and the compound became ineffective in some areas where it had been regularly used. Resistance was overcome, at least in the short term, by a second generation of ARs sometimes called 'superwarfarins'. Examples include brodifacoum, difenacoum, flocoumafen and bromodiolone. The superwarfarins are more hydrophobic, persistent and toxic than warfarin itself. They have usually been effective in overcoming resistance to warfarin and, consequently, have come into more wide-scale use in the last two decades. Although knowledge about their environmental fate and effects is at present very limited, enough is known to raise questions about the risks that might be associated with their increasing use. The combination of persistence with very high vertebrate toxicity has set the alarm bells ringing. The ensuing account will be principally concerned with the second-generation rodenticides.

11.2 *Chemical properties*

The formulae of some ARs are given in Figure 11.1, where it can be seen that they have some structural resemblance both to dicoumarol and to vitamin K in its quinone form. All possess quinone rings linked to unsubstituted phenyl rings. The phenyl rings of the rodenticides confer hydrophobicity, especially in the relatively large and

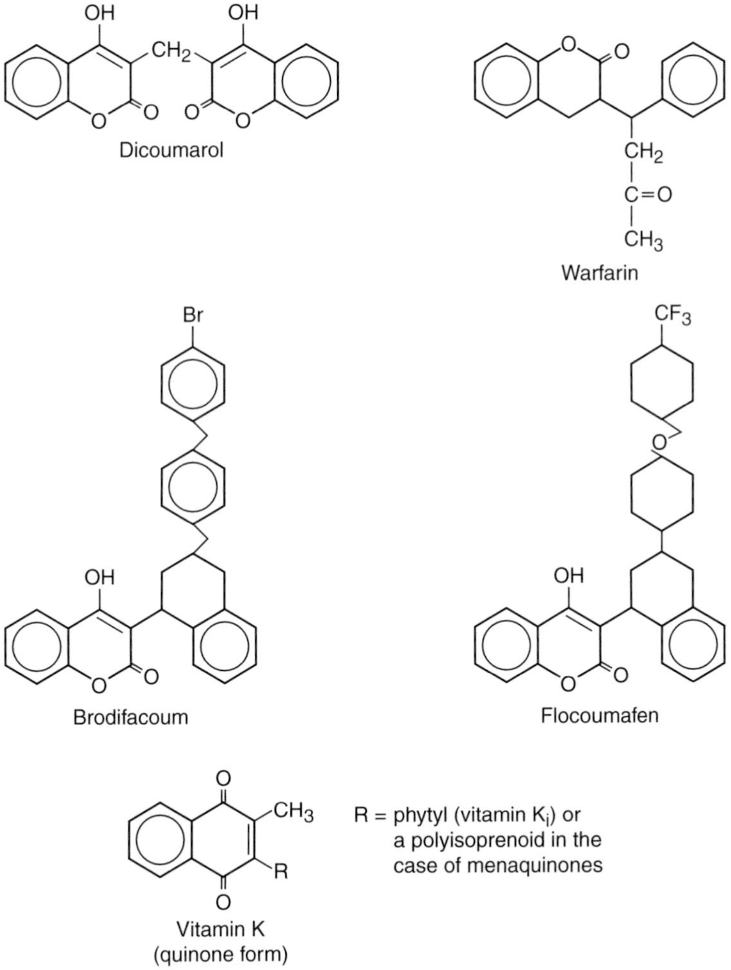

FIGURE 11.1 *Anticoagulant rodenticides.*

complex molecules of brodifacoum and flocoumafen. The chemical properties of some anticoagulant rodenticides are given in Table 11.1. All the compounds listed in Table 11.1 are solids. Flocoumafen and brodifacoum have particularly low vapour pressures. The hydrophobicity of brodifacoum and flocoumafen is reflected in their low water solubility. It should be remembered, however, that because they possess ionisable hydroxyl groups water solubility is pH dependent. With brodifacoum, for example, water solubilities are 3.8×10^{-3} and 10.0 mg/L at pH values of 5.2 and 9.3 respectively. An increase in pH encourages some ionisation, and solubility increases accordingly.

11.3 *Metabolism of anticoagulant rodenticides*

Metabolism has been studied in more detail for warfarin than for other related rodenticides. The metabolism of warfarin appears to be essentially similar to that of related compounds and will be taken as a model for the group. Two main types of primary metabolic attack have been recognised: (1) monooxygenase attack upon diverse positions on the molecule to yield hydroxy metabolites, and (2) conjugation of the hydroxyl group to yield, for example, glucuronides. More than one form of P450 is involved in warfarin metabolism. In a study of metabolism of [14C]flocoumafen by the Japanese quail (Huckle *et al.*, 1989), biotransformation was extensive and rapid, with eight metabolites detected in excreta. The elimination of radioactivity from the liver of Japanese quail was biphasic (Figure 11.2). After an initial period of rapid decline, there followed a period of slow exponential decline, indicating a half-life of more than 100 days. Slow elimination in the second phase has also been demonstrated in rodents and is attributed to very strong binding to one or more proteins of the hepatic endoplasmic reticulum. Other highly toxic 'superwarfarins' such as brodifacoum and difenacoum also show very slow 'second-phase' elimination from the liver, which is associated with strong protein binding. Depending on the compound and the species, < 2 μg/g (2 ppm by weight) of the residue of superwarfarins can be strongly bound in the liver. Most other tissues show less capacity for protein binding.

Some of the strong protein binding in liver is to the target site of the rodenticide, the reductase of vitamin K 2,3-epoxide (see section 11.5). There is also evidence that flocoumafen can bind strongly to cytochrome P4501A1. In the Japanese quail, the

TABLE 11.1 *Properties of anticoagulant rodenticides*

Compound	Water solubility (mg/L)	Log K_{ow}	Vapour pressure (mPa)
Warfarin	17		1.5×10^{-3}
Flocoumafen	1.1	4.7	1.3×10^{-7}
Brodifacoum	0.24 (pH 7.4)	8.5	$<< 1 \times 10^{-3}$
Difenacoum	2.5 (pH 7.3)		

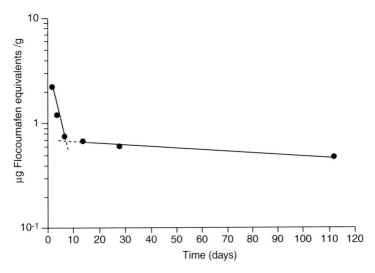

FIGURE 11.2 *Loss of flocoumafen residues from quail liver. Depletion of radioactivity from Japanese quail after a single oral dose (14 mg/kg) of {^{14}C}flocoumafen. Data are presented as microgram equivalents of {^{14}C}flocoumafen per gram of tissue and are mean values of two animals. Data collected at day 7 and at day 12 were from four animals and three animals respectively. After Huckle* et al. *(1989).*

binding of flocoumafen to hepatic microsomes was linked to strong inhibition of EROD-ase activity, both *in vivo* and *in vitro* (Fergusson, 1994). Strong binding to two sites in the liver could explain the biphasic elimination of flocoumafen by this species (Figure 11.2).

The marked persistence of the readily biodegradable superwarfarins in liver is something of a paradox. Ready biodegradability does not guarantee a short half-life. Superwarfarins are much more biodegradable than refractory pollutants such as p,p'-DDE, dioxin and higher chlorinated PCBs. Yet they can still be highly persistent at low concentrations in certain tissues because tenacious binding ensures that they are virtually unavailable to the P450s that can degrade them. Such long-term storage brings the risk of long-term toxic effects in individuals and in food chains.

11.4 *Environmental fate of anticoagulant rodenticides*

Warfarin and related rodenticides are usually incorporated into baits and placed in locations where they will be found by rats, mice and certain other vertebrate pests. There is a potential risk that baits will be eaten by pets, farm animals and other non-target species. To guard against this, baits are made inaccessible to vertebrates other than the target species. For example, baits may be put in short lengths of piping, wide enough for rats and mice to enter but narrow enough to exclude cats and dogs. In Western Europe, warfarin is used both inside and outside farm buildings. The use of more toxic rodenticides such as brodifacoum and flocoumafen is often restricted to the interior of buildings, to reduce the risk to non-target species.

The main concern about these compounds is that they can be transferred via rodents to terrestrial predators and scavengers that feed upon them. Species at risk include members of the crow family, e.g. raven (*Corvus corax*), magpie (*Pica pica*), carrion crow (*Corvus corone*) and rook (*Corvus frugilegus*), owls such as barn owl (*Tyto alba*) and tawny owl (*Strix aluca*), and large predators which are also carrion feeders such as buzzard (*B. buteo*) and red kite (*M. milvus*).

The long half-lives of residues of the superwarfarins in the livers of rodents give particular cause for concern. Owls and other predators may acquire these compounds from their live prey. Individual rats and mice live for several days after consuming lethal doses of rodenticide before they die from haemorrhaging. Furthermore, some resistant strains of rodents can tolerate relatively high levels of rodenticide and so act as more efficient vectors of the pesticide than susceptible strains. Also, rodents in the later stages of poisoning may be more vulnerable to predation than normal healthy individuals; in other words, there may be selective predation, working in favour of the most contaminated members of the prey population. Apart from predation, there is the obvious concern that scavengers will feed upon poisoned individuals if they die above ground.

A number of reports have established the presence of rodenticides in predators and scavengers found dead in the field [see, for example, reports by the UK Wildlife Incident Investigation Scheme (WIIS)]. Brodifacoum, difenacoum, bromodiolone and flocoumafen have all been found, albeit at low levels in most cases (< 1 ppm in liver). Sometimes, more than one type of rodenticide has been found in one individual. The toxicological significance of these residues will be discussed in the next section. One study conducted in Britain between 1983 and 1989 was of barn owls found dead in the field. Ten per cent of the sample of 145 birds contained AR residues in their livers, difenacoum and brodifacoum were prominent among them (Newton *et al.*, 1990). In another study, barn owls that had been fed rats dosed with flocoumafen showed that a substantial proportion of the rodenticide ingested by owls was eliminated in pellets (Eadsforth *et al.*, 1991). The authors suggest that the exposure of owls to rodenticides in the field may be monitored by analysis of pellets dropped at roosts or regular perching places.

11.5 *The toxicity of anticoagulant rodenticides*

The mode of action of warfarin and related rodenticides is illustrated in Figure 11.3. Vitamin K in its reduced hydroquinone form is a cofactor for a carboxylase located in the rough endoplasmic reticulum of hepatocytes. It is converted into an epoxide when it participates in the conversion of glutamate residues of certain proteins to γ-carboxy-glutamate (Gla). The regeneration of the hydroquinone form of vitamin K from the

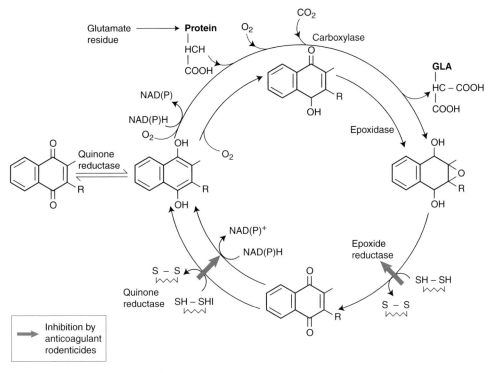

FIGURE 11.3 *Action of warfarin and related rodenticides on the vitamin K cycle.*

2,3 epoxide is dependent on the function of a reductase. ARs bind to the reductase and prevent the conversion of the epoxide to the quinone. They also inhibit the subsequent reduction of the quinone to the hydroquinone. Thus, ARs can prevent the cyclic regeneration of the hydroquinone form so that carboxylation of glutamate slows down or ceases.

In the case of prothrombin and related clotting factors, interruption of the vitamin K cycle leads to the production of non-functional, undercarboxylated proteins, which are duly exported from hepatocytes into blood (Thijssen, 1995). They are non-functional because there is a requirement for the additional carboxyl residues in the clotting process. Ionised carboxyl groups can establish links with negatively charged sites on neighbouring phospholipid molecules of cell surfaces via calcium bridges, and these links are necessary for the formation of blood clots. Because clotting proteins turn over slowly (half-life of prothrombin in the rat is approximately 10 h), several days will elapse after inhibition of the vitamin K cycle before the level of functional clotting proteins is sufficiently low to allow severe haemorrhaging. Typically, rats and mice begin to die from haemorrhaging 5 days or more after exposure to ARs. Owing to the strong affinity of superwarfarins for the reductase, the available binding sites may be progressively occupied by ARs over a period of weeks or even months of continuing exposure to low levels of the compounds. Evidently all the superwarfarins bind to the same site, and it is to be expected that they will have an additive toxic effect. What is unclear, and can make interpretation of residue data difficult, is what degree of occupancy of the reductase binding sites by ARs will lead to serious haemorrhaging. Some individuals appear perfectly healthy when carrying liver residues that are high enough to cause haemorrhaging in others.

On account of the uncertainties associated with the interpretation of residue data, it is important to have other evidence to establish toxic effects in the field. In the investigation of deaths in the field, haemorrhaging is usually easy to identify in a post-mortem examination if carcasses are in a reasonable condition. If early toxic effects are to be identified in live vertebrates, however, a biomarker assay is needed (Fergusson, 1994). The detection of undercarboxylated Gla proteins in blood has already been used to monitor human exposure to warfarin and other ARs (Knapen *et al.*, 1993). The development of such an assay, e.g. an ELISA (enzyme-linked immunosorbent assay), that could be used to assay blood samples from predators/ scavengers exposed to rodenticides in the field has obvious attractions. It should then be possible to establish when levels of exposure in the field are high enough to begin to inhibit the vitamin K cycle and increase the blood level of undercarboxylated clotting proteins. Toxic effects could be identified at an early stage before the occurrence of deaths due to haemorrhaging.

Because of the delay in the appearance of haemorrhaging after exposure to warfarin and related anticoagulant rodenticides, a suitable interval must elapse between exposure of experimental animals to the chemical and the assessment of mortality in toxicity testing. Typically this period is at least 5 days. Some values of acute oral LD_{50} of rodenticides to vertebrates are given in Table 11.2. Looking at the data overall, brodifacoum and flocoumafen are more toxic than warfarin to mammals. Toxicity of the former two compounds to birds varies considerably between species. On the available evidence, the galliform birds chicken (*Gallus domesticus*) and Japanese quail (*C. coturnix japonica*) are much less sensitive to flocoumafen than are mammals. The chicken is less sensitive to brodifacoum than mammals. The mallard duck (*Anas platyrhynchus*), however, is just as sensitive as mammals to brodifacoum; studies with the barn owl (*T. alba*) indicate that it is of similar sensitivity to the rat or mouse. Newton *et al.* (1990) fed mice containing brodifacoum to the owls and estimated that birds lethally poisoned by the rodenticide (*n* = 4) had consumed 0.150–0.182 mg/kg of the compound. The birds died within 6–17 days of receiving a single dose of brodifacoum, and the concentration of rodenticide in the liver was 0.63–1.25 mg/kg. Owls were also dosed with difenacoum in this study, which was found to be less toxic than brodifacoum.

A number of studies have shown that predatory and scavenging birds can acquire liver residues of second-generation ARs when these compounds are used in the field (Newton *et al.*, 1990; and Annual Reports of WIIS), and that the levels are sometimes high enough to cause death by haemorrhaging (Merson *et al.*, 1984). In a field trial with brodifacoum conducted in Virginia, USA, five screech owls (*Otus asio*) that were found dead 5–37 days after treatment contained 0.4–0.8 µg/g of the rodenticide in

TABLE 11.2 *Acute oral LD$_{50}$ values for some anticoagulant rodenticides*

Compound	Species	Acute oral LD$_{50}$ (mg/kg)
Warfarin	Rat	1
Warfarin	Pig	1
Brodifacoum	Rat (male)	0.27
Bridifacoum	Rabbit (male)	0.3
Brodifacoum	Mouse (male)	0.4
Brodifacoum	Chicken	4.5
Brodifacoum	Mallard duck	0.31
Flocoumafen	Rat	0.25
Flocoumafen	Mouse	0.8
Flocoumafen	Chicken	> 100
Flocoumafen	Japanese quail	> 300
Flocoumafen	Mallard duck	~ 24

Data from Tomlin (1997).

liver (Hegdal and Colvin, 1988). These residues are of a similar magnitude to the levels found in poisoned barn owls in the study mentioned above. Field deaths of red kites and goshawks have also been attributed to secondary poisoning caused by bromodiolone (WIIS scheme). In another study, ravens died of brodifacoum poisoning during a rat control programme on Langara Island, British Columbia, Canada (Howald *et al.*, 1999). They had acquired the rodenticide directly from bait and indirectly by predating or scavenging poisoned rats. Post-mortem examination established that the birds had died of severe haemorrhaging, and contained 0.98–2.52 mg/kg brodifacoum in the liver, mean value 1.35 mg/kg ($n = 13$). In New Zealand, primary and secondary brodifacoum poisoning of wekas (*Gallirallus australis*), flightless rails which are predators and scavengers, has been reported (Eason and Spurr, 1995).

As discussed earlier (p. 210), a problem with these field incidents is that the low levels of rodenticides found in many of the poisoned birds are of similar magnitude to those in birds that survive exposure. A low residue level may signify everything or nothing. Additional evidence is needed to establish that the concentrations of rodenticide present in the livers of birds (or mammals) found in the field are sufficient to have caused death, e.g. the presence of haemorrhaging in the carcasses.

11.6 *Ecological effects of anticoagulant rodenticides*

The demonstration that owls and other birds can acquire lethal doses of superwarfarins in the field following normal patterns of use has raised questions about the possibility of these compounds causing population declines. In one case reported in Malaysia, a population decline in owls was related to the use of superwarfarins (see Newton *et al.*, 1990). In New Zealand, the entire population of wekas on one island was wiped out by primary and secondary brodifacoum poisoning (Eason and Spurr, 1995). In Britain, however, a widespread decline in the barn owl population during the 1980s could not be explained in terms of rodenticide use. Only a small proportion (2%) of the barn owls found dead during the period 1983–9 contained residues of brodifacoum + difenacoum of 0.1 ppm or above in the liver. Thus, no more than 2% of the dead owls contained residues of rodenticides high enough to suggest poisoning. However, it should also be pointed out that the use of superwarfarins was restricted at the time of the survey. The recommended use of brodifacoum, for example, was (and still is) restricted to the interior of buildings. Superwarfarins have mainly been used in areas where resistance has developed to warfarin. The critical question is whether increasing use of superwarfarins will bring a significant risk to owls and other species exposed to them. During the 1990s, cases have been reported of red kites and other birds of prey dying as a consequence of poisoning by superwarfarins (see reports of WIIS). This is a controversial issue because red kites have been reintroduced into England in recent years. The population is still small, but growing. It has been suggested that mortality due to rodenticide poisoning may prevent the re-establishment of the species in certain areas of the country where ARs are more widely used.

In summary, there is as yet no clear evidence that superwarfarins have caused any widespread declines in predators/scavengers that feed upon rodents. However, the persistence and very high cumulative toxicity of these compounds suggest that they could pose a serious hazard to such species if they are more widely used. The situation should be kept under close review.

Resistance to warfarin has developed in populations of rats after repeated exposure to the rodenticide (Thijssen, 1995). One resistant strain discovered in Wales was found to have a much reduced capacity to bind warfarin to liver microsomes in comparison with susceptible rats. Resistance was due to a gene which encoded for a form of vitamin K epoxide reductase that was far less sensitive to warfarin inhibition than the form found in susceptible rats. Another resistant strain, which arose in the area of Glasgow, was found to differ from the resistant Welsh strain. Resistance was again due to an altered form of vitamin K epoxide reductase. However, the strain from the Glasgow area contained a form of the enzyme which bound warfarin just as strongly as that from susceptible rats! The difference was that the binding was readily reversible (Thijssen, 1995). In addition to these strains, another from the Andover area may owe its resistance to enhanced detoxication by a P450-based monooxygenase.

11.7 *Summary*

Warfarin and the second-generation superwarfarins are anticoagulant rodenticides that have a structural resemblance to dicoumarol and vitamin K. They act as vitamin

K antagonists, thereby retarding or stopping the carboxylation of clotting proteins in the hepatic endoplasmic reticulum. The build-up of non-functional, undercarboxylated clotting proteins in the blood leads eventually to death by haemorrhaging.

Brodifacoum, difenacoum, flocoumafen and other superwarfarins bind strongly to proteins of the hepatic endoplasmic reticulum and consequently have long half-lives in vertebrates, often exceeding 100 days. Thus, they present a hazard to predators and scavengers which feed upon rodents that have been exposed to them. A number of species of predatory and scavenging birds have died as a consequence of secondary poisoning by superwarfarins in field incidents, and questions have been asked about the long-term risks associated with expanding use of these compounds.

11.8 *Further reading*

There is a shortage of appropriate texts on the ARs. Buckle and Smith (1994) describe the use of ARs in rodent control. Thijssen (1995) gives a concise account of mode of action and resistance mechanisms. For effects on non-target species, reference should be made to the individual citations given in the foregoing text.

CHAPTER **12**

Pyrethroid insecticides

12.1 *Background*

The insecticidal properties of pyrethrum, a product prepared from the dried and powdered heads of flowers belonging to the genus *Chrysanthemum*, have long been recognised. First introduced into Europe in the middle of the nineteenth century, early sources were the region of the Caucasus and the Adriatic coast. Subsequently, the major source of commercial pyrethrum was the species *Chrysanthemum cinerariaefolium* grown in East Africa. In the course of time the insecticidal ingredients of pyrethrum, the pyrethrins, were chemically characterised. Six pyrethrins were identified, all of them lipophilic esters (Figure 12.1). They are formed from two acids, chrysanthemic acid and pyrethric acid, in combination with three bases, pyrethrolone, cinerolone and jasmolone.

A serious limitation of pyrethrins as commercial insecticides is their instability. On the one hand, they are photolabile so have only limited life when applied to surfaces, e.g. plant leaves exposed to direct sunlight. On the other, they are readily biodegradable, and often have only a short 'knock down' effect on target insects unless they are synergised with compounds (e.g. piperonyl butoxide) that will repress their oxidative metabolism. The important point is that they have served as models for the development of the synthetic pyrethroids, one of the most widely used types of insecticide at the present time. The first synthetic pyrethroids, compounds such as allethrin and bioallethrin, were not sufficiently photostable to have great commercial potential (Leahey, 1985). Subsequently, a series of compounds were discovered that had greater stability, and these achieved great commercial success. Included among these are permethrin, cypermethrin, deltamethrin, fenvalerate, cyfluthrin, cyhalothrin and others. Their widespread introduction during the 1970s came on the heels of the environmental problems associated with the persistent organochlorine insecticides. Although the synthetic pyrethroids have sufficient metabolic stability to be effective insecticides, they are, nevertheless, readily biodegradable by vertebrates and do not tend to be biomagnified in food chains. At the time of their introduction they were seen to be environmentally friendly insecticides that, for some purposes, were effective alternatives to organochlorine insecticides.

12.2 *Chemical properties*

The structures of some pyrethroid insecticides are shown in Figure 12.1. They are all lipophilic esters showing some structural resemblance to the natural pyrethrins. They can all exist in a number of different enantiomeric forms. Permethrin, cypermethrin and deltamethrin, for example, all have three asymmetric carbon atoms, and consequently eight possible enantiomers (Leahey, 1985; EHC 94, 95, 97). Their enantiomers fall into two categories, *cis* or *trans*, depending on the stereochemistry of the 1 relative to the 3 position of the three-membered ring of the acid moiety (see, for

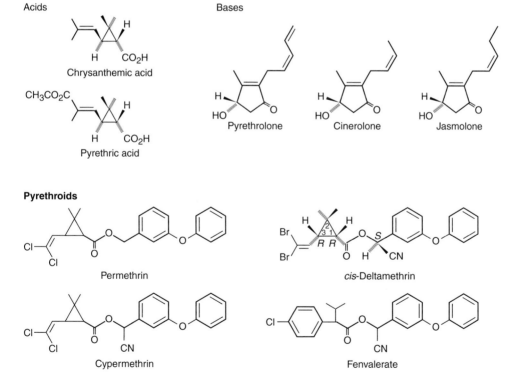

FIGURE 12.1 *Structure of pyrethrins and pyrethroids.*

example, *cis*-deltamethrin in Figure 12.1). Thus, there are four possible *cis* enantiomers and four possible *trans* enantiomers for each of these three compounds. Commercial products are usually racemic mixtures of different enantiomers. A notable exception is deltamethrin, which is marketed as a single cis isomer (EHC 97). Fenvalerate differs from the other pyrethroids featured in Figure 12.1 on account of the structure of its acid moiety. Nevertheless, it has similar biological properties to the other pyrethroids. The properties of some pyrethroids are given in Table 12.1.

The pyrethroids apart from fenvalerate are solids with low water solubility, marked lipophilicity and low vapour pressure. Fenvalerate is a viscous liquid with an appreciable vapour pressure. As they are esters, they are subject to hydrolysis at high pH. They are sufficiently stable to heat and light to be effective insecticides in the field.

12.3 *Metabolism of pyrethroids*

The metabolism of permethrin will be taken as an example of the metabolism of pyrethroids in general (Figure 12.2). The two types of primary metabolic attack are by microsomal monooxygenases and by esterases. Monooxygenase attack involves different forms of cytochrome P450 and yields metabolites with hydroxyl groups substituted in both the acidic and basic moieties. The principal metabolites formed by primary oxidation are compounds 1 and 2 in Figure 12.2. Hydroxylation occurs on a methyl group of the acid moiety and on a free *para*-ring position in the basic moiety. Hydrolysis yields free acids and bases, with or without hydroxyl introduced by oxidative attack. Thus, metabolites 4 and 5 are the esteratic products of permethrin; metabolites 3 and 6 are also hydrolytic products, but they have additional hydroxy groups introduced by oxidative attack. The hydroxyl groups are then available for

TABLE 12.1 *Properties of some pyrethroid insecticides*

Compound	Water solubility µg/mL at 20 or 25 °C	Log K_{ow}	Vapour pressure Pa at 20 or 25 °C
Permethrin (racemate)	0.2	6.5	1.3×10^{-6}
Cypermethrin (racemate)	0.009	6.3	1.9×10^{-7}
α-Cypermethrin (two *cis* isomers)	0.005–0.01	5.16	1.7×10^{-7}
Deltamethrin	< 0.002	5.43	2×10^{-6}
Fenvalerate	0.002	6.2	3.7×10^{-5}

1 pascal (Pa) = 0.0075 torr (i.e. mmHg).
Data from EHC 82, EHC 94, EHC 95, EHC 97 and EHC 142.

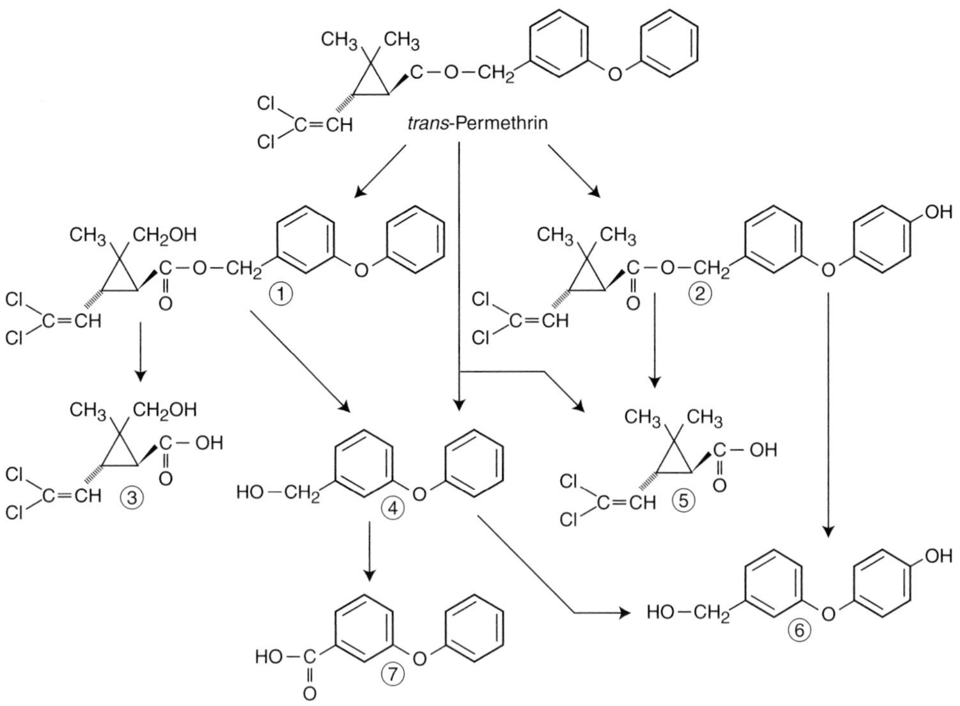

FIGURE 12.2 *The metabolism of* trans-*permethrin.*

conjugation with glucuronide, sulphate, peptide, etc., depending on species. In both insects and vertebrates the excreted products are mainly conjugates.

There has been some controversy over the relative importance of oxidation and esteratic hydrolysis in primary metabolic attack. The strong potentiation of toxicity to insects of certain pyrethroids by PBO and other P450 inhibitors (see section 2.6) suggests the dominance of oxidation over hydrolysis as a detoxication mechanism. A problem of interpretation of metabolic studies has been the shortage, sometimes the apparent absence, of primary oxidative metabolites such as those shown in Figure 12.2. Related to this is the question of the mechanism by which hydroxylated metabolites such as compounds 3 and 6 were formed. Did hydroxylation occur before or after hydrolytic cleavage of the ester bond? In most cases available evidence strongly suggests that oxidation predominates over hydrolysis as a primary mode of oxidative attack. In insects, the marked synergistic action of P450 inhibitors such as PBO and EBI fungicides (see section 12.5) is not consistent with esterase attack being important in primary metabolism (and consequent detoxication) of pyrethroids. Further, the products of esteratic cleavage are polar in character and hardly ideal substrates for the hydrophobic active centres of cytochrome P450s! It should also be mentioned that the primary products of oxidative attack are more polar than the original insecticides and are likely, on that account, to be better substrates for hydrolytic attack by esterases [cf. the esteratic hydrolysis of OP oxons but not OP thions (Chapter 10)]. Primary attack by oxidation, followed by hydrolysis (Figure 12.2), can explain an observation made by several workers studying microsomal metabolism of pyrethroids – that switching on P450 oxidation by addition of NADPH can increase the rate of hydrolysis (Lee *et al.*, 1989).

12.4 *Environmental fate of pyrethroids*

Pyrethroids are extensively used in agriculture, so agricultural land is commonly contaminated by them. They can also reach field margins and hedgerows through spray drift. Because of their high toxicity to aquatic organisms, precautions are taken so that they do not enter surface waters, which might happen as a consequence of spray drift or soil run off. Their use in sheep dips, to control ectoparasites, has raised concern over the safe disposal of unused dipping liquids; dips should not be discharged into adjacent water courses.

Two major factors determining the environmental fate of pyrethroids are marked lipophilicity and relatively rapid biodegradation by many species (Leahey, 1985). When pyrethroids reach soils or aquatic systems, they become strongly adsorbed to the colloidal fraction – mineral particles and associated organic matter. Consequently, their levels fall rapidly in water and in most organisms. Bioconcentration studies with fish have shown BCF factors in the steady state from 50 to several thousand, depending on species, age, etc., often considerably below the values predicted by the high K_{ow} values (EHC 97, 142). Major factors responsible for this are believed to be the relatively rapid metabolism (see Chapter 4) and also the strong adsorption to colloidal material in certain test systems. In some studies, the pyrethroid concentrations measured in water by chemical analysis included considerable amounts of insecticide in the adsorbed state, which was not readily available to the fish. This represented an overestimate of the concentration that organisms were actually exposed to in water, and consequently an underestimate of the BCF (Leahey, 1985). In a laboratory study of the persistence of five pyrethroids in soil, the rates of loss followed the order fenpropathrin $>$ permethrin $>$ cypermethrin $>$ fenvalerate $>$ deltamethrin (Chapman and Harris, 1981). Microbial degradation was an important factor in their disappearance. Half-lives of deltamethrin determined in two German soils were found to be 35 days in a sandy soil but 60 days in a sandy loam. Hill and Schaalje (1985) showed that deltamethrin applied in the field underwent a biphasic pattern of loss: an initial rapid loss being succeeded by a slower first-order degradation. This is essentially similar to the pattern of loss of another group of hydrophobic insecticides – the organochlorine compounds – except that the organochlorine compounds are eliminated much more slowly, especially in the later stages of the process (see section 4.3 and Chapter 5). Both types of insecticide have log K_{ow} values in the range 5–7, but organochlorine insecticides are metabolised much more slowly than pyrethroids by soil microorganisms.

Pyrethroids can also persist in sediments. In one study, α-cypermethrin was applied to a pond as an emulsifiable concentrate (EHC 142). Sixteen days after application, 5% of the applied dose was still present in sediment, falling to 3% after a further 17 days. This suggests a half-life of the order of 20–25 days – similar in magnitude to half-lives measured in temperate soils.

The general picture, then, is that pyrethroids are reasonably persistent in soils and sediments but, because of their ready biodegradability, do not undergo biomagnification with movement up food chains. There is, however, concern that residues in sediments may continue to be available to certain bottom feeders long after initial contamination, and that some aquatic invertebrates in lower trophic levels, which are deficient in detoxifying enzymes, may bioconcentrate them.

12.5 *Toxicity of pyrethroids*

Pyrethroids, like p,p'-DDT, are toxic because they interact with Na^+ channels of the axonal membrane, thereby disturbing the transmission of nerve action potential (Eldefrawi and Eldefrawi, 1990; section 5.2.4 of this book). In both cases marked hydrophobicity leads to a build-up in concentration of the insecticides in the axonal membrane and reversible association with the Na^+ channel. Consequently both DDT and pyrethroids show negative temperature coefficients in arthropods; increasing temperature brings decreasing toxicity, because it favours desorption of insecticide from the site of action.

Pyrethroids show very marked selective toxicity (Table 12.2). They are highly toxic to terrestrial and aquatic arthropods and somewhat less toxic to fish. They are only moderately toxic to rodents, and less toxic still to birds. The selectivity ratio between bees and rodents is of the order 10 000- to 100 000-fold with topical application of insecticide! They therefore appear to be environmentally safe insecticides – so far as terrestrial vertebrates are concerned. There are, inevitably, concerns about their possible side-effects in aquatic systems – especially on invertebrates. A field problem that has emerged relatively recently is the synergistic action of certain EBI fungicides upon pyrethroids. Some combinations of EBIs with pyrethroids are highly toxic to bees,

TABLE 12.2 *Toxicity of some pyrethroids*

Compound	LD$_{50}$ rat (mg/kg)	LD$_{50}$ birds (mg/kg)	96 h LC$_{50}$ fish (μg/L)	LC$_{50}$ aquatic invertebrates (μg/L)
Permethrin	500	> 13 000 (4)	0.6–314	0.018–1.2
Cypermethrin	250	> 10 000 (1)	0.4–2.8	0.01–5
Fenvalerate	451	> 4000 (3)	0.3–200	0.008–1
Deltamethrin	129	4000 (1)	0.4–2.0	5 (*Daphnia*)

Mean values given for birds; the number of species tested is given in brackets.
Data from EHC 82, EHC 94, EHC 95 and EHC 97.

with synergistic ratios of the order 10–20 (Colin and Belzunces, 1992; Pilling, 1993; Meled *et al.*, 1998). There have been reports from France and Germany of deaths of bees in the field that are attributable to synergistic effects of this kind after the use of tank mixes by spray operators. The enhancement of the toxicity of l-cyhalothrin to bees by the EBI fungicide prochloraz has been demonstrated in a semi-field trial (Bromley-Challenor, 1992). The synergistic action of EBI fungicides has been attributed, largely or entirely, to inhibition of detoxication by cytochrome P450 (see, for example, Pilling *et al.*, 1995). Questions are now being asked about possible hazards to wild bees and other pollinators posed by pyrethroid/EBI mixtures.

12.6 *Ecological effects of pyrethroids*

Contamination of surface waters by pyrethroids might reasonably be expected to have effects, at the population level and above, on aquatic invertebrates because of their very high toxicity. In one study of a farm pond, cypermethrin was applied aerially, adjacent to the water body (Kedwards *et al.*, 1999a). Changes were observed in the composition of the macroinvertebrate community of the pond, which were related to levels of the pyrethroid in the hydrosoil. Diptera were most affected, showing a decline in abundance with increasing cypermethrin concentration. Chironomidae larvae first declined and later recovered. Harmful effects on macroinvertebrate communities were also demonstrated in mesocosm studies. Cypermethrin and λ-cyhalothrin were individually applied to experimental ponds at the rates of 0.7 and 1.7 g.a.i./ha, and the results subjected to multivariate analysis (Kedwards *et al.*, 1999b). The treatments caused a decrease in abundance of Gammaridae and Asellidae, but a concomitant increase in Planorbidae and Chironomidae, Hirudinae and Lymnaeidae. Gammaridae were found to be more sensitive to the chemicals than Asellidae, their numbers remaining depressed until the termination of the experiment (15 weeks) with both treatments. This may have been because they inhabit the sediment surface, where there will be relatively high levels of recently adsorbed pyrethroid, whereas the asellidae are epibenthic and burrow into the hydrosoil, where lower levels of insecticide should exist – at least in the short term. In general, it should be possible to pick up effects of this kind in natural waters using ecological profiling, e.g. RIVPACS. There is a need here for combining ecological profiling with chemical analysis to facilitate the detection of chemicals that cause changes in the community structure in natural waters.

The continuing use of pyrethroids in agriculture has led to the emergence of resistant strains of pests. One of the best-studied examples is the tobacco budworm (*H. virescens*), a very serious pest of cotton in the southern states of the USA (McCaffery, 1998). Indeed, the resistance problem has become so severe in certain areas that there is a danger of losing control of the pest. A study of a number of resistant strains from the field has revealed two major types of resistance mechanism. Some individuals possess aberrant forms of the target site, the Na^+ channel. At least two forms are known which confer either kdr (< 100-fold) or super-kdr (>100-fold) resistance, which is the consequence of the presence of an insensitive form of the Na^+ channel protein (McCaffery, 1998; section 4.5 of this book).

This type of resistance has been found in a number of species of insects, including *M. domestica, H. virescens, P. xylostella, B. germanica, Anopheles gambiae* and *M. persicae*. Kdr has been attributed to three different changes in single amino acids of the voltage-dependent sodium channel and super-kdr to changes in pairs of amino acids also located in the sodium channel (Salgado, 1999). Interestingly, it appears that earlier selective pressure by DDT raised the frequency of *kdr* genes in the population before pyrethroids came to be used! Thus, some 'pyrethroid resistance' already existed before the insecticides were applied in the field.

The other major mechanism of pyrethroid resistance found in some field strains of *H. virescens* was enhanced detoxication due to a high rate of oxidative metabolism, mediated by a form of cytochrome P450 (McCaffery, 1998). Some strains, such as PEG 87, which was subject to a high level of field and laboratory selection, possessed both mechanisms. Other examples of pyrethroid resistance due to enhanced detoxication may be found in the literature on pesticides.

12.7 *Summary*

Pyrethroid insecticides were modelled upon naturally occurring pyrethrins, which were once quite widely use as insecticides but had the disadvantages of being photochemically unstable and susceptible to rapid metabolic detoxication. Pyrethroids are more stable than pyrethrins and, like DDT, act upon the voltage-dependent sodium channel of the nerve axon. Although lipophilic, they are readily biodegradable and do not undergo biomagnification along food chains to any important extent. They are, however, strongly adsorbed in soils and sediments, where they can be persistent.

Pyrethroids are much more toxic to invertebrates than to most vertebrates. They can have serious effects upon aquatic invertebrates, at least in the short term. They can be synergised by inhibitors of cytochrome P450, such as EBI fungicides, so there are potential hazards associated with the use of mixtures of these two types of pesticides. Resistance to pyrethroids has developed in a number of pest species as a result of both insensitive forms of the target site (sodium channel) and enhanced metabolic detoxication.

12.8 *Further reading*

EHC 82 (*Cypermethrin*), EHC 94 (*Permethrin*), EHC 95 (*Fenvalerate*), EHC 97 (*Deltamethrin*) and EHC 142 (*Alphacypermethrin*) are all valuable sources of information on the environmental toxicology of pyrethroids.

Leahey, J. P. (ed.) (1985) *The Pyrethroid Insecticides*. A multiauthor work that covers many aspects of the toxicology and ecotoxicology of the earlier pyrethroids.

PART 3

Further issues and
future prospects

CHAPTER 13

The ecotoxicological effects of herbicides

13.1 Introduction

Chapters 5–12 deal with groups of pollutants that have been studied in some depth and detail, largely because they have appreciable – sometimes very high – mammalian toxicity and are perceived as human health hazards. Some of them are markedly persistent and undergo biomagnification with passage along food chains. As has been explained, individual compounds, or mixtures of related compounds, have sometimes been shown to cause adverse ecological effects. Although it was necessary to take such pollutants as examples, it is important now to consider the more complex situation in which organisms are exposed to mixtures of compounds differing in their chemistry and/or mode of action.

The present chapter is the first of the final part of the text, in which the emphasis moves on from the detailed descriptions of particular types of pollutants given in the foregoing chapters to address some wider issues. Apart from the question of the complexity of pollution in the real world, certain other issues arise directly from the findings of ecotoxicological studies on individual pollutants reported in the earlier text.

Herbicides constitute a large and diverse class of pesticides that, with a few exceptions, have very low mammalian toxicity and have received relatively little attention as environmental pollutants. Paraquat and other bipyridyl herbicides have appreciable mammalian toxicity and will be discussed in Chapter 14. Dinoseb and related dinitrophenols, which act as uncouplers of oxidative phosphorylation in mitochondria, are general biocides that are little used today because of their hazardous nature. They will be mentioned, briefly, in section 14.2. Herbicides are, in general, readily biodegradable by vertebrates and are not known to undergo substantial biomagnification in food chains. Their principal use has been weed control in agriculture and horticulture, although they have also been used as defoliants in forests (e.g. in the Vietnam war), for controlling weeds on roadside verges and in water courses and as management tools on estates and nature reserves. This chapter will be mainly concerned with their impact on the agricultural environment. A brief mention will be made of their wider dispersal in the aquatic environment.

13.2 *Some major groups of herbicides*

The following brief account identifies only major groups of herbicides not mentioned elsewhere in the text and is far from comprehensive. For a more detailed account see Hassall (1990).

The phenoxyalkane carboxylic acids are among the most successful and widely used herbicides. They act as plant growth regulators and produce distorted growth patterns in treated plants. Compounds such as 2,4-D, MCPA, and mecoprop (Figure 13.1) are used as selective herbicides to control dicotyledenous weeds in monocotyledenous crops such as cereals and grass. They are formulated as water-soluble potassium or sodium salts, or as lipophilic esters, and they are frequently sprayed in combination with other types of herbicides that have different modes of action and patterns of weed control. They are applied to foliage and are not soil acting.

Ureides (e.g. diuron, linuron, isoproturon) and triazines (e.g. atrazine, simazine, ametryne) all act as inhibitors of photosynthesis and are applied to soil (see Figure 13.1 for structures). They are toxic to seedling weeds, which can absorb them from soil. Some of them (e.g. simazine) have very low water solubility and, consequently, are persistent and relatively immobile in soil (section 4.3).

Sulphonylurea herbicides such as chlorsulfuron and sulfometuron are also soil acting, affect cell division and have very high phytotoxicity. Indeed, they can be toxic to plants when present in soil at levels low enough to make chemical analysis difficult. Carbamate herbicides constitute a relatively diverse group. Some, like barban (Figure 13.1), are applied to foliage, whereas others (e.g. chlorpropham) are soil acting. The latter type affect cell division. Other important herbicides, or groups of herbicides, include glyphosate, aminotriazole, chlorinated benzoic acids (e.g. dicamba) and phenolic nitriles (e.g. ioxynil, bromoxynil).

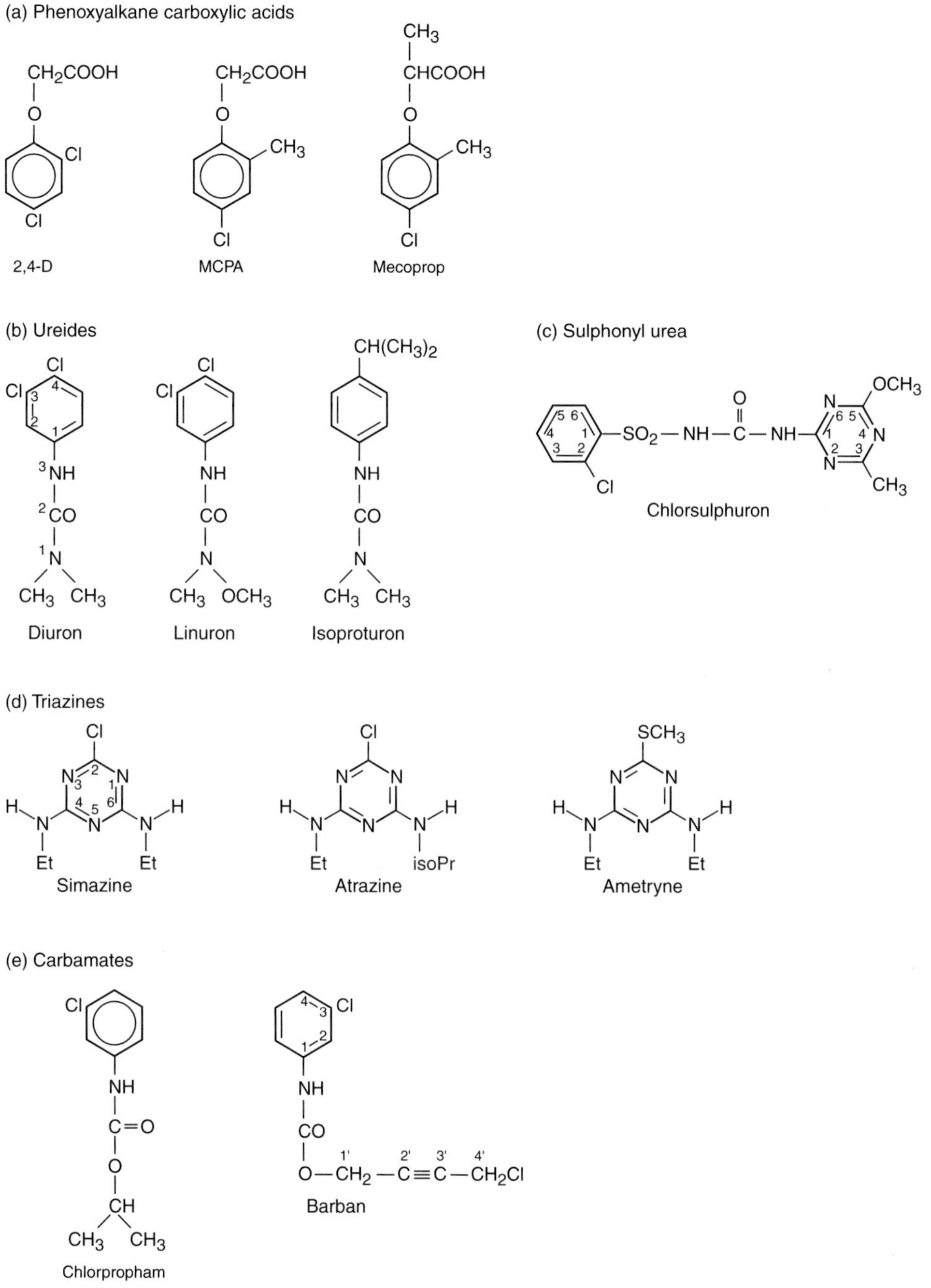

FIGURE 13.1 *Structures of some herbicides.*

13.3 *Agricultural impact of herbicides*

Since the Second World War herbicides have come to be widely used in agriculture and horticulture in the developed world. Frequently, they have been used in 'cocktails' containing several ingredients of contrasting modes of action, thus giving control over a wide range of weed species. The effectiveness of the application of herbicides together with cultivation of the land is evident in many agricultural areas in Western Europe and North America, where few weeds are seen. It is easier to control plants, which are stationary, than to control mobile insect or vertebrate pests. Weed species have been very effectively controlled over large areas of agricultural land. In Britain concern has been expressed over the near extinction of certain once common farmland species that are of botanical interest, e.g. corn cockle and pheasants eye.

Ecologically, such a large reduction in weed species represents a major change to farmland ecosystems, and may be expected to have knock-on effects upon other species. Certain problems have come to light with the investigation of the status of birds on farmland. In one study, the Game Conservancy Council investigated the reasons for the severe and continuing decline of the grey partridge (*Perdix perdix*) on farmland in Britain. The study commenced in the late 1960s, and established that the decline was closely related to increased chick mortality (Potts, 1986, 2000; also Chapter 12 in Walker *et al.*, 2000). The chick mortality was largely explained by a shortage of their insect food (e.g. sawflies) due, in turn, to the absence of the weeds upon which the insects themselves feed. An effect at the bottom of the food chain led to a population decline further up. It is worth reflecting that such an effect by herbicides could not have been forecasted by normal risk assessment (see Chapters 14 and 15). The herbicides responsible are, in general, of very low avian toxicity, and ordinary risk assessment would have declared them perfectly safe to use, so far as partridges and other birds are concerned! Subsequent work has shown that partridge populations can continue to survive on agricultural land if headlands are left unsprayed, thereby allowing weeds to survive, which will support the insects upon which young partridges feed.

This study helped to ring the alarm bells about possible other indirect effects of the wide variety of herbicides used in agriculture. More recently, further evidence has been gained of the reduction in populations of insects and other arthropods on farmland that may relate, at least in part, to the removal of weeds by the use of herbicides. A survey of farmland birds in Britain has established the marked decline of several species in addition to the grey partridge, which may be the consequence of indirect effects of herbicides and other pesticides (Crick *et al.*, 1998; also Chapter 12 of Walker *et al.*, 2000). Species affected include tree sparrow (*Passer montanus*), turtle dove (*Streptopelia purpur*), spotted flycatcher (*Musciapa striata*) and skylark (*Alauda arvensis*). A study is currently in progress to attempt to establish the cause of these declines. Recently, concern about the side-effects of herbicides used on agricultural land has intensified with the development of genetically manipulated (GM) crops. Some GM crops are relatively insensitive to the action of herbicides, thus permitting the application to them of unusually high levels of certain herbicides. The advantage of increasing dose, from the agricultural point of view, is the control of certain difficult weeds. From an ecotoxicological point of view, increasing dose rates of herbicides above the currently approved levels may cause undesirable ecological side-effects. It is very important that any such change in practice is rigorously tested in field trials, as part of environmental risk assessment, before approval is given by regulatory authorities. Such new technology, based on GM crops, should only be introduced if it is shown to be environmentally safe.

One problem that has arisen with the use of herbicides in agriculture is spray or vapour drift. When fine-spray droplets are released, especially if applied aerially, they may be deposited outside the target area because of air movements and cause damage there. In the first place, this is a question of application technique. Herbicides, like other pesticides, should not be applied as sprays under windy conditions. In most situations, herbicides are not applied aerially because of the danger of drift. Where herbicides have appreciable vapour pressure, there may be problems with vapour drift. Under hot conditions, volatile herbicides may go into the vapour state, and the vapour may drift further than the spray droplets. Such was the case with early volatile ester formulations of phenoxyalkanecarboxylic acids (Hassall, 1990). Nowadays, formulations are of less volatile esters or of aqueous concentrates of sodium or potassium salts (which are of low volatility). Spray drift of herbicides can cause damage to crops and wild plants outside the spray area. The cause of such damage can be hard to establish with highly active herbicides (e.g. sulphonylureas), where the phytotoxic concentrations are low enough to make chemical detection difficult.

13.4 Movement of herbicides into surface waters and drinking water

As discussed earlier (section 4.3), pesticides have a very limited tendency to move through soil profiles into drainage water because of the combined effects of adsorption by soil colloids (important for herbicides such as simazine, which have relatively high K_{ow} values), metabolism (important for water-soluble and readily biodegradable herbicides such as 2,4-D and MCPA) and in some cases volatilisation. In reality, however, there are complications. In the first place there may be run-off from agricultural land into neighbouring water courses after heavy rainfall. Soil colloids, with adsorbed herbicides, can be washed into drainage ditches and streams. There is an additional problem with certain soils high in clay minerals (Williams *et al.*, 1996; and section 4.3). During dry periods these soils shrink and develop deep cracks. If heavy rains follow, free herbicides located near the soil surface, and colloids with adsorbed herbicides, can be quickly washed down into the drainage system without passing through the soil profile. In the Rosemaund experiment, the herbicides atrazine, simazine, isoproturon, trifluralin, and MCPA were all detected in drainage water after heavy rain. The respective maximum concentrations in micrograms per litre (ppb) were – 81, 68, 16, 14 and 47 (Williams *et al.*, 1996). These levels were reached after normal approved use of the herbicides and raise questions about possible effects on aquatic plants growing in receiving waters. As mentioned elsewhere (section 10.3.4) the level of carbofuran found during the same study was sometimes high enough to kill freshwater shrimps (*G. pulex*) used as a bioindicator (Matthiessen *et al.*, 1995).

Recent surveys have been providing more information on the levels of herbicides in rivers. In one study a number of different herbicides were detected in the River Humber, UK (House *et al.*, 1997). A number of triazines were detected in the Rivers Aire, Calder, Trent, Don and Ouse, the most abundant of them being atrazine and simazine. The results for simazine showed peaks in the spring and again in the early autumn of 1994 for some rivers, the latter coinciding with the first major storm of the year (Figure 13.2). The maximum level of simazine recorded was 8 μg/L. This was high enough to be toxic to phytoplankton and algae, but was not sustained. Phenyl ureas and phenoxyalkanoic acids were also detected. Concentrations were generally low, but levels of the following herbicides were detected up to the maximum value (μg/L) given in brackets: diuron (< 8.7), chlortoluron (< 0.67), mecoprop (< 8.2).

These high levels were sporadic and transitory. However, they were sometimes high enough to cause phytotoxicity, and more work needs to be carried out to establish whether herbicides in contaminated streams and rivers are having adverse effects upon populations of aquatic plants.

With the acceptable concentrations of herbicides in drinking water being taken to very low levels by some regulatory authorities (e.g. the EC), there has been interest in low levels of atrazine present in groundwater and in drinking water. This finding illustrates the point that mobility of pesticides becomes increasingly evident as the sensitivity of analysis improves.

13.5 *Summary*

As the first chapter in the final part of the book, contamination by herbicides is taken as an example of the complexity of pollution in the real world. A wide variety of compounds of diverse structure, chemical properties and mechanism of action are used as herbicides. Very few of them have appreciable toxicity to animals, and they do not usually undergo significant biomagnification with movement along food chains. Important groups of herbicides are phenoxyalkane carboxylic acids, ureides, triazines and carbamates. Herbicides are often applied as mixtures of compounds with contrasting properties.

The successful use of herbicides and associated cultivation procedures has greatly reduced the populations of weed species in many agricultural areas, sometimes bringing species of botanical interest to near extinction. Intensive weed control in cereal farming has been shown to cause the reduction of certain insect populations and the reduction

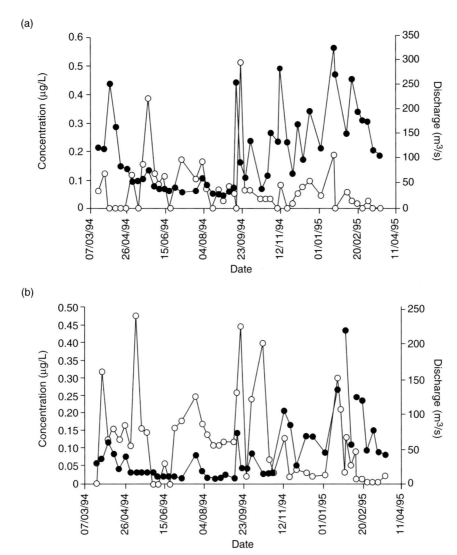

FIGURE 13.2 *Atrazine levels in the Humber River area. Comparison of the concentration of simazine and river discharge over one annual cycle for (a) River Trent at Cromwell Lock and (b) River Aire at Beale. ○, Simazine concentration; ●, river discharge. From House* et al. *(1997) with permission.*

of the grey partridge. The decline of some other insectivorous birds on agricultural land may have a similar cause. The introduction of GM crops with high tolerance to herbicides may lead to increases in dose rates of herbicides on agricultural land with attendant ecotoxicological risks.

Significant levels of herbicides have also been detected in rivers, although these are usually transitory. Heavy rainfall can move herbicides from agricultural land to nearby ditches and streams because of run off and percolation of water through deep fissures in certain soils that are high in clay.

13.6 *Further reading*

Ashton, F. M. and Crafts, A. S. (1973) *Mode of Action of Herbicides*. This book describes the mode of action of major types of herbicides.

Hassall, K. A. (1990) Includes a readable account of the biochemistry of herbicides.

Potts (1986) *The Biochemistry and Use of Pesticides*. An authoritative account of the factors responsible for the decline of the grey partridge on agricultural land.

CHAPTER 14

Dealing with complexity: the toxicity of mixtures

14.1 *Introduction*

In Part 2, attention was focused on particular pollutants. Their chemical and biochemical properties were related to their ecotoxicological effects. Sometimes, with the aid of biomarker assays, their effects upon individuals were related to consequent effects at the level of population and above. The biomarker assays provided the essential evidence that adverse effects on populations, communities and ecosystems were being caused by environmental levels of particular chemicals. The examples given included population declines in raptors due to eggshell thinning caused by p,p'-DDE, and the decline or extinction of dog whelk populations due to imposex caused by TBTs. These were relatively straightforward situations in which most of the adverse change was attributable to a single chemical. In environmental studies, however, one is usually dealing with mixtures of pollutants, and adverse effects may be due to combinations of chemicals.

There are many cases where adverse effects at the level of population or above have been shown to correlate with levels of either individual pollutants or combinations of pollutants, e.g. the studies on pollution by polychlorinated aromatic compounds in the Great Lakes (Chapters 5–7). The problem is establishing where pollutant levels are *causally* related to adverse effects. Many factors other than the pollutants actually determined by chemical analysis may cause populations to decline, including shortage of food, habitat change, disease and climatic change – and other pollutants that have not been analysed! Such factors may very well correlate with the measured pollutant levels, especially where comparison is simply being made between the populations in one or two polluted areas and the population in a 'clean' area. In badly polluted areas there may be elevated levels of other pollutants in addition to those determined by chemical analysis, and these may have direct effects upon the species being studied – or indirect ones, e.g. by causing changes in food supply.

The following account will be concerned with methods used for measuring the toxicity of mixtures, both in laboratory studies and in the field, and the related question of the development of predictive models.

14.2 *Measuring the toxicity of mixtures*

As explained earlier (section 2.6), the toxicity testing of pesticides and industrial chemicals for the purposes of statutory environmental risk assessment is usually carried out on single compounds. For reasons of practicality and cost, only a minute proportion of the combinations of pollutants that may occur in the natural environment can be tested for their toxicity. This dilemma will be discussed further in section 14.3. In the real world, however, mixtures of pollutants are found in contaminated ecosystems, in effluents discharged into surface waters, e.g. sewage, industrial effluents, and in waste waters from pulp mills. The tests or bioassays used here measure the toxicity of mixtures, and investigators are presented with the problem of identifying the individual components of the mixture that, singly or collectively, are bioactive. The issue is even further complicated by the possibility that naturally occurring xenobiotics, such as phytoestrogens taken up by fish, may contribute significantly to the toxicity that is measured.

In the simplest situation, chemicals in a mixture will show additive toxicity. If environmental samples are submitted for both toxicity testing and chemical analysis, the toxicity of the mixture may be estimated from the chemical data and this compared with the actual measured toxicity. As explained earlier for the estimation of dioxin equivalents (section 7.2.4), the toxicity of each component of a mixture may be expressed relative to that of the most toxic component (toxic equivalency factor or TEF). Using TEFs as conversion factors, the concentration of each component can then be converted into toxicity units (toxic equivalents or TEQs), the summation of which gives the predicted toxicity for the whole mixture. Often, the estimated toxicity of mixtures of chemicals in environmental samples falls short of the measured toxicity. Both failure to detect certain toxic molecules (including natural xenobiotics), and limited availability of pollutants (for example, when adsorbed to sediments), can contribute to this underestimation of toxicity. Potentiation between pollutants can also contribute to the underestimation of toxicity (see Doi, Chapter 12 in vol. 2 of Calow, 1994). Sometimes mixtures of pollutants present in environmental samples are subjected to a fractionation procedure in an attempt to identify the main toxic components. By a process of elimination, toxicity can be tracked down to one or two fractions and compounds.

The advantages of combining toxicity testing with chemical analysis when dealing with complex mixtures of environmental chemicals are clearly evident. More useful information can be obtained than would be possible if one or the other were to be used alone. However, chemical analysis can be a very expensive matter, which places a limitation on the extent to which it can be used. There has been a growing interest in the development of new cost-effective biomarker assays for assessing the toxicity of mixtures. Of particular interest are bioassays that incorporate biomarker responses and are inexpensive, rapid and simple to use. These can be used alone or in combination with standard toxicity tests, and some of them give an indication of the types of pollutants responsible for toxic effects.

14.3 Shared mode of action – an integrated biomarker approach to measuring the toxicity of mixtures

A very large number of toxic organic pollutants – both man-made and naturally occurring – exist in the living environment. However, they express their toxicity through a much smaller number of mechanisms. Some of the more important sites of action of pollutants were described earlier (section 2.4). Thus, a logical approach to measuring or estimating the toxicity of mixtures of pollutants is to use appropriate biomarker assays for monitoring the operation of mechanisms of toxic action, and to relate this to the levels of individual chemicals of mixtures to which organisms are exposed (Peakall, 1992; Peakall and Shugart, 1993). Such an approach can provide an index of additive toxicity that takes into account potentiation of toxicity at the toxicokinetic level (Walker, 1998c). Biomarker assays of this type are both qualitative and quantitative; they identify a mechanism of toxic action and the degree to which it operates. Thus, they can provide an integrated measure of the overall effect of a group of compounds that operate through the same mode of action. When the mechanism of action is specific to a particular class of chemical, then it can be related to particular components of a mixture.

Four examples will now be given of mechanistic biomarker assays that can give integrative measures of toxic action by pollutants, all of which have been described earlier in the text. Where groups of pollutants share a common mode of action and their effects are additive, TEQs can, in principle, be estimated from concentrations. In these examples toxicity was thought to be simply related to the percentage of the total number sites of action that were occupied by the pollutants, and the toxic effects additive where two or more compounds of the same type were attached to the binding site.

1 The inhibition of brain cholinesterase is a biomarker assay for OPs and carbamate insecticides (section 10.2.4). Organophosphorous insecticides inhibit the enzyme by forming covalent bonds with a serine residue at the active centre. Inhibition is, at best, slowly reversible. The degree of the toxic effect depends upon the extent of cholinesterase inhibition caused by one or more OP or carbamate insecticides. In the case of OPs administered to vertebrates a typical scenario is as follows: sublethal symptoms begin to appear at 40–50% inhibition of cholinesterase, lethal toxicity above 70% inhibition.

2 The anticoagulant rodenticides warfarin and super warfarins are toxic because they have high affinity for a vitamin K binding site of hepatic microsomes (section 11.5). In theory, an ideal biomarker would measure the percentage of vitamin K binding sites occupied by rodenticides. However, the technology is not currently available to do that. On the other hand, the measurement of increases in plasma levels of undercarboxylated clotting proteins some time after exposure to rodenticide provides a good biomarker for this toxic mechanism.

3 Some hydroxy metabolites of coplanar PCBs, such as 4-OH-3,3′,4,5′-tetrachlorobiphenyl, act as antagonists of thyroxine (section 6.2.4). They have high affinity for the thyroxine binding site on TTR in plasma. Toxic effects include vitamin A deficiency. Biomarker assays for this toxic mechanism include the percentage of thyroxine binding sites to which rodenticide is bound, the plasma levels of thyroxine and the plasma levels of vitamin A.

4 Coplanar PCBs, PCDDs and PCDFs express Ah receptor-mediated toxicity (sections 6.2.4 and 7.2.4). Binding to the receptor leads to induction of cytochrome P4501 and a number of associated toxic effects. Again, toxic effects are related to the extent of binding to this receptor and appear to be additive, even with complex mixtures of planar polychlorinated compounds. Induction of P4501A1/2 has been widely used as the basis of a biomarker assay. Residue data can be used to estimate TEQs for dioxin (see section 7.2.4).

In addition to the foregoing, three further examples of toxic mechanisms (numbers 5–7) deserve consideration. These are (5) the interaction of endocrine disruptors with the oestrogen receptor, (6) the action of uncouplers of oxidative phosphorylation and (7) the mechanisms of oxidative stress. Until now only the first is well represented by biomarker assays that have been used in ecotoxicology.

5 The oestrogen receptor has been exploited in the development of biomarker assays for endocrine disruptors (EDs). The considerable range of biomarker assays (including bioassays) already developed is reviewed by Janssen *et al.* (1998). A surprisingly diverse range of chemicals can act as agonists or antagonists for the oestrogen receptor, producing 'feminising' or 'masculinising' effects. These include *o,p′*-DDE, certain PAHs, PCBs, PCDDS, PCDFs, alkylphenols and naturally occurring phyto- and myco-oestrogens. However, it should be borne in mind that (1) some EDs (e.g. *o,p′*-DDE, PCBs) probably act through their hydroxy-

metabolites, which bear a closer resemblance to natural oestrogens than do the parent compounds, and (2) others (e.g. alkyl phenols) are only very weak oestrogens.

   A number of biomarker assays have been developed for fish. Apart from a variety of non-specific end points such as organ weight, histochemical change, etc., vitellogenin synthesis has provided a specific and sensitive end point, which has been very useful for detecting the presence of environmental oestrogens at low concentrations. A number of different cell lines have been developed as bioassays for rapid screening of environmental samples. These include fish and bird hepatocytes, mouse hepatocytes, human mammary tumour cells and yeast cells (Janssen *et al.*, 1998). The end points include vitellogenin production, ED binding, the activation of glucosidase and the generation of light through the intermediacy of reporter genes and the elevation of mRNA levels. The diversity of the available bioassays reflects the high profile that endocrine disruptors have been given in recent years. Some of these assays are described in more detail in section 14.5.

6 Oxidative phosphorylation of ADP to generate ATP is a function of the mitochondrial inner membrane of animals and plants. Compounds that uncouple the process are general biocides, showing toxicity to animals and plants alike. For oxidative phosphorylation to proceed, a proton gradient must be built up across the inner mitochondrial membrane to provide the energy required to run the process (for a more detailed account see Nicholls, 1982). The maintenance of a proton gradient depends on the inner mitochondrial membrane remaining impermeable to protons. Most uncouplers of oxidative phosphorylation are weak acids that are lipophilic when in the undissociated state. Examples include the herbicides DNOC and dinitro secondary butyl phenol (dinoseb) and the fungicide pentachlorophenol (PCP). The proton gradients across inner mitochondrial membranes are built up by active transport, utilising energy from the electron transport chain that operates within the membrane. The gradient falls from outside the membrane to the inside (Figure 14.1). The dissociated forms ('conjugate bases') of the weak acids combine with protons on the outside of the membrane to form the undissociated lipophilic acids, which then dissolve in the membrane and diffuse across to the inside. Here, where the H^+ concentration is lower than on the outside of the membrane, they dissociate to release protons, and so act as proton translocators. They run down proton gradients, and hence 'uncouple' oxidative phosphorylation, dissipating the energy that would otherwise have driven ATP synthesis. The action of uncouplers can be studied in isolated mitochondria utilising an oxygen electrode to follow the rate of oxygen consumption in relation to the rate of NADH consumption (see Nicholls, 1982). Thus, the combined toxic action of mixtures of 'uncouplers' can be studied in isolated mitochondria. Such studies can be used to investigate the significance of tissue levels of mixtures of, for example, substituted phenols, which are found after animals have been exposed to them *in vivo*.

7 It has become increasingly apparent that the toxicity of certain compounds is due

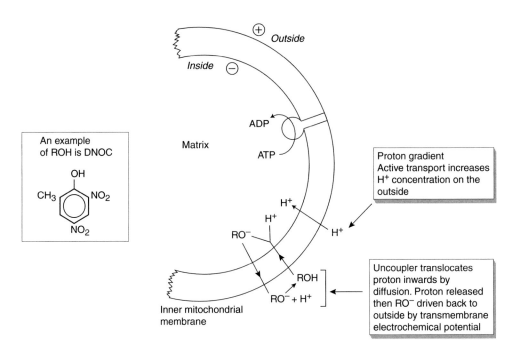

FIGURE 14.1 *Uncouplers of oxidative phosphorylation.*

to their ability to generate highly unstable oxyradicals such as the superoxide anion, $O_2\cdot^-$, and the hydroxyl radical, $OH\cdot$, as well as hydrogen peroxide, H_2O_2. These reactive species can cause cellular damage, e.g. lipid peroxidation and DNA damage, and have been implicated in certain disease states, e.g. atherosclerosis and certain forms of cancer (Halliwell and Gutteridge, 1986). Because they are so unstable, they are difficult or impossible to detect. Proof of their existence depends upon indirect evidence. The characteristic products of oxyradical attack (e.g. oxidised lipids, malonaldehyde from lipid peroxidation and oxidative adducts of DNA) and the induction of enzymes involved in their destruction (e.g. superoxide dismutase, catalase and peroxidase) can all provide evidence for the presence of oxyradicals and give some indication of their cellular concentrations.

These highly reactive species can be generated as a consequence of the presence of certain organic pollutants, e.g. bipyridyl herbicides and aromatic nitro compounds (Figure 14.2). Taking the examples of the herbicide paraquat (Hathway, 1984) and nitropyrene (Hetherington *et al.*, 1996), both can receive single electrons from reductive sources to form unstable free radicals. These radicals can then pass the electrons on to molecular oxygen to form the superoxide anion, with regeneration of the original molecule. Thus, a cyclic process is established, the net effect being to transfer electrons from a reductive source to oxygen to generate an oxyradical. Once formed, superoxide can undergo further reactions to form hydrogen peroxide and the highly reactive hydroxy radical. The toxicity of paraquat to plants and to animals is believed to be due, largely or entirely, to

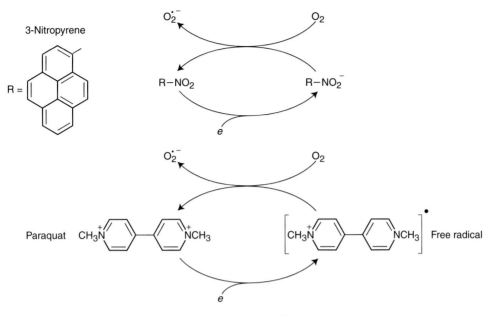

Figure 14.2 *Superoxide generation by 3-nitropyrene and paraquat.*

cellular damage caused by oxyradicals. In the case of plants, these radicals attack the photosynthetic system (see Hassall, 1990). In animals, toxic action is mainly against type I and type II alveolar cells, which take up the herbicide by a selective active transport system (see Timbrell, 1999).

There is evidence that mechanisms other than the production of free radicals of nitrogen-containing aromatic compounds are important in the generation of oxyradicals by pollutants. Refractory substrates for cytochrome P450, such as higher chlorinated PCBs, may cause the generation of the superoxide anion by occupying substrate binding sites of the haem protein not being oxidised by the activated oxygen produced by the haem nucleus. The unused activated oxygen is believed to escape from the domain of the cytochrome P450 in the form of superoxide to cause oxidative damage elsewhere in the cell.

At the time of writing, the toxicity of oxyradicals generated by the action of pollutants is highly topical because of the relevance to human diseases. It is not an easy subject to investigate because of the instability of the radicals and the different mechanisms by which they may be generated. It is hoped that rapid progress will be made, so that monitoring the effects of oxyradicals will make an important contribution to the growing armoury of mechanistic biomarkers for the study of environmental effects of organic pollutants.

Taking an overall view of the foregoing examples, the first five all involve interaction between organic pollutants and well-defined sites on proteins, one of which is the active site of an enzyme, the others 'receptors' to which chemicals bind to produce toxicological effects. Knowledge of the structures and properties of receptors facilitates the development of QSAR models for pollutants, where toxicity can be predicted from chemical parameters (Box 15.1). Indeed, new pesticides are sometimes designed on the basis of such models, e.g. some EBI fungicides that can lock into the catalytic site of P450 have been discovered in this way. Interactions such as these are essentially similar to the interaction of agonists and/or antagonists with receptors in pharmacology.

The last two examples do not belong in the same category, there being no clearly defined single binding site on a protein. Uncouplers of oxidative phosphorylation operate across the inner mitochondrial membranes, their critical properties being the ability to reversibly interact with protons, and their existence in the uncharged lipophilic state after protons are bound. Oxyradicals can, in principle, be generated by a variety of redox systems that are able to transfer single electrons to oxygen under cellular conditions. The systems that carry out one-electron reduction of nitroaromatic compounds and aromatic amines have yet to be properly elucidated. Neither of these mechanisms of toxic action is susceptible to the kind of QSAR analysis described above, in which the detailed stereo structure of particular binding sites are known.

When chemicals have toxic effects, the initial molecular interaction between the chemical and its site of action (receptor, membrane, redox system, etc.) is followed by a sequence of changes at the cellular and whole-organism levels that lead to the appearance of overt symptoms of intoxication. Biomarkers that measure the changes at higher levels of organisation, e.g. the release of stress proteins, damage to cellular organelles, disturbances to the nervous system or endocrine system can, in principle, provide integrated measures of the effects of diverse chemicals in a mixture. They can measure the combined effects of chemicals working through different modes of action, if these modes of action have the same higher-level effect. For example, two chemicals may act on different receptors in the nervous system, but they may both produce similar disturbances, e.g. tremors, hyperexcitability, even certain changes in the EEG pattern. Moving from the primary toxic lesion to the knock-on effects at higher levels of organisation, the higher one goes, the harder it becomes to relate effects to particular mechanisms of toxic action. This is one advantage of using combinations of biomarkers rather than single biomarker assays when investigating toxic effects of mixtures of dissimilar compounds. A higher-level biomarker (e.g. scope for growth in molluscs, or behavioural effects in vertebrates) can give an integrated measure of the toxicity of the mixture of chemicals, without providing any evidence of causation; the mechanistic biomarker (e.g. any of the first five listed above) gives a measure of a specific biochemical effect that can be related to a particular type of compound. Taken together, these two types of biomarker can give an 'in-depth' picture of the sequence of adverse events following exposure to the mixture.

The account thus far has been concerned primarily with mechanistic biomarker responses in living organisms. In the next section the discussion will move on to the exploitation of this principle in the development of bioassay systems which can be used in environmental risk assessment.

14.4 *Bioassays for toxicity of mixtures*

Both cellular systems and genetically manipulated microorganisms have been used to measure the toxicity of individual compounds and of mixtures present in environmental samples, e.g. of water, soil and sediment. Such bioassays can have the advantages of being simple, rapid and inexpensive to run. They can provide evidence for the existence in environmental samples of chemicals with toxic properties, acting either singly or in combination. Some of them provide measures of the operation of certain modes of action, thus giving evidence of the types of compounds responsible for toxic effects; simple bioassays that use broad indications of toxicity such as lethality or reduction of growth rate as end points do not do this. The shortcoming of bioassay systems is the difficulty of relating the toxic responses that they measure to the toxic effects that would be experienced by free-living organisms if exposed to the same concentrations of chemicals in the field. These simple systems do not reproduce the complex toxicokinetics of living vertebrates and invertebrates. As explained in Chapter 2, toxicokinetic factors are determinants of toxicity, and there can be very large differences in toxicity between species because of metabolic differences. With persistent pollutants this problem may be partially overcome by conducting bioassays upon tissue extracts, but even here there are complications. How closely does the use of an extract reproduce the actual cellular concentrations at the site of action in the living animal? How similar are the toxicodynamic processes between the test system and the living animal? The site of action may very well differ in the two cases. It is clear from many examples of resistance to pesticides that a difference of just one amino acid residue of a target protein can profoundly change the affinity for the pesticide, and consequently the toxicity (see section 2.5 and various examples in Chapters 5–14).

Notwithstanding these complications, bioassay systems have considerable potential for biomonitoring and environmental risk assessment. By giving a rapid indication of where toxicity exists, they can pave the way for the use of more sophisticated methods of establishing cause and effect, including chemical analysis and biomarker assays on living organisms. In the context of biomonitoring, they are useful for checking the quality of surface waters and effluents, and for giving early warning of pollution problems. In these respects they have considerable advantages over chemical analysis. They can be very much cheaper and, because chemical analysis is not comprehensive, they can measure the toxicity of compounds that escape detection in the chemical laboratory.

A number of bioassays utilise microorganisms. Some, such as the Microtox test system, give a non-specific measure of toxicity. This system utilises the bioluminescent marine organism *Vibrio fischeri*, which emits light because of the action of the enzyme luciferase (see Calow, 1994). Toxicity is measured by the degree of inhibition of light. A more specific type of test is the bacterial mutagenicity assay, the best-known example of which is the Ames test (Maron and Ames, 1983). This type of test has been widely used in the chemical industry to screen pesticides and drugs for mutagenic properties. In the Ames test, histidine-dependent strains of the bacterium *Salmonella typhimurium* are exposed to individual chemicals or mixtures. Mutation is shown by a loss of histidine dependence, and the mutation rates are related to dose. An important feature of the Ames test is that it incorporates a metabolic activation system; usually a preparation of mammalian hepatic microsomes with high monooxygenase activity. Thus, a distinction can be made between pollutants that are themselves mutagenic, and others that require metabolic activation by the P450 system.

A number of mammalian and fish cell lines have been used to test for toxicity, some of them measuring particular mechanisms. Bioassay systems have been developed which test for Ah receptor-mediated toxicity (section 7.2.4). Some cell hepatoma lines, e.g. from mice, contain the Ah receptor, and cells of this type have been transfected with reporter genes (Garrison *et al.*, 1996). An example is the CALUX system, where interaction of coplanar PCBs, dioxins, etc., with the Ah receptor of hepatoma cells triggers the synthesis of luciferase and consequent light emission. The degree of occupancy of the Ah receptor by these compounds determines the quantity of light that is emitted. Thus, the CALUX system can give an integrated measure of the effects of mixtures of polyhalogenated compounds on the Ah receptor, and an indication, therefore, of the potential of such mixtures to cause Ah receptor-mediated toxicity.

In another example, fish hepatocyte lines have been used to detect the presence of environmental oestrogens. Primary cultures of rainbow trout hepatocytes containing the oestrogen receptor can show elevated levels of vitellogenin when exposed to environmental oestrogens (Sumpter and Jopling, 1995). Assays with this system, together with assays for vitellogenin production in caged male fish, have demonstrated the presence of oestrogenic activity at sewage outfalls. Subsequent investigation established that much of the oestrogenic activity was due to natural oestrogens in sewage, but there was also evidence that nonyl phenols derived from detergents had an oestrogenic effect in a highly polluted stretch of river. The oestrogen receptor is responsive to a number of environmental compounds, including organochlorine compounds such as dicofol and o,p'-DDT, nonyl phenols (rather weak) and naturally occurring phytoestrogens (IEH Assessment, 1995). Once again, an assay system that is mechanistically based can give an integrated measure of the adverse effects of mixtures of environmental chemicals.

Fish hepatocyte lines have also been developed that can show cytochrome P4501A1 induction due to PAHs and planar polychlorinated aromatic compounds binding to the Ah receptor (Vaillant *et al.*, 1989; Pesonen *et al.*, 1992).

Apart from the scientific advantages offered by this new technology, it has also been welcomed by organisations seeking a reduction in the number of animals used in toxicity testing (see section 15.6 and Walker, 1998b).

14.5 *Potentiation of toxicity in mixtures*

The problem of potentiation is discussed in section 2.6. Potentiation is often the consequence of interactions at the toxicokinetic level, especially inhibition of detoxication or increased activation. The consequences of such potentiation may be evident not only at the whole animal level but also in enhanced responses of biomarker assays that measure toxicity (Figure 14.3). Biomarkers of exposure alone should not reflect potentiation at the toxicokinetic level.

The real problem about potentiation is anticipating where it may occur, when the only available toxicity data are for the individual compounds that will constitute the mixture. This is a frequent issue in the regulation of pesticides. When should the use of new mixtures of old pesticides be approved? When considering the toxicity of mixtures, it tends to be assumed that chemicals will interact in an additive fashion, unless there is clear evidence to the contrary, an approach that has worked out reasonably well in practice. However, there are important exceptions (section 2.6 and Chapters 10 and 12). Full consideration should be given to known mechanisms of potentiation when questions are raised about the possible toxicity of mixtures. Where there is an apparent risk on good mechanistic grounds, tests should be carried out to establish the toxicity of the mixtures in question. In this way the very limited resources available for testing mixtures should be targeted on the most important cases. With the very rapid growth of understanding of the mechanistic basis of toxicity, it should become increasingly possible to anticipate where substantial potentiation of toxicity will occur. In this field, there is no substitute for expert knowledge. The resources do not exist for any general statutory requirement for the toxicity testing of mixtures of industrial chemicals that may be released into the environment. Even if such resources did exist, such an exercise would be very largely a waste of time, because substantial potentiation of toxicity in mixtures is a rare event.

As understanding grows of biochemical mechanisms that lead to strong potentiation of toxicity, more attention should be given to the release of compounds which are not toxic in themselves, but which may increase the toxicity of pollutants already present in the environment. Recognition of the synergistic action of the P450 inhibitor

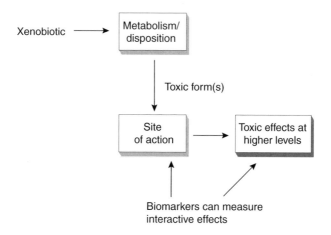

FIGURE 14.3 *Biomarkers of toxic effect.*

piperonyl butoxide has made it unlikely that insecticide formulations containing it would be approved for use on food crops. Another example is the EBI fungicides, which can potentiate the toxicity of pyrethroid and phosphorothionate insecticides (section 2.6). There may be situations in which their use in agriculture increases the hazards presented by certain commonly applied insecticides to wildlife.

14.6 *Summary*

Most statutory toxicity testing is carried out on individual compounds. In the natural environment, however, organisms are exposed to complex mixtures of pollutants. Toxicity testing procedures are described for environmental samples that contain mixtures of different chemicals.

Particular attention is given to the development of new mechanistic biomarker assays and bioassays that can be used as indices of the toxicity of mixtures. These biomarker assays are based upon toxic mechanisms such as brain AChE inhibition, vitamin K antagonism, thyroxine antagonism, Ah receptor-mediated toxicity and interaction with the oestrogenic receptor. They can give integrative measures of the toxicity of mixtures of compounds that share the same mode of action, and they can give evidence of potentiation as well as additive toxicity. Bioassays can be used for rapid screening of environmental samples to detect the presence of mixtures of toxic chemicals.

14.7 *Further reading*

Fossi, M. C. and Leonzio, C. (1994) *Non-destructive Biomarkers in Vertebrates*. A collection of detailed reviews of non-destructive biomarkers.

Janssen, P. A. H., *et al.* (1998) The authors review biomarker assays and bioassays for EDs in reasonable detail.

Peakall, D. B. (1992) *Animal Biomarkers as Pollution Indicators*. An authoritative text on biomarkers.

Walker, C. (1998b) gives a broad review of biomarker assays in ecotoxicology.

CHAPTER 15

The environmental impact of organic pollutants: future prospects

15.1 *Introduction*

During the second half of the twentieth century it was discovered that a number of organic pollutants had harmful side-effects in natural ecosystems, prominent among them chemicals that combined high toxicity (lethal or sublethal) with marked biological persistence. Examples such as dieldrin, DDT, TBT and methyl mercury constitute a major part of the second section of the present text. After these discoveries, restrictions and bans upon their release into the environment were introduced in many countries. Persistent organochlorine insecticides, for example, were withdrawn from many uses and replaced by less persistent OP and carbamate insecticides. More stringent legislation was brought in to control the production and marketing of new chemicals, with clearer guidelines for environmental risk assessment. Particularly strong rules were applied to new pesticides – something of a double-edged weapon. More stringent regulations may be expected to reduce the risk of new persticides causing further problems, but they also make the discovery and registration of newer, more environmentally friendly compounds a more costly and time-consuming business. Thus, they can discourage the development of more environmentally safe pesticides than those currently in use.

Looking ahead, two major issues present themselves.

1 What changes may be expected in environmental risk assessment practices to evaluate new pesticides and industrial chemicals?
2 What improvements are likely in the techniques and strategies used to investigate existing complex pollution problems?

As explained earlier in the text (section 4.5), the central concern is about effects at the level of population or above, but this can be very difficult to establish, let alone to predict. These issues will be discussed in the sections that follow.

15.2 *The design of new pesticides*

It is not surprising that many of the organic pollutants that have caused serious environmental problems have been pesticides. Pesticides, after all, are designed with a view to causing damage to pests, and selectivity between pests and other organisms can only be achieved to a limited degree. In designing new pesticides, manufacturers seek to produce compounds of greater efficacy, cost effectiveness and environmental safety than are offered by existing products (for an account of the issues involved in the development of new safer insecticides see Hodgson and Kuhr, 1990). Sometimes the driving force behind pesticide innovation is to overcome a developing resistance problem, where existing products are becoming ineffective against major pests. It may also be to provide a product that is more 'environmentally safe' than those currently on the market. Innovation, however, is to some extent hampered by escalating costs, not least the cost associated with ecotoxicity testing and environmental risk assessment.

With the rapid growth of knowledge in the field of biochemical toxicology, it is becoming increasingly possible to design new pesticides based upon structural models of the site of action – the QSAR approach. Sophisticated computer graphic systems make life easier for the molecular modeller. The discovery and development of EBI fungicides as inhibitors of certain forms of P450 provide an example of the successful application of this approach.

There has also been rapid growth in understanding of the enzyme systems that metabolise pesticides and other xenobiotics (see Chapter 2 and Hutson and Roberts, 1999). As more is discovered about the mechanisms of catalysis by P450-based monooxygenases, esterases, glutathione-S-transferases, etc., so it becomes easier to predict the routes and rates of metabolism of pesticides. In theory, it should become easier to design readily biodegradable pesticides that have better selectivity than existing products (there are large species differences in metabolism that can be exploited!). It should also be possible to design pesticides that are selectively toxic towards resistant strains.

Another ongoing interest is to identify more naturally occurring compounds that

BOX 15.1 *Quantitative structure–activity relationships.*

There has long been an interest in mathematical relationships between chemical structure and toxicity, and the development of models from them that can be used to predict the toxicity of chemicals (see Donkin, Chapter 14, in vol. 2, Calow 1994). If considering groups of compounds that share the same mode of action, much of the variation in toxicity between different molecules is related to differences in cellular concentration when the same dose is given. In other words, toxicokinetic differences are of primary importance in determining selective toxicity (see section 2.3). The simplest situation is represented by non-specific narcotics, which include general anaesthetics. Toxicity here is related to the (relatively high) concentrations that the compounds reach in biological membranes and is not due to any specific interaction with cellular 'receptors' (see section 2.4). Simple models can relate chemical properties to both cellular concentration and toxicity. Good QSARs have been found for narcotics when using descriptors for lipophilicity such as log K_{ow} values. For example, the following equation relates the hydrophobicity of members of a group of aliphatic, aromatic and alicyclic narcotics to their toxicity to fish.

$$\log 1/LC_{50} = 0.871 \log K_{ow} - 4.87 \text{ (Könemann, 1981)}$$

Other much more toxic compounds operating through specific biochemical mechanisms (e.g. OP anticholinesterases) cannot be modelled in this way. If toxicity were to be plotted against log K_{ow}, such compounds would be represented as 'outliers' in relation to the straight line provided by the data for the narcotics (Lipnick, 1991). Their toxicity would be much greater than predicted by the simple 'hydrophobicity' model for the narcotics. For such compounds more sophisticated QSAR equations are required which bring in descriptors for chemical properties relating, for example, to their ability to interact with a site of action. An example of such an equation relates the properties of OPs to their toxicity to bees (Vighi *et al.*, 1991).

$$\log 1/LD_{50} = 1.14 \log K_{ow} - 0.28 (\log K_{ow})^2 + 0.28^2 \Xi - 0.76^2 \Xi_{ox}$$
$$- 1.09 \Psi 3 + 0.096 (\Psi 3)^2 + 12.29$$

where Ξ and Ψ are chemical descriptors for reactivity with the active site of cholinesterase. The K_{ow} is for the active 'oxon' form of an OP.

In general, it is easier to use models such as these to predict the distribution of chemicals (i.e. relation between exposure and tissue concentration) than it is to predict their toxic action. The relationship between tissue concentrations and toxicity is not straightforward for a diverse group of compounds, and depends on the mode of action. Even with distribution models, however, the picture can be complicated by species differences in metabolism, as in the case of models for bioconcentration and bioaccumulation (see Chapter 4). Rapid metabolism can lead to lower tissue concentrations than would be predicted from a simple model based on K_{ow} values. Thus, such models need to be used with caution when dealing with different species.

act as pesticides (Hodgson and Kuhr, 1990). These may be useful as pesticides in their own right, or they may serve as models for the design of new products. Examples of natural products that have already been of interest from this point of view include pyrethrins, nicotine, rotenone, plant growth regulators, insect juvenile hormones, precocene and extracts of the seed of the neem tree (*Azadirachta indica*) (see Hodgson and Kuhr, 1990; Otto and Weber, 1992). It is probable that natural products will continue to be a rich source of new pesticides or models for new insecticides in the years ahead. A vast array of natural chemical weapons have been produced during the evolutionary history of the planet, and many are still waiting to be discovered (Chapter 1).

15.3 *The adoption of more ecologically relevant practices in ecotoxicity testing*

Currently, the environmental risk assessment of chemicals for registration purposes depends on the comparison of two things: (1) an estimate (sometimes a measure) of environmental concentration of the chemical; and (2) an estimate of 'environmental toxicity'. Environmental concentration is difficult to estimate, especially for mobile species of terrestrial ecosystems. The estimation of environmental toxicity may be based on a no observable effect concentration (NOEC) or LC_{50} for the most sensitive organism found in a series of ecotoxicity tests. For further information on these issues see Chapter 6 in Walker *et al.* (2000), Calow (1994) and Walker (1998b). Because of the high levels of uncertainty involved, the estimate of environmental toxicity is divided by a large safety factor, commonly 1000. If the estimate of environmental toxicity (2) is larger than the estimate of environmental concentration of the chemical (1) there is perceived to be a risk.

The limitations of this approach are not difficult to see (see, for example, Kapustka *et al.*, 1996). It is based on the approach to risk assessment used in human toxicology and has been regarded as the best that can be done with existing resources. It is concerned with estimating the likelihood that there will be a toxic effect upon a sensitive species after the release of a chemical into the environment. With the very large safety factors that are used, it may well seriously overestimate the risks presented by some chemicals. More fundamentally, it does not address the basic issue of effects upon populations, communities or ecosystems. Small toxic effects may be of no significance at these higher levels of biological organisation, where population numbers are often controlled by density-dependent factors (Chapter 4). Also, it does not deal with the question of indirect effects. As mentioned earlier (Chapter 13), standard environmental risk assessment of herbicides would have given no indication that they could be the indirect cause of the decline of the grey partridge on agricultural land.

There has been growing pressure from biologists for the development of more ecologically relevant end points when carrying out toxicity testing for the purposes of environmental risk assessment (see Walker *et al.*, 1998; and Chapter 12 in Walker *et al.*, 2000). In concept, populations will decline when pollutants, directly or indirectly, have a sufficiently large effect on rates of mortality and/or rates of recruitment to reduce population growth rate (section 4.4). Thus, sublethal effects, e.g. on reproduction or behaviour, can be more important than lethal ones. If pollutant effects can be quantified in this way, for example through the use of biomarker assays for toxic effect (see section 15.4), then better risk assessment is made possible by including them in appropriate population models. In practice, this approach is still at an early stage of development; it is a research strategy that can only be used in a few cases and cannot yet deal with the large numbers of compounds submitted for risk assessment. Nevertheless, looking at the problem from this more fundamental point of view does suggest certain improvements that could quickly be made in the protocols for environmental risk assessment.

First, a large proportion of the resources currently being spent on the precise determination of LD_{50} values for birds or LC_{50} values for fish could be diverted to more relevant testing procedures. At best, these values give only a rough indication of lethal toxicity in a small number of species. A rough ranking of compounds with respect to toxicity, e.g. low toxicity, moderate toxicity, etc., is good enough for such a crude and empirical approach; knowing particular values a little more precisely does practically nothing to improve the quality of environmental risk assessment. In the first place, greater consideration of ecological aspects before embarking on testing should lead to the selection of more appropriate species, life stages and end points in the testing protocol. It might be sensible, for example, to include tests on behavioural effects if testing neurotoxic pesticides, or of reproductive effects if testing a compound that can disturb steroid metabolism. These are mechanisms that, on the basis of experience, might be expected to have adverse ecological effects. In a number of instances, population declines have been the consequence of reproductive failure (e.g. the effects of p,p'-DDE in shell thickness of raptors, the effects of PCBs and other polychlorinated compounds on reproduction of fish-eating birds in the Great Lakes and the effects of TBT on the dog whelk). Effects on behaviour may affect breeding and feeding.

In some species there may be good reasons for looking at the toxicity of certain types of compounds to early developmental stages (e.g. avian embryos, larval stages of amphibians) rather than adults. In short, testing protocols should be more flexible, so that there can be a greater opportunity for expert judgement, rather than following a rigid set of rules. Knowledge of the metabolism and the mechanism of action of a new chemical may suggest the most appropriate end points in toxicity testing. Indeed, mechanistic biomarkers can provide better and more informative end points than lethality; they can be used to monitor progression through sublethal (including subclinical) effects before lethal tissue concentrations are reached.

An approach that has gained much interest recently is the use of model ecosystems, microcosms, mesocosms and macrocosms, for testing chemicals (section 4.6). In these, replicated and controlled tests can be carried out to establish the effects of chemicals upon the structure and function of the (artificial) communities that they contain. The major problem is relating effects produced in mesocosms to events in the real world (see Crossland, 1994). Nevertheless, it can be argued that mesocosms do incorporate certain relationships (e.g. predator/prey) and processes (e.g. carbon cycle) that are found in the outside world, and they test the effects of chemicals on these things. Once again, the judicious use of biomarker assays during the course of mesocosm studies may help to relate effects of chemicals in them to similar effects in the natural environment.

15.4 *The development of more sophisticated methods of toxicity testing: mechanistic biomarkers*

Mechanistic biomarkers can, in theory, overcome many of the basic problems associated with establishing causality. In the field they can be used to measure the extent to which pollutants act upon wild species through defined toxic mechanisms, thus giving more insight into the sublethal as well as the lethal effects of chemicals. Most importantly, they can provide measures of the integrated effects of mixtures of compounds operating through the same mechanism, measures that take into account potentiation at the toxicokinetic level (section 14.4). In theory, they can provide the vital link between known levels of exposure and changes in mortality rates or recruitment rates; estimates of mortality rates or recruitment rates can then be incorporated into population models (section 4.4). Graphs can be generated that link a biomarker response to a population parameter, as has already been achieved with eggshell thinning in the sparrowhawk induced by p,p'-DDE and imposex in the dog whelk caused by TBT (Figure 4.4). The current problem is that there are too few suitable biomarker assays. At present, this approach lies in the realm of research and cannot be applied to most problems with environmental chemicals.

At the practical level, an ideal mechanistic biomarker should be simple to use, sensitive, relatively specific, stable and useable on material (e.g. blood, skin biopsies) that can be obtained by non-destructive sampling. A tall order! And no biomarker yet developed has all of these attributes. However, the judicious use of combinations of biomarkers can overcome the shortcomings of individual assays. The main point to emphasise is that the resources so far invested in the development of biomarker technology for environmental risk assessment has been very small (cf. the investment in biomarkers for use in medicine). Knowledge of toxic mechanisms of organic pollutants is already substantial (especially of pesticides), and it grows apace. The scientific basis is already there for technological advance; it comes down to a question of investment.

As mentioned earlier, the development of bioassay techniques is one important aspect of biomarker technology. Cell lines have been developed for species of interest in ecotoxicology, e.g. of birds and fish, and have sometimes been genetically manipulated (e.g. with incorporation of receptors and reporter genes) to facilitate their employment as biomarker assays (Walker, 1998b). In principle, it should be possible to conserve the activities of enzymes concerned with detoxication and activation in these cellular systems, so that the toxicokinetics of the *in vitro* assay bear some resemblance to those in the living animal. Bioassays with such cellular systems could be developed for species of ecotoxicological interest that are not available for ordinary toxicity testing. Also, they could go some way to overcoming the fundamental problem of interspecies differences in toxicity. One difficulty encountered with cell lines has been that of gene expression. Enzymes concerned with detoxication or activation have sometimes not been expressed in cell systems. However, recent work with genetically manipulated cell lines has begun to overcome this problem (Glatt *et al.*, 1997).

15.5 *Field studies*

Ecotoxicology is primarily concerned with effects of chemicals on populations, communities and ecosystems, but the trouble is that field studies are expensive and difficult to perform, and can only be used to a limited degree. To a large extent, the risk assessment of chemicals has to be accomplished by other means. With the registration of pesticides, field studies are occasionally carried out to resolve questions that turn up in normal risk assessment (Somerville and Walker, 1990), but are far too expensive and time-consuming to be used with any regularity. Lack of control of variables and the difficulty of achieving adequate replication are fundamental problems. However, the development of new strategies, and the development of new biomarker assays could pave the way for more informative and cost-effective investigations of the effects of pollutants in the field.

The use of biotic indices in environmental monitoring is one way of identifying existing/developing pollution problems in the field (see Chapter 11 in Walker *et al.*, 2000). Such ecological profiling can flag up structural changes in communities that may be the consequence of pollution. For example, the RIVPACS system can identify changes in the macroinvertebrate communities of freshwater systems (Wright, 1995). It is important that adverse changes found during biomonitoring are followed up by the use of biomarker assays (indicator organisms and/or bioassays) and chemical analysis to identify the cause. As noted above, improvements in biomarker technology should make this task easier and cheaper to perform.

Biomarker assays can be used to establish the relationship between the levels of chemicals present and consequent biological effects both in controlled field studies (e.g. field trials with pesticides) and in the investigation of the biological consequences of existing or developing pollution problems in the field. In the latter case, clean organisms can be deployed to both 'clean' and polluted sites in the field and biomarker responses can be measured in them. Organisms can be deployed along pollution gradients, so that dose–response curves can be obtained for the field as well as in the laboratory, and the two compared. An example of the adoption this approach was the deployment of *M. edulis* along PAH gradients in the marine environment and the measurement of scope for growth (section 9.6). The challenge here is to take the further step and relate biomarker responses to population parameters, so that predictions of population effects can be made using mathematical models. The predictions from the models can then be compared with the actual state of the populations in the field. The validation of such an approach should lead to its wider use in the general field of environmental risk assessment.

15.6 *Ethical questions*

There has been growing opposition to the use of vertebrate animals for toxicity testing. This has ranged from the extremism of some animal rights organisations to the reasoned approach of the Fund for the Replacement of Animals in Medical Experiments (FRAME), and the European Centre for the Validation of Alternative Methods (ECVAM) (see Balls *et al.*, 1991; issues of the journal ATLA and publications of ECVAM at the Joint Research Centre, Ispra, Italy). FRAME, ECVAM and related organisations advocate the adoption of the principles of the three Rs (Van Zutphen and Balls, 1997), namely the reduction, refinement and replacement of testing procedures that cause suffering to animals.

Regarding ecotoxicity testing, these proposals gain some strength from the criticisms raised earlier to existing practices in environmental risk assessment. There is a case for making testing procedures more ecologically relevant, and this goes hand in hand with attaching less importance to crude measures of lethal toxicity in a few species of birds and fish (Walker, 1998b). The savings made by a substantial reduction in the numbers of vertebrates used for 'lethal' toxicity testing could be used for the development and subsequent use of testing procedures that do not cause suffering to animals and are more ecologically relevant. Examples include sublethal tests (e.g. on behaviour or reproduction), tests involving the use of non-destructive biomarkers, the use of eggs for testing certain chemicals and the improvement of tests with mesocosms. Rigid adherence to fixed rules would prolong the use of unscientific and outdated practices, and slow down much needed improvements in techniques and strategies for ecotoxicity testing. Better science should, for the most part, further the requirements of the three Rs.

15.7 *Summary*

With improvements in scientific knowledge and related technology there is an expectation that more environmentally friendly pesticides will continue to be introduced, and that ecotoxicity testing procedures will become more sophisticated. There is much interest in the introduction of better testing procedures that work to more ecologically relevant end points than the lethal toxicity tests that are still widely used. Such a development should be consistent with the aims of organisations such as FRAME and ECVAM, which seek to reduce toxicity testing with animals. Mechanistic biomarker assays would be an important part of this approach. They have potential for use in field studies, providing the vital link between exposure to chemicals and consequent toxicological and ecotoxicological effects.

15.8 *Further reading*

New developments are best followed by reading current issues of the leading journals in the field, which include *Environmental Toxicology and Chemistry, Ecotoxicology, Environmental Pollution, Environmental Health Perspectives, Bulletin of Environmental Contamination and Toxicology, Archives of Environmental Contamination and Toxicology, Functional Ecology, Applied Ecology* and *Biomarkers*.

AChE: Acetylcholinesterase.

Adducts: In the context of toxicology, products of stable linkages between xenobiotics and endogenous molecules, e.g. between PAH metabolites and DNA.

Ah receptor (aryl hydrocarbon receptor): A receptor located on a cytoplasmic protein to which bind planar compounds such as PAHs, coplanar PCBs and PCDDs. Binding initiates the induction of cytochrome P4501A1/2.

Ah receptor-mediated toxicity: Toxic effects associated with the binding of polychlorinated aromatic compounds, such as coplanar PCBs and PCDDs, to the Ah receptor.

Alkaloids: A diverse group of nitrogen-containing organic compounds synthesised by plants, many of which show biological activity.

Antagonism: With reference to toxicity, when the toxicity of a mixture is less than the sum of the toxicities of its components.

Anthropogenic: Generated by the activities of man.

Anticoagulant rodenticides (ARs): Rodenticides that cause haemorrhaging, usually through disturbing the synthesis of clotting proteins, e.g. warfarin, brodifacoum, difenacoum.

Aryl: Aromatic moiety.

ATPases: Adenosine triphosphatases.

Bioaccumulation factor (BAF): Concentration of a chemical in an animal/concentration of same chemical in its food.

Bioconcentration factor (BCF): Concentration of a chemical in an organism/concentration of same chemical in the ambient medium.

Biomagnification: Increase in concentration of a chemical in living organisms with passage along a food chain.

Biomarker: A biological response to an environmental chemical at the individual level or below demonstrating a departure from normal status.

Biotransformation: Conversion of a chemical into one or more products by a biological mechanism (predominantly by enzyme action).

Carbanion: Chemical moiety bearing a negative charge on a carbon atom.

Carbene: Free radical with two unpaired electrons on a carbon atom.

Carbonium ion: Chemical moiety bearing a positive charge on a carbon atom.

Carboxylesterases: Esterases that hydrolyse organic compounds with carboxylester bonds. Carboxylesterases that are inhibited by OPs belong to the category EC 3.1.1.1 in the IUB classification of enzymes.

Carcinogen: A substance able to cause cancer.

ChE (cholinesterase): A general term for esterases that hydrolyse cholinesters.

Cholinergic: Associated with the neurotransmitter acetylcholine.

Congener: A member of a group of structurally related compounds.

Conjugate: In biochemical toxicology, a structure (often an anion) formed by the combination of a xenobiotic (usually a phase I metabolite) with an endogenous component, e.g. glucuronate sulphate or glutathione.

Coordination: In chemistry, the donation of electrons by one atom to another in bond formation.

Cyclodienes: A group of organochlorine insecticides some of which are highly toxic and persistent, e.g. aldrin, dieldrin and heptachlor.

Cytochrome P450: A haem protein that catalyses many biological oxidations (see also microsomal monooxygenases).

p,p'-DDE: p,p'-Dichlorodiphenyldichloroethylene (stable metabolite of p,p'-DDT).

p,p'-DDT: p,p'-Dichlorodiphenyltrichloroethane (main insecticidal component of the insecticide DDT).

EBI fungicides: Ergosterol biosynthesis inhibitors used as fungicides.

$EC(D)_{50}$: Concentration (dose) that has an effect upon 50% of a population. Also known as median effect concentration (dose).

Electrophile: An electron-seeking atom or group.

Endocrine disruptors: Chemicals that cause disturbances of the endocrine system, e.g. by acting as agonists or antagonists at the oestrogen receptor.

Endoplasmic reticulum: Membranous network within cells which contains many enzymes that metabolise xenobiotics. Hepatic microsomes consist mainly of vesicles derived from the endoplasmic reticulum of liver.

Epoxide hydrolase: A type of enzyme that converts epoxides to diols by the addition of water.

Ester: An organic salt which yields an acid and a base when hydrolysed.

Esterases: Enzymes that hydrolyse esters.

Eukaryotes: Organisms that contain DNA within their nuclei.

Free radical: A molecule or atom possessing an unpaired electron.

Fugacity: Tendency of a chemical to escape from the phase in which it is located into another phase (e.g. from liquid into gas).

GABA: Gamma-amino butyric acid, a neurotransmitter. Acts upon GABA receptors located especially in the nervous system.

Genotoxic: Toxic by acting upon genetic material, especially DNA.

Gla proteins: Proteins containing residues of gamma carboxy glutamate, including clotting proteins of the blood. They are able to bind calcium ions.

Glucuronyl transferases: A group of enzymes that catalyse the formation of conjugates between glucuronide and a xenobiotic (usually a phase I metabolite).

Glutathione-S-transferases: A group of enzymes that catalyse the formation of conjugates between reduced glutathione and xenobiotics.

Hydrophilic: 'Water-loving'. Polar organic compounds tend to be hydrophilic.

Hydrophobic: 'Water hating'. Non-polar organic compounds are hydrophobic.

Immunotoxicity: Toxicity to the immune system.

Imposex: The imposition of male characteristics upon females in prosobranch molluscs, e.g. the dog whelk (*Nucellus lapillus*).

Induction: With reference to enzymes, an increase in activity due to an increase in their cellular concentrations. This may be a response to a xenobiotic and often involves an increased rate of synthesis of the enzyme.

Isoenzymes (isozymes): Enzymes that are very similar to one another in size and structure but with differences in catalytic ability.

K_{ow}: Octanol/water partition coefficient.

Ligand: A compound with specific binding properties.

Ligandin: A form of glutathione-S-transferase with a marked capacity for binding certain lipophilic xenobiotics.

Lipophilic: 'Lipid-loving'. Such organic compounds tend to be of low polarity and are hydrophobic.

Lipoproteins: Macromolecules that are associations of lipids with proteins. Involved in the transport of both lipids and lipophilic xenobiotics in the blood.

Microcosm, mesocosm and macrocosm: 'Small', 'medium' or 'large' multispecies system in which effects of chemicals can be studied.

Microsomes: Vesicles obtained from homogenised tissues by ultracentrifugation. They are derived mainly from the endoplasmic reticulum in the case of the liver ('hepatic microsomes').

Mitochondrion: A subcellular organelle in which oxidative phosphorylation occurs, leading to the generation of ATP.

Monooxygenases (MOs): Enzyme systems of the endoplasmic reticulum of many cell types which can catalyse the oxidation of a great diversity of lipophilic xenobiotics, and are particularly well developed in hepatocytes. Forms of cytochrome P450 constitute the catalytic centres of monooxygenases.

Neuropathy target esterase (NTE): An esterase of the nervous system whose inhibition by certain OPs (e.g. mipafox, leptophos) can lead to the development of delayed neuropathy.

Neurotransmitter: Endogenous substance involved in the transmission of nerve impulses.

Nucleophile: An atom or group that seeks a positive charge.

NOE(C)D: No observed effect concentration or dose.

Oxyradical: An unstable form of oxygen possessing an unpaired electron, e.g. superoxide anion.

PAH: Polycyclic aromatic hydrocarbon.

PCB: Polychlorinated biphenyl.

PCDD: Polychlorinated dibenzodioxin.

PCDF: Polychlorinated dibenzofuran.

Phosphorothionates: OP compounds containing thion groups (cf. organophosphates which contain oxon groups).

Phytotoxic: Toxic to plants.

Poikilotherms: Organisms that are unable to regulate their body temperatures.

Polarity: Possessing electrical charge.

Population growth rate (r): Per capita rate of increase of population.

Potentiation: With reference to toxicity; the situation where the toxicity of a combination of compounds is greater than the summation of the toxicities of its individual components.

Pyrethrins: Naturally occurring lipophilic esters which are toxic to many insects.

Pyrethroids: Synthetic insecticides having a strong resemblance to pyrethrins.

QSARs (quantitative structure–activity relationships): Relationships between structural parameters of chemicals and their toxicity.

Recalcitrant: see Refractory.

Reductase: An enzyme catalysing reductions.

Refractory: With reference to environmental chemicals, those that are unreactive ('difficult to manage').

Resistance: Reduced susceptibility to a chemical that is genetically determined.

RIVPACS: River Invertebrate Prediction and Classification.

Rotenone: A complex flavonoid produced by the plant *Derris ellyptica*. It has insecticidal activity due to its ability to inhibit electron transport in the mitochondrion.

Selective toxicity (Selectivity): Difference in toxicity of a chemical towards different species, strains, sexes, age groups, etc.

Superoxide anion ($O_2$$^{.-}$): Reactive oxyradical implicated in oxidative stress.

Superwarfarins: Second-generation anticoagulant rodenticides related to warfarin.

Synergism: Similar to potentiation (see below), but some authors use the term in a more restricted way, e.g. where one component of a mixture, the synergist, would not cause toxicity if applied alone at the dose in question.

TBT: Tributyltin.

TCDD: Tetrachlorodibenzodioxin.

Toxic equivalent (TEQ): A value that expresses the toxicity of a mixture of chemicals relative to that of a reference compound.

Toxicodynamics: Relating to the toxic action of chemicals upon living organisms.

Toxicokinetics: Relating to the fate of toxic chemicals within living organisms, i.e. questions of uptake, distribution, metabolism, storage and excretion; factors that determine how much of a toxic form reaches the site of action.

Transthyretin (TTR): A protein complex found in blood that binds both retinol (vitamin A) and thyroxine.

Uncouplers of oxidative phosphorylation: Compounds that uncouple oxidative phosphorylation from electron transport in the inner mitochondrial membrane. Most are weak lipophilic acids that can run down the proton gradient across this membrane.

Vitamin K: A cofactor for the carboxylase of the hepatic endoplasmic reticulum which is responsible for completing the synthesis of blood clotting proteins.

Vitellogenin: A protein that forms part of the yolk of egg-laying vertebrates.

Xenobiotic: A 'foreign compound' that has no role in the normal biochemistry of a living organism. A 'normal' endogenous compound to one species can be a xenobiotic to another species.

References

AARTS, J. M. M. J. G., DENISON, M. S., DE HAAN, L. H. J., COX, M. A., SCHALK, J. A. C. and BROUWER, A. (1993) Antagonistic effects of di ortho PCBs on Ah receptor-mediated induction of luciferase activity by 3,4,3′,4′TCB in mouse hepatocyte 1c1c7 cells. *Organohalogen Compounds* **14**, 69–72.

AGOSTA, W. (1996) *Bombardier Beetles and Fever Trees – A Close up Look at Chemical Warfare and Signals in Animals and Plants*. Reading, MA: Addison Wesley.

AHLBORG, U. G., BECKING, G. C., BIRNBAUM, L. S., BROUWER, A., *et al.* (1994) Toxic equivalency factors for dioxin-like PCBs. *Chemosphere* **28**, 1049–67.

ALDRIDGE, W. N. (1953) Serum esterases I. *Biochemical Journal* **53**, 110–17.

ALDRIDGE, W. N. and STREET, B. W, (1964) Oxidative phosphorylation; biochemical effects and properties of trialkyl tin. *Biochemical Journal* **91**, 287–97.

ALZIEU, C., HERAL, M., THIBAUD, Y., DARDIGNAC, M.-J. and FEUILLET, M. (1982) Influence des peintures antisalissures a base d'organostanniques sur la calcification de la coquille de l'huitre Crassostrea gigas. *Rev. Trav. Inst. Peches Marit.* **45**, 101–16.

ASHTON, F. M. and CRAFTS, A. S. (1973) *Mode of Action of Herbicides*. New York: Wiley.

BACCI, E. (1994) *Ecotoxicology of Organic Pollutants*. Boca Raton, FL: Lewis.

BAILEY, S., BUNYAN, P. J., JENNINGS, D. M., NORRIS, J. D., STANLEY, P. I. and WILLIAMS, J. H. (1974) Hazards to wildlife from the use of DDT in orchards. II. A further study. *Agro-Ecosystems* **1**, 323–38.

BAKKE, J. E., BERGMAN, A. and LARSEN, G. L. (1982) Metabolism of 2,4′,5 trichlorobiphenyl by the mercapturic acid pathway. *Science* **217**, 645–7.

BAKKE, J. E., FEIL, V. J. and BERGMAN, A. (1983) Metabolites of 2,4′,5 trichlorobiphenyl in rats. *Xenobiotica* **13**, 555–64.

BALLANTYNE, B. and MARRS, T. C. (eds) (1992) *Clinical and Experimental Toxicology of Organophosphates and Carbamates*. Oxford: Butterworth/Heinemann.

BALLS, M., BRIDGES, J. and SOUTHEE, J. (eds) (1991) *Animals and Alternatives in Toxicology*. Basingstoke, UK: Macmillan.

BALLSCHMITTER, K. and ZELL, M. (1989) Analysis of PCB by glass capillary gas chromatography. Composition of Aroclor and Clophen PCB mixtures. *Zeitung Analytische Chemie* **302**, 20–31.

BATTEN, P. L. and HUTSON, D. H. (1995) Species differences and other factors affecting metabolism and extrapolation to man. In HUTSON, D. H. and PAULSON, G. D. (eds) *Progress in Pesticide Biochemistry and Toxicology*, vol. 8. *The Metabolism of Agrochemicals,* pp. 267–308. Chichester: J. Wiley.

BEAUVAIS, S. L., JONES, S. B., BREWER, S. K. and LITTLE, E. E. (2000) Physiological measures of neurotoxicity of diazinon and malathion to larval rainbow trout and their correlation with behavioural measures. *Environmental Toxicology and Chemistry* **19**, 1875–80.

BERNARD, R. F. (1966) DDT residues in avian tissues. *Journal of Applied Ecology* 3 (supplement), 193–8.

BOON, J. P., VAN ARNHEM, E. V., JANSEN, S., KANNEN, N., *et al.* (1992) The toxicokinetics of PCBs in marine mammals with special reference to possible interactions of individual congeners with cytochrome P450 dependent monooxygenase systems: an overview. In WALKER, C. H. and LIVINGSTONE, D. R. (eds) *Persistent Pollutants in Marine Ecosystems*, pp. 119–60. Oxford: Pergamon Press.

BORG, K., WANNTORP, H., ERNE, K. and HANKO, E. (1969) Alkyl mercury poisoning in terrestrial Swedish wildlife. *Viltrevy* 6, 301–79.

BORG, K., ERNE, K., HANKO, E. and WANNTORP, H. (1970) Experimental secondary mercury poisoning in the goshawk. *Environmental Pollution* 1, 91–104.

BORLAKOGLU, J. T., WILKINS, J. P. G. and WALKER, C. H. (1988) Polychlorinated biphenyls in sea birds – molecular features and metabolic interpretations. *Marine Environmental Research* 24, 15–19.

BOSVELD, A. T. C., NIEBOER, R., DE BONT, A., MENNEN, J., *et al.* (2000) Biochemical and developmental effects of dietary exposure to PCBs 126 and 153 in common tern chicks. *Environmental Toxicology and Chemistry* 19, 719–30.

BOYD, I. L., MYHILL, D. G. and MITCHELL-JONES, A. J. (1988) Uptake of lindane by pipistrelle bats and its effect on survival. *Environmental Pollution* 51, 95–111.

BREALEY, C. J. (1980) *Comparative metabolism of Pirimiphos-methyl in rat and Japanese Quail*. PhD Thesis. University of Reading.

BREALEY, C. J., WALKER, C. H. and BALDWIN, B. C. (1980) 'A' Esterase activities in relation to differential toxicity of pirimiphos-methyl. *Pesticide Science* 11, 546–54.

BROLEY, C. L. (1958) Plight of the American bald eagle. *Audubon Magazine* 60, 162–71.

BROMLEY-CHALLENOR, K. C. A. (1992) *Synergistic mechanisms of synthetic pyrethroids and fungicides in Apis mellifera*. MSc Thesis. University of Reading.

BROOKS, G. T. (1972) Pathways of enzymatic degradation of pesticides. *Environ. Quality Safety* 1, 106–63.

BROOKS, G. T. (1974) *The Chlorinated Insecticides*, vols 1 and 2. Cleveland, OH: CRC Press.

BROOKS, G. T. (1992) Progress in structure–activity studies on cage convulsants and related GABA receptor chloride ionophore antagonists. In OTTO, D. and WEBER, B. (eds) *Insecticides: Mechanism of Action and Resistance*, pp. 237–42. Andover: Intercept.

BROOKS, G. T., PRATT, G. E. and JENNINGS, R. C. (1979) The action of precocenes in milkweed bugs and locusts. *Nature* 281, 570–2.

BROUWER, A. (1991) Role of biotransformation in PCB-induced alterations in vitamin A and thyroid hormone metabolism in laboratory and wildlife species. *Biochemical Society Transactions* 19, 731–7.

BROUWER, A. (1996) Biomarkers for exposure and effect assessment of dioxins and PCBs. In IEH Report on 'The use of Biomarkers in Environmental Exposure Assessment', pp. 51–8. Leicester: Institute of Environmental Health.

BROUWER, A., KLASSON-WEHLER, E., BOKDAM, M., MORSE, D. C. and TRAAG, W. A. (1990) Competitive inhibition of thyroxine binding to transthyretin by monohydroxy metabolites of 3,4,3',4 TCB. *Chemosphere* 20, 1257–62.

BROUWER, A., MORSE, D. C., LANS, M. C., SCHUUR, G., *et al.* (1998) Interactions of persistent environmental organohalogens with the thyroid hormone system: mechanisms and possible consequences for animal and human health. *Toxicology and Industrial Health* 14, 59–84.

BROWN, A. W. A. (1971) Pest resistance to pesticides. In WHITE-STEVENS, R. (ed.) *Pesticides in the Environment*, pp. 437–551. New York: Dekker.

BRUGGERS, R. L. and ELLIOTT, C. C. H. (1989) *Quelea quelea: Africa's Bird Pest*. Oxford: Oxford University Press.

BUCKLE, A. P. and SMITH, R. H. (eds) (1994) *Rodent Pests and their Control*. CAN International.

BULL, K. R., EVERY, W. J., FREESTONE, P., HALL, J. R., *et al.* (1983) Alkyl lead pollution and bird mortalities on the Mersey Estuary, UK 1979–1981. *Environmental Pollution* (Series A) 31, 239–59.

BURTON, G. A. Jr (ed.) (1992) *Sediment Toxicity Assessment*. Boca Raton, FL: Lewis.

CALOW, P. (ed.) (1994) *Handbook of Ecotoxicology*, vols 1 and 2. Oxford: Blackwell Science.

CHAPMAN, R. A. and HARRIS, C. R. (1981) Persistence of pyrethroid insecticides in a mineral and an organic soil. *Journal of Environmental Science and Health B* 16, 605–15.

CHENG, Z. and JENSEN, A. (1989) Accumulation of organic and inorganic tin in the blue mussel, Mytilus edulis, under natural conditions. *Marine Pollution Bulletin* 20, 281–6.

CHIPMAN, J. K. and WALKER, C. H. (1979) The metabolism of dieldrin and two of its analogues: the relationship between rates of microsomal metabolism and rates of excretion of metabolites in the male rat. *Biochemistry and Pharmacology* 28, 1337–45.

CLARK, R. B. (1992) *Marine Pollution*, 3rd edn. Oxford: Oxford Scientific Publications.

CLARKSON, T. W. (1987) Metal toxicity in the central nervous system. *Environmental Health Perspectives* 75, 59–64.

COLIN, M. E. and BELZUNCES, L. P. (1992) Evidence of synergy between prochloraz and deltamethrin: a convenient biological approach. *Pesticide Science* 36, 115–19.

CONNELL, D. W. (1994) The octanol–water partition coefficient. In CALOW, P. (ed.) *Handbook of Ecotoxicology*, vol. 2, pp. 775–84. Oxford: Blackwell Science.

CONNOR, M. S. (1983) Fish/sediment concentration ratios for organic compounds. *Environmental Science and Technology* 18, 31–5.

COPPING, L. G. and MENN, J. J. (2000) Biopesticides: a review of their action, application and efficacy. *Pesticide Management Science* 56, 651–76.

CRAIG, P. J. (ed.) (1986) *Organometallic Compounds in the Environment: Principles and Reactions*. Longmans.

CRANE, M., DELANEY, P., WATSON, S., PARKER, P. and WALKER, C. H. (1995) The effect of malathion 60 on *Gammarus pulex* below watercress beds. *Environmental Toxicology and Chemistry* 14, 1181–8.

CRICK, H. Q. P., BAILLIE, S. R., BALMER, D. E., BASHFORD, R. I., *et al.* (1998) Breeding birds in the wider countryside: their conservation status. Research Report 198. Thetford, UK: British Trust for Ornithology.

CROSBY, D. G. (1998) *Environmental Toxicology and Chemistry*. New York: Oxford University Press.

CROSSLAND, N. O. (1994) Extrapolation from mesocosms to the real world. *Toxicology and Ecotoxicology News* 1, 15–22.

DANERUD, P. O., MORSE, D. C., KLASSSON-WEHLER, E. AND BROUWER, A. (1996) Binding of 3,3',4,4' TCB metabolite to foetal transthyretin and effects on foetal thyroid hormone levels in mice. *Toxicology* 106, 105–14.

DAVILA, D. R., MOUNHO, B. J. and BURCHIEL, S. W. (1997) Toxicity of PAH to the human immune system; models and mechanisms. *Toxicology and Ecotoxicology News* 4, 5–9.

DAVIS, D. and SAFE, S. H. (1990) Immunosuppressive activities of PCBs in C57BL/6N mice: structure–activity relationships as Ah receptor agonists and partial agonists. *Toxicology* 63, 97–111.

DE MATTEIS, F. (1974) Covalent binding of sulphur to microsomes and loss of cytochrome P450 during oxidative desulphuration of several chemicals. *Molecular Pharmacology* 10, 849.

DEVONSHIRE, A. L. (1991) Role of esterases in resistance of insects to insecticides. *Biochemical Society Transactions* **19**, 755–9.

DEVONSHIRE, A. L. and SAWICKI, R. M. (1979) Insecticide resistant *Myzus persicae* as an example of evolution by gene duplication. *Nature* **280**, 140–1.

DEVONSHIRE, A. L., BYRNE, G. D., MOORES, G. D. and WILLIAMSON, M. S. (1998) Biochemical and molecular characterization of insecticide insensitive acetylcholinesterase in resistant insects. In DOCTOR, B. P., QUINN, D. M., ROTUNDO, R. L. and TAYLOR, P. (eds) *Structure and Function of Cholinesterases and Related Proteins*. New York: Plenum Press.

DE VOOGT, P. (1996) Ecotoxicology of chlorinated aromatic hydrocarbons. In HESTER, R. E. and HARRISON, R. M. (eds) *Chlorinated Organic Micropollutants* no. 6 in series 'Issues in Environmental Science and Technology', pp. 89–112. London: Royal Society of Chemistry.

DRABEK, J. and NEUMANN, R. (1985) Proinsecticides. In HUTSON, D. H. and ROBERTS, T. R. (eds) *Progress in Pesticide Biochemistry and Toxicology*, vol. 5. *Insecticides*, pp. 35–86. Chichester: J. Wiley.

EADSFORTH, C. V., DUTTON, A. J., HARRISON, E. G. and VAUGHAN, J. A. (1991) A barn owl feeding study with (14C) flocoumafen-dosed mice. *Pesticide Science* **32**, 105–19.

EASON, C. T. and SPURR, E. B. (1995) Review of toxicity and impacts of brodifacoum on non target wildlife in New Zealand. *New Zealand Journal of Zoology* **22**, 371–9.

EDSON, E. F., SANDERSON, D. M. and NOAKES, D. N. (1966) Acute toxicity data for pesticides. *World Review of Pest Control* **5** (3), 143–51.

EDWARDS, C. A. (1973) *Persistent Pesticides in the Environment*, 2nd edn. Cleveland, OH: CRC Press.

EHRLICH, P. R. and RAVEN, P. H. (1964) Butterflies and plants: a study in co-evolution. *Evolution* **18**, 586–608.

ELDEFRAWI, M. E. and ELDEFRAWI, A. T. (1990) Nervous-system-based insecticides. In HODGSON, E. and KUHR, R. J. M. *Safer Insecticides – Development and Use*. New York: Dekker Inc.

ELLIOTT, J. E., NORSTROM, R. J. and J. A. (1988) Organochlorines and eggshell thinning in Northern gannets (Sula bassanus) from Keith, Eastern Canada. *Environmental Pollution* **52**, 81–102.

ENVIRONMENTAL HEALTH CRITERIA NO. 9 (1979) *DDT and its Derivatives*. Geneva: WHO.

ENVIRONMENTAL HEALTH CRITERIA NO. 18 (1981) *Arsenic*. Geneva: WHO.

ENVIRONMENTAL HEALTH CRITERIA NO. 38 (1984) *Heptachlor*. Geneva: WHO.

ENVIRONMENTAL HEALTH CRITERIA NO. 63 (1986) *Organophosphorous Insecticides: A General Introduction*. Geneva: WHO.

ENVIRONMENTAL HEALTH CRITERIA NO. 64 (1986) *Carbamate Pesticides: A General Introduction*. Geneva: WHO.

ENVIRONMENTAL HEALTH CRITERIA NO. 83 (1989) *DDT and its Derivatives – Environmental Aspects*. Geneva: WHO.

ENVIRONMENTAL HEALTH CRITERIA NO. 82 (1989) *Cypermethrin*. Geneva: WHO.

ENVIRONMENTAL HEALTH CRITERIA NO. 85 (1989) *Lead – Environmental Aspects*. Geneva: WHO.

ENVIRONMENTAL HEALTH CRITERIA NO. 86 (1989) *Mercury – Environmental Aspects*. Geneva: WHO.

ENVIRONMENTAL HEALTH CRITERIA NO. 88 (1989) *PCDDs and PCDFs*. Geneva: WHO.

ENVIRONMENTAL HEALTH CRITERIA NO. 91 (1989) *Aldrin and Dieldrin*. Geneva: WHO.

ENVIRONMENTAL HEALTH CRITERIA NO. 94 (1990) *Permethrin*. Geneva: WHO.

ENVIRONMENTAL HEALTH CRITERIA NO. 95 (1990) *Fenvalerate*. Geneva: WHO.

ENVIRONMENTAL HEALTH CRITERIA NO. 97 (1990) *Deltamethrin*. Geneva: WHO.

ENVIRONMENTAL HEALTH CRITERIA NO. 101 (1990) *Methylmercury*. Geneva: WHO.

ENVIRONMENTAL HEALTH CRITERIA NO. 116 (1990) *Tributyltin Compounds*. Geneva: WHO.

ENVIRONMENTAL HEALTH CRITERIA NO. 121 (1991) *Aldicarb*. Geneva: WHO.

ENVIRONMENTAL HEALTH CRITERIA NO. 123 (1992) *Alpha and Beta Hexachlorocyclohexanes*. Geneva: WHO.

ENVIRONMENTAL HEALTH CRITERIA NO. 124 (1991) *Lindane*. Geneva: WHO.

ENVIRONMENTAL HEALTH CRITERIA NO. 130 (1992) *Endrin*. Geneva: WHO.

ENVIRONMENTAL HEALTH CRITERIA NO. 140 (1993) *Polychlorinated Biphenyls and Terphenyls*. Geneva: WHO.

ENVIRONMENTAL HEALTH CRITERIA NO. 142 (1992) *Alpha-cypermethrin*. Geneva: WHO.

ENVIRONMENTAL HEALTH CRITERIA NO. 152 (1994) *Polybrominated Biphenyls*. Geneva: WHO.

ENVIRONMENTAL HEALTH CRITERIA NO. 153 (1994) *Carbaryl*. Geneva: WHO.

ENVIRONMENTAL HEALTH CRITERIA NO. 197 (1997) *Demeton-S-methyl*. Geneva: WHO.

ENVIRONMENTAL HEALTH CRITERIA NO. 198 (1998) *Diazinon*. Geneva: WHO.

ENVIRONMENTAL HEALTH CRITERIA NO. 202 (1998) *Non Heterocyclic Polycyclic Aromatic Hydrocarbons*. Geneva: WHO.

ERNST, W. (1977) Determination of the bioconcentration potential of marine organisms – a steady state approach. I. Bioconcentration data for 7 chlorinated pesticides in mussels and their relation to solubility data. *Chemosphere* **13**, 731–40.

ERNST, W. R., PEARCE, P. A. and POLLOCK, T. L. (eds) (1989) Environmental Effects of Fenitrothion Use in Forestry. Environment Canada, Atlantic Region Report.

ETO, M. (1974) *Organophosphorous Insecticides: Organic and Biological Chemistry*. Cleveland, OH: CRC Press.

FENT, K. (1996) Ecotoxicology of organotin compounds. *Critical Reviews in Toxicology* **26**, 1–117.

FENT, K., WOODIN, B. R. and STEGEMAN, J. J. (1998) Effects of triphenyl tin and other organotins on hepatic monooxygenase system in fish. *Comparative Biochemistry and Physiology* **121C** (special issue), 277–88.

FERGUSSON, D. (1994) *The effects of 4-hydroxycoumarin anticoagulant rodenticides on birds and the development of techniques for non-destructively monitoring their ecological effects*. PhD Thesis, University of Reading.

FEST, C. and SCHMIDT, K.-J. (1982) *Chemistry of Organophosphorous Compounds*, 2nd edn. Berlin: Springer-Verlag.

FFRENCH-CONSTANT, T. A., ROCHELAU, J. C., STEICHEN, J. C. and CHALMERS, A. E. (1993) A point mutation in a *Drosophila* GABA receptor confers insecticide resistance. *Nature* **363**, 449–51.

FORGUE, S. T., *et al.* (1980) Direct evidence that an arene oxide is a metabolic intermediate of 2,2',5,5' TCB. In Proceedings of the 19th Meeting of the Society of Toxicology. Abstract no. 383.

FOSSI, M. C. and LEONZIO, C. (1994) *Non-Destructive Biomarkers in Vertebrates*. Boca Raton, FL: Lewis.

FRY, D. M. and TOONE, C. K. (1981) DDT-induced feminisation of gull embryos. *Science* **2132**, 922–4.

FUCHS, P. (1967) Death of birds caused by application of seed dressings in the Netherlands. *Mededelingen Rijksfaculteit Landbouwweetenschappen Gent* **32**, 855–9.

GAGE, J. C. and HOLM, S. (1976) The influence of molecular structure on the retention and excretion of PCBs by the mouse. *Toxicology and Applied Pharmacology* **36**, 555–60.

GARRISON, P. M., TULLIS, K., AARTS, J. M. M. J. G. and BROUWER, A., *et al.* (1996) Species specific recombinant cell lines as bioassay systems for the detection of dioxin-like chemicals. *Fundamentals of Applied Toxicology* **30**, 194–203.

GEORGHIOU, G. P. and SAITO, T. (eds) (1983) *Pest Resistance to Pesticides*. New York: Plenum Press.

GIBBS, P. E. (1993) A male genital defect in the dog whelk favouring survival in a polluted area. *Journal of the Marine Biological Association of the United Kingdom* **73**, 667–8.

GIBBS, P. E. and BRYAN, G. W. (1986) Reproductive failure in populations of the dog whelk caused by imposex induced by TBT from antifouling paints. *Journal of the Marine Biological Association of the United Kingdom* **66**, 767–77.

GIESY, J. P. (1997) PCDDs, PCDFs, PCBs and 2,3,7,8-TCDD equivalents in fish from Saginaw Bay, Michigan. *Environmental Toxicology and Chemistry* **16**, 713–24.

GILBERTSON, M., FOX, G. A. and BOWERMAN, W. W. (eds) (1998) *Trends in Levels and Effects of Persistent Toxic Substances in the Great Lakes*. Dordrecht: Kluwer Academic Publishers.

GINGELL, R. (1976) Metabolism of ^{14}C-DDT mouse and hamster. *Xenobiotica* **6**, 15–20.

GLATT, H., *et al.* (1997) The use of cell lines genetically engineered for human xenobiotic metabolising enzymes. In VAN ZUTPHEN, L. F. M. and BALLS, M. (eds) *Animal Alternatives, Welfare and Ethics*, pp. 81–94. Amsterdam: Elsevier.

GOTELLI, N. J. (1998) *A Primer of Ecology*, 2nd edn. Sunderland, MA: Sinauer Associates.

GREIG-SMITH, P. W., WALKER, C. H. and THOMPSON, H. M. (1992a) Ecotoxicological consequences of interactions between avian esterases and organophosphorous compound. In BALLANTYNE, B. and MARRS, T. C. (eds) *Clinical and Experimental Toxicology of Organophosphates and Carbamates*, pp. 295–304. Oxford: Butterworth/Heinemann.

GREIG-SMITH, P. W., FRAMPTON, G. and HARDY, A. R. (1992b) *Pesticides, Cereal Farming, and the Environment: the Boxworth Project*. London: HMSO.

GRUE, C. E., HART, A. D. M. and MINEAU, P. (1991) Biological consequences of depressed brain cholinesterase activity in wildlife. In MINEAU, P. (ed.) *Cholinesterase-inhibiting Insecticides – their Impact on Wildlife and the Environment*, pp. 151–210. Amsterdam: Elsevier.

HALL, A. T. and URIS, J. T. (1991) Anthracene reduces reproductive potential and is maternally transferred during long term exposure in fathead minnows. *Aquatic Toxicology* **19**, 249–64.

HALLIWELL, B. and GUTTERIDGE, J. M. C. (1986) Oxygen free radicals and iron in relation to biology and medicine – some problems and concepts. *Archives of Biochemistry and Biophysics* **246**, 501–14.

HAMILTON, G. A., HUNTER, K., RITCHIE, A. S., RUTHVEN, A. D., BROWN, P. M. and STANLEY, P. I. (1976) Poisoning of wild geese by carbophenothion treated winter wheat. *Pesticide Science* **7**, 175–83.

HARBORNE, J. B. (1993) *Introduction to Ecological Biochemistry*, 4th edn. London: Academic Press.

HARBORNE, J. B. and BAXTER, H. (eds) (1993) *Phytochemical Dictionary*. London: Taylor & Francis.

HARDY, A. R. (1990) Estimating exposure: the identification of species at risk and routes of exposure. In SOMERVILLE, L. and WALKER, C. H. (eds) *Pesticide Effects on Terrestrial Wildlife*, pp. 81–98. London: Taylor & Francis.

HASSALL, K. A. (1990) *The Biochemistry and Uses of Pesticides*, 2nd edn. Basingstoke: Macmillan.

HATHWAY, D. E. (1984) *Molecular Aspects of Toxicology*. London: Royal Society of Chemistry.

HAYES, W. J. and LAWS, E. R. (1991) *Handbook of Pesticide Toxicology*, vol. 2. *Classes of Pesticides*. San Diego: Academic Press.

HEBERT, C. E., NORSTROM, R. J., SIMON, M., BRAUNE, B., M., WESELOH, D. V. and MACDONALD, C. R. (1994) Temporal trends and sources of PCDDs and PCDFs in the Great Lakes; herring gull monitoring. *Environmental Science and Technology* **28**, 1266–77.

HEGDAL, P. L. and COLVIN, B. A. (1988) Potential hazard to Eastern Screech-owls and other raptors of brodifacoum bait used for vole control in orchards. *Environmental Toxicology and Chemistry* **7**, 245–60.

HEINZ, G. H. and HOFFMAN, D. J. (1998) Methylmercury chloride and selenomethionine interactions on health and reproduction in mallards. *Environmental Toxicology and Chemistry* **17**, 139–45.

HETHERINGTON, L. H., LIVINGSTONE, D. R. and WALKER, C. H. (1996) Two and one-electron dependent reductive metabolism of nitroaromatics by Mytilus edulis, Carcinus maenas and Asterias rubens. *Comparative Biochemistry and Physiology* 113C, 231–9.

HILL, B. D. and SCHAALJE, G. B. (1985) A two compartment model for the dissipation of deltamethrin in soil. *J. Agric. Fd Chem.* 33, 1001–6.

HILL, E. F. (1992) Avian toxicology of anticholinesterases. In BALLANTYNE, B. and MARRS, T. C. (eds) *Clinical and Experimental Toxicology of Organophosphates and Carbamates*, pp. 272–94. Oxford: Butterworth/Heinemann.

HILL, I. R., MATTHIESSEN, P. and HEIMBACH, F. (eds) (1993) Guidance document on sediment toxicity tests and bioassays for freshwater and marine environments. *SETAC Europe Workshop on Sediment Toxicity Assessment.* 8–10 November, 1993, Renesse, The Netherlands.

HODGSON, E. and KUHR, R. J. (eds) (1990) *Safer Insecticides: Development and Use*. New York: Marcel Dekker.

HODGSON, E. and LEVI, P. (eds) (1994) *Introduction to Biochemical Toxicology*, 2nd edn. Norwalk, CT: Appleton and Lange.

HOLDEN, A. V. (1973) International cooperation on organochlorine and mercury residues in wildlife. *Pesticide Monitoring* 7, 37–52.

HOOPER, M. J., *et al.* (1989) Organophosphate exposure in hawks inhabiting orchards during winter dormant spraying. *Bulletin of Environmental Contaminants and Toxicology* 42, 651–60.

HOSOKAWA, M., MAKI, T. and SATOH, T. (1987) Multiplicity and regulation of hepatic microsomal carboxylesterases in rats. *Molecular Pharmacology* 31, 579–84.

HOUSE, W. A., LEACH, D., LONG, J. L. A., CRANWELL, P., *et al.* (1997) Micro-organic compounds in the Humber rivers. *The Science of the Total Environment* 194/195 (Special issue), 357–72.

HOWALD, G. R., MINEAU, P., ELLIOTT, J. E. and CHENG, K. M. (1999) Brodifacoum poisoning of avian scavengers during rat control at a seabird colony. *Ecotoxicology* 8, 431–7.

HUCKLE, K. R., WARBURTON, P. A., FORBES, S. and LOGAN, C. J. (1989) Studies on the fate of flocoumafen in the Japanese quail. *Xenobiotica* 19, 51–62.

HUGGETT, R. J., KIMERLE, R. H., MEHRLE, P. M. Jr and BERGMAN, H. (eds) (1992) *Biomarkers. Biochemical, Physiological, and Histological Markers of Anthropogenic Stress*. Boca Raton, FL: Lewis.

HUNT, E. G. and BISCHOFF, A. I. (1960) Inimical effects on wildlife of periodic DDD applications to Clear Lake California. *Fish and Game* 46, 91–106.

HUTSON, D. H. (1976) Comparative metabolism of dieldrin in the CFE rat and in two strains of mice (CF1 and LACG). *Fd Cosmet. Toxicol.* 14, 577–91.

HUTSON, D. H. and PAULSON, G. D. (eds) (1995) *The Mammalian Metabolism of Agrochemicals. Progress in Pesticide Biochemistry and Toxicology*, vol. 8. Chichester, UK: J. Wiley.

HUTSON, D. H. and ROBERTS, T. (eds) (1999) *Metabolic Pathways of Agrochemicals*, part 2. *Insecticides and Fungicides*. London: Royal Society of Chemistry.

INSTITUTE FOR ENVIRONMENTAL HEALTH (1995) *Environmental Oestrogens: Consequences to Human Health and Wildlife*. Report A1. Leicester, UK: Institute for Environmental Health (MRC).

INTERNATIONAL ATOMIC ENERGY AGENCY (1972) *Mercury Contamination in Man and his Environment*. Technical Report Series No 137. Vienna: IAEA.

JAGER, K. (1970) *Aldrin, Dieldrin, Endrin and Telodrin*. Amsterdam: Elsevier.

JAKOBY, W. B. (ed.) (1980) *Enzymatic Basis of Detoxication*. New York: Academic Press.

JANSSEN, P. A. H., FABER, J. H. and BOSVELD, A. T. C. (1998) *(Fe)male?* IBN Scientific Contributions DLO. Wageningen, The Netherlands: Institute for Forestry and Nature Research.

JEFFERIES, D. J. (1975) The role of the thyroid in the production of sublethal effects by organochlorine insecticides and PCBs. In MORIARTY, F. M. *Organochlorine Insecticides: Persistent Organic Pollutants*, pp. 132–230. London: Academic Press.

JEFFERIES, D. J. and PARSLOW, J. L. F. (1972) Thyroid changes in PCB-dosed guillemots and their indication of one of the mechanisms of action of these materials. *Environmental Pollution* **10**, 293–311.

JOHNSON, M. K. (1992) Molecular events in delayed neuropathy: experimental aspects of neuropathy target esterase. In BALLANTYNE, B. and MARRS, T. C. (eds) *Clinical and Experimental Toxicology of Organophosphates and Carbamates*, pp. 90–113. Oxford: Butterworth/Heinemann.

JOHNSTON, G. O., COLLETT, G., WALKER, C. H., DAWSON, A., BOYD, I. and OSBORN, D. (1989) Enhancement of malathion toxicity in the hybrid red-legged partridge following exposure to prochloraz. *Pesticide Biochemistry and Physiology* **35**, 107–18.

JOHNSTON, G. O., WALKER, C. H. and DAWSON, A. (1994a) Interactive effects between EBI fungicides and OP insecticides in the hybrid red-legged partridge. *Environmental Toxicology and Chemistry* **13**, 615–20.

JOHNSTON, G. O., WALKER, C. H. and DAWSON, A. (1994b) Interactive effects of prochloraz and malathion in the pigeon, starling and hybrid red-legged partridge. *Environmental Toxicology and Chemistry* **13**, 115–20.

JOHNSTON, G. O., WALKER, C. H. and DAWSON, A. (1994c) Potentiation of carbaryl toxicity to the hybrid red-legged partridge following exposure to malathion. *Pesticide Biochemistry and Physiology* **49**, 198–208.

JONES, D. M., BENNETT, D. and ELGAR, K. E. (1978) Deaths of owls traced to insecticide treated timber. *Nature*, **272**, 52.

JORGENSEN, S. E. (ed.) (1990) *Modelling in Ecotoxicology*. Amsterdam: Elsevier.

KAISER, T. E., REICHEL, W. L., LOCKE, L. H., Cromartie, E., *et al*. (1980) Organochlorine pesticides, PCBs and PBB residues and necropsy data for bald eagles from 29 states. *Pesticides Monitoring Journal* **13**, 145–9.

KAPUSTKA, L. A., WILLIAMS, B. A. and FAIRBROTHER, A. (1996) Evaluating risk predictions at population and community levels in pesticide registration. *Environmental Toxicology and Chemistry* **15**, 427–31.

KARR, J. R. (1981) Assessment of biotic integrity using fish communities. *Fisheries* **6**, 21–7.

KATO, T. (1986) Sterol biosynthesis in Fungi; a target for broad spectrum fungicides. In HAUG, G. and HOFFMANN, H. (eds) *Chemistry of Plant Protection*, no. 1, *Sterol Biosynthesis Inhibitors and Anti-feeding Compounds*, pp. 1–24. Berlin: Springer-Verlag.

KEDWARDS, T. J., MAUND, S. J. and CHAPMAN, P. F. (1999a) Community level analysis of ecotoxicological field studies. I. Biological monitoring. *Environmental Toxicology and Chemistry* **18**, 149–57.

KEDWARDS, T. J., MAUND, S. J. and CHAPMAN, P. F. (1999b) Community level analysis of ecotoxicological field studies. II. Replicated design studies. *Environmental Toxicology and Chemistry* **18**, 158–66.

KLASSON-WEHLER, E. (1989) *Synthesis of Some Radiolabelled Organochlorines and Metabolism Studies in vivo of two PCBs*. Doctoral Dissertation, University of Stockholm, Sweden.

KLASSON-WEHLER, E., KUROKI, H., ATHANASIADOU, M. and BERGMAN, A. (1992) Selective retention of hydroxylated PCBs in blood. In *Organohalogen Compounds*, vol. 10. *Toxicology, Epidemiology, Risk Assessment, and Management*, pp. 121–2. Helsinki: Finnish Institute of Occupational Health.

KNAPEN, M. H. J., KON-SIONG, G. J., HAMULYAK, K. and VERMEER, C. (1993) Vitamin K-induced changes in markers for osteoblast activity in urinary calcium loss. *Calcif. Tissue International* **53**, 81–5.

KOEMAN, J. H. (ed.) (1972) Side effects of persistent pesticides and other chemicals on birds and mammals. In The Netherlands Report by the Working Group on Birds and Mammals of the Committee TNO for research on Side effects of pesticides. *TNO-nieuws* **27**, 527–632.

KOEMAN, J. H. and PENNINGS, J. H. (1970) An orientational survey on the side effects and environmental distribution of dieldrin in a tsetse control area in S. W. Kenya. *Bulletin of Environmental Contaminants and Toxicology* **5**, 164–70.

KOEMAN, J. H. and VAN GENDEREN, H. (1970) Tissue levels in animals and effects caused by chlorinated hydrocarbon insecticides, chlorinated biphenyls, and mercury in the marine environment along the Netherlands coast. FAO Technical Conference on Marine Pollution. Rome, December, 1970.

KOEMAN, J. H., OSKAMP, A. A. G., VEEN, J., BROUWER, E., *et al.* (1967) Insecticides as a factor in the mortality of the sandwich tern. A preliminary communication. *Mededelingen Rijksfaculteit Landbouwweetenschappen Gent* **32**, 841–54.

KOISTINEN, J. (1997) 2,3,7,8-TCDD equivalent in extracts of Baltic white-tailed sea eagles. *Environment, Toxicology and Chemistry* **16**, 1533–44.

KÖNEMANN, H. (1981) QSAR relationships in fish toxicity studies. Part 1: relationship of 50 industrial pollutants. *Toxicology* **19**, 209–21.

KORTE, F. and ARENT, H. (1965) Metabolism of insecticides. IX. Isolation and identification of dieldrin metabolites from urine of rabbits after oral administration of C-14 dieldrin. *Life Science* **4**, 2017–26.

KUHR, R. J. and DOROUGH, H. W (1976) *Carbamate Insecticides*. Cleveland: CRC Press.

KURELEC, B., SMITAL, P., PIVCEVIC, B., EUFEMIA, N. and EPEL, D. (2000) Multixenobiotic resistance, P-glycoprotein and chemosensitizers. *Ecotoxicology* **9**, 307–27.

LANDIS, W. G., MOORE, D. R. J. and NORTON, S. B. (1998) Ecological risk assessment; looking in, looking out. In DOUBEN, P. E. T. (ed.) *Pollution Risk Assessment and Management*, pp. 273–310. Chichester: Wiley.

LANS, M. C., KLASSON-WEHLER, E., WILLEMSEN, M., MEUSSEN, E., *et al.* (1993) Structure-dependent competitive interaction of hydroxy-PCB, PCDD and PCDF with transthyretin. *Chemistry and Biology International* **88**, 7–21.

LEAHEY, J. P. (ed.) (1985) *The Pyrethroid Insecticides*. London: Taylor & Francis.

LEE, K.-S., *et al.* (1989) Metabolism of trans cypermethrin in *H. armigera* and *H. virescens*. *Pesticide Biochemistry and Physiology* **34**, 49–57.

LEWIS, D. F. V. (1996) *Cytochromes P450. Structure, Function and Mechanism*. London: Taylor & Francis,

LEWIS, D. F. V. and LAKE, B. G. (1996) Molecular modelling of CYP1A subfamily members based on an alignment with CYP102. *Xenobiotica* **26**, 723–53.

LIPNICK, R. L. (1991) Outliers: their origin and use in the classification of molecular mechanisms of toxicity. *The Science of the Total Environment* **109/110**, 131–53.

LIVINGSTONE, D. R. (1985) Responses of the detoxication/toxication enzyme system of molluscs to organic pollutants and xenobiotics. *Marine Pollution Bulletin* **16**, 158–64.

LIVINGSTONE, D. R. (1991) Organic xenobiotic metabolism in marine invertebrates. In GILLES, R. (ed.) *Advances in Comparative and Environmental Physiology*, vol. 7, pp. 46–185. Heidelberg: Springer-Verlag.

LIVINGSTONE, D. R. and STEGEMAN, J. J. (eds) (1998) Forms and functions of cytochrome P450. *Comparative Biochemistry and Physiology* **121C** (special issue).

LIVINGSTONE, D. R., MOORE, M. N. and WIDDOWS, J. (1988) Ecotoxicology: Biological effects measurements on molluscs and their use in impact assessment. In SALOMANS, W., BAYNE, B. L., DUURSMA, E. K. and FORSTNER U. (eds) *Pollution of the North Sea An Assessment*, pp. 624–37. Berlin: Springer-Verlag.

LUDWIG, J. P., *et al.* (1993) Caspian tern reproduction in the Saginaw Bay ecosystem following a 100 year flood event. *Journal of the Great Lakes Research* **19**, 96–108.

LUDWIG, J. P., KURITA-MATSUBA, H. J., AUMAN, M. E., LUDWIG, C. L., *et al.* (1996) Deformities, PCDDs and TCDD-equivalents in double crested cormorants and Caspian terns of the Upper Great Lakes 1986–1991; testing a cause–effect hypothesis. *Journal of the Great Lakes Research* **22**, 172–97.

LUNDHOLM, E. (1987) Thinning of eggshells in birds by DDE: mode of action of the eggshell gland. *Comparative Biochemistry and Physiology (part C)* **88**, 1–22.

LUTZ, R. J., DEDRICK, R. L., MATTHEWS, H. B., ELING, T. and ANDERSON, M. W. (1977) A preliminary pharmacokinetic model for several chlorinated biphenyls in rat. *Drug Metab. Dispos.* **5**, 386–96.

MCCAFFERY, A. R. (1998) Resistance to insecticides in heliothine Lepidoptera: a global view. *Philosophical Transactions of the Royal Society, London, B* **353**, 1735–50.

MCCAFFERY, A. R., GLADWELL, R. T., EL-NAYIR, H., WALKER, C. H., PERRY, J. N. and MILES, M. J. (1991) Mechanisms of resistance to pyrethroids in laboratory and field strains of *Heliothis virescens*. *Southwestern Entomologist Supplement* **15**, 143–58.

MCCARTY, J. P. and SECORD, A. L. (1999) Reproductive ecology of tree swallows with high levels of PCB contamination. *Environmental Toxicology and Chemistry* **18**, 1433–9.

MACHIN, A. F., *et al.* (1975) Metabolic aspects of the toxicology of diazinon. 1. Hepatic metabolism in the sheep, cow, guinea-pig, rat, turkey, chicken and duck. *Pesticide Science* **6**, 461–73.

MACKAY, D. (1991) *Multimedia Environmental Models: The Fugacity Approach.* Chelsea, MI: Lewis.

MACKAY, D., SHIU, W. Y. and SUTHERLAND, R. P. (1979) Determination of air–water Henry's law constants for hydrophobic pollutants. *Environmental Science and Technology* **13**, 333–7.

MACKNESS, M. I., WALKER, C. H., ROWLANDS, D. G. and PRICE, N. R. (1982) Esterase activity in homogenates of 3 strains of rust red flour beetle. *Comparative Biochemistry and Physiology* **74C**, 65–68.

MACKNESS, M. I., THOMPSON, H. M. and WALKER, C. H. (1987) Distinction between A esterases and arylesterases and implications for esterase classification. *Biochemical Journal* **245**, 293–6.

MARON, D. M. and AMES, B. N. (1983) Revised methods for the Salmonella mutagenicity test. *Mutation Research* **113**, 173–215.

MATTHIESSEN, P. and GIBBS, P. E (1998) Critical appraisal of the evidence for TBT-mediated endocrine disruption in mussels. *Environmental Toxicology and Chemistry* **17**, 37–43.

MATTHIESSEN, P., *et al.* (1995) Use of a Gammarus pulex bioassay to measure the effects of transient carbofuran runoff from farmland. *Ecotoxicology and Environmental Safety* **30**, 111–19.

MEEHAN, A. P. (1986) *Rats and Mice: Their Biology and Control.* East Grinstead, UK: Rentokil.

MELED, M., THRASYVOULOU, A. and BELZUNCES, L. P. (1998) Seasonal variation in susceptibility of Apis mellifera to the synergistic action of prochloraz and deltamethrin. *Environmental Toxicology and Chemistry* **17**, 2517–20.

MELLANBY, K. M. (1967) *Pesticides and Pollution.* London: Collins.

MENSINK, B. P. (1997) *Tributyl Tin causes Imposex in the Common Whelk Mechanism and Occurrence.* NIOZ RAPPORT 1997–6. Institut voor Onderzoek der Zee. Den Burg: The Netherlands.

MENTLEIN, R., RONAI, A., ROBBI, M., HEYMANN, E., *et al.* (1987) Genetic identification of rat liver carboxylesterases isolated in different laboratories. *Biochimica et Biophysica Acta* **913**, 27–38.

MERSON, M. H., BYERS, R. E. and KAUKEINEN D. E. (1984) Residues of the rodenticide brodifacoum in voles and raptors after orchard treatment. *Journal of Wildlife Management* **48**, 212–16.

MEYER, M. W. (1998) Ecological risk of mercury in the environment: the inadequacy of 'the best available science' (Editorial). *Environmental Toxicology and Chemistry* **17**, 138.

MIZUTANI, T., HIDAKA, K., OHE, T. and MATSUMOTO, M. (1977) A comparative study on accumulation and elimination of tetrachlorobiphenyl isomers in mice. *Bulletin of Environmental Contaminants and Toxicology* **18**, 452–61.

MONOD, G. (1997) L'induction du cytochrome P4501A1. In *Biomarquers en Ecotoxicologie: Aspects Fondamentaux*, pp. 33–51. Paris: Masson.

MOORE, M. N., LIVINGSTONE, D. R., WIDDOWS, J., LOWE, D. M. AND PIPE, R. K. (1987) Molecular, cellular, and physiological effects of oil-derived hydrocarbons in molluscs and their use in impact assessment. *Philosophical Transactions of the Royal Society, London, B* 316, 603–23.

MOORE, N. W. and WALKER, C. H. (1964) Organic chlorine insecticide residues in wild birds *Nature* 201, 1072–3.

MORIARTY, F. M. (1968) The toxicity and sublethal effects of *p,p'*-DDT and dieldrin to *Aglais urticae* and *Chorthippus brunneus*. *Annals of Applied Biology* 62, 371–93.

MORIARTY, F. M. (ed.) (1975) *Organochlorine Insecticides: Persistent Organic Pollutants*. London: Academic Press.

MORIARTY, F. M. (1999) *Ecotoxicology*, 3rd edn. London: Academic Press.

MORSE, D. C., KLASSON-WHELER, E., VAN DE PAS, M., DE BIE, A. T. H. J., VAN BLADEREN, P. J. and BROWER, A. (1995) Metabolism and biochemical effects of 3,3′,4,4′ TCB in pregnant and foetal rats. *Chemical and Biological Interactions* 95, 42–56.

MORSE, D. C., *et al.* (1996) Alterations in rat brain thyroid hormone status following pre- and post-natal exposures to PCBs. *Toxicology of Applied Pharmacology* 139, 269–79.

MUIR, D. C. G., *et al.* (1992) Arctic marine ecosystem contamination. *The Science of the Total Environment* 122, 75–134.

NATURAL ENVIRONMENT RESEARCH COUNCIL (1971) The sea bird wreck in the Irish Sea. Autumn, 1969. NERC Publications C No. 4.

NEBERT, D. W. and GONZALEZ, F. J. (1987) P450 genes: evolution, structure, and regulation. *Annual Review of Biochemistry* 56, 945–93.

NELSON, D. R. (1998) Metazoan P450 evolution. *Comparative Biochemistry and Physiology* 121C (special issue), 15–22.

NELSON, D. R. and STROBEL, H. W. (1987) The evolution of cytochrome P 450 proteins. *Molecular Biology and Evolution* 4, 572–93.

NEWMAN, M. C. (1996) *Fundamentals of Ecotoxicology*. Chelsea, MI: Ann Arbor Press.

NEWTON, I. (1986) *The Sparrowhawk*. Calton: Poyser.

NEWTON, I. and HAAS, M. B. (1984) The return of the sparrowhawk. *British Birds* 77, 47–70.

NEWTON, I. and WYLLIE, I. (1992) Recovery of a sparrowhawk population in relation to declining pesticide contamination. *Journal of Applied Ecology* 29, 476–84.

NEWTON, I., MEEK, E. and LITTLE, B. (1978) Breeding ecology of the merlin in Northumberland. *British Birds* 71, 376–98.

NEWTON, I., WYLLIE, I. and FREESTONE, P. (1990) Rodenticides in British Barn Owls. *Environmental Pollution* 68, 101–17.

NICHOLLS, D. G. (1982) *Bioenergetics: An Introduction to Chemiosmotic Theory*. London: Academic Press.

NISBET, I. C. T. (1989) Organochlorines: reproductive impairment and declines in bald eagle populations, mechanisms and dose–response relationships. In MEYBURG, B.-U. and CHANCELLOR, R. D. (eds) *Raptors in the Modern World*. Proceedings of the Third World Conference on Birds of Prey and Owls. Berlin, pp. 483–89. Berlin: WWGBP.

NORSTROM, R. J. (1988) Bioaccumulation of polychlorinated biphenyls in Canadian wildlife. In CRINE, J.-P. (ed.) *Hazards, Decontamination and Replacement of PCBs*. New York: Plenum.

NORSTROM, R. J., McKINNON, A. E. and de FREITAS, A. S. W. (1976) A bioenergetic based model for pollutant accumulation by fish. *Journal of the Fisheries Research Board of Canada* 33, 248–67.

NORSTROM, R. J., SIMON, M. and WESELOH, D. V. (1986) Long term trends of PCDD and PCDF contamination in the great Lakes. Proceedings of Dioxin 86, the Sixth International Symposium on Chlorinated Dioxins and Related Compounds, Fukuoka, Japan, September.

O'CONNOR, D. J. and NIELSEN, S. W. (1981) Environmental survey of methylmercury levels in wild mink and otter from the North Eastern USA and experimental pathology of methylmecurialism in the otter. Proceedings, Worldwide Furbearer Conference, Frostburg, MD, USA, 3–11, August 1981, pp. 1728–45.

ODA, J. and MULLER, W. (1972) Identification of a mammalian breakdown product of dieldrin. In COULSTON, F. AND KORTE, F. (eds) *Environmental Quality and Safety*, 1, pp. 248–9. Stuttgart: Georg Thieme; New York: Academic Press

OPPENOORTH, F. J. and WELLING, W. (1976) Biochemistry and physiology of resistance. In WILKINSON, C. F. (ed.) *Pesticide Biochemistry and Physiology*, pp. 507–54. London: Heyden.

ORIS, J. T. and GIESY, J. P. (1986) Photoinduced toxicity of antracene to juvenile bluegill sunfish; photoperiod effects and predictive hazard evaluation. *Environmental Toxicology and Chemistry* 5, 761–8.

ORIS, J. T. and GIESY, J. P. (1987) The photoinduced toxicity of PAH to larvae of the fathead minnow. *Chemosphere* 16, 1395–404.

OSBORN, D., EVERY, W. J. and BULL, K. R. (1983) The toxicity of trialkyl lead compounds to birds. *Environmental Pollution* (Series A) 31, 261–75.

OTTO, D. and WEBER, B. (1992) *Insecticides: Mechanism of Action and Resistance*. Andover: Intercept.

PARKER, P. J.-A. and CALLAGHAN, A. (1997) Esterase activity and allele frequency in field populations of Simulium equinum exposed to organophosphate pollution. *Environmental Toxicology and Chemistry* 16, 2550–5.

PEAKALL, D. B. (1992) *Animal Biomarkers as Pollution Indicators*. London: Chapman & Hall.

PEAKALL, D. B. (1993) DDE-induced eggshell thinning; an environmental detective story. *Environmental Reviews* 1, 13–20.

PEAKALL, D. B. and FAIRBROTHER, A. (1998) Biomarkers for monitoring and measuring effects. In DOUBEN, P. E. T. (ed.) *Pollution Risk Assessment and Management*, pp. 351–6. Chichester: Wiley.

PEAKALL, D. B. and SHUGART, L. R. (eds) (1993) *Biomarker Research and Application in the Assessment of Environmental Health*. Berlin: Springer-Verlag.

PEAKALL, D. B., LINCER, J. L., RISEBROUGH, R. W., PRITCHARD, J. B. and KINTER, W. B. (1973) DDE-induced eggshell thinning: structural and physiological effects in three species. *Comparative and General Pharmacology* 4, 305–13.

PESONEN, M., GOKSOYR, A. and ANDERSSON, T. (1992) Expression of P4501A1 in a primary culture of rainbow trout microsomes exposed to B-naphthoflavone or 2,3,7,8-TCDD. *Archives of Biochemistry and Biophysics* 292, 228–33.

PILLING, E. D. (1992) Evidence for pesticide synergism in the honeybee. *Aspects of Applied Biology* 31, 43–7.

PILLING, E. D. (1993) *Synergism between EBI fungicides and a pyrethroid insecticide in the honeybee*. PhD Thesis, University of Southampton.

PILLING, E. D., BROMELEY-CHALLENOR, K. A. C., WALKER, C. H. and JEPSON, P. C. (1995) Mechanism of synergism between the pyrethroid insecticide lamda cyhalothrin and the imidazole fungicide prochloraz in the honeybee. *Pesticide Biochemistry and Physiology* 51, 1–11.

POIGER, H. and BUSER, H. R. (1983) Structure elucidation of mammalian TCDD metabolites. In TUCKER, R. E., YOUNG, A. L. and GRAY, A. P. (eds) *Human and Environmental Risks of Chlorinated Dioxins and Related Compounds*, pp. 483–92. New York: Plenum Press.

POTTS, G. R. (1986) *The Partridge*. London: Collins.

POTTS, G. R. (2000) *The Grey Partridge.* In PAIN, D. and DIXON, J. (eds) *Bird Conservation and Farming Policy in the European Union.* London: Academic Press.

PURCHASE, I. F. H. (1994) Current knowledge of mechanisms of carcinogenicity: genotoxic vs non-genotoxic. *Human and Experimental Toxicology* 13, 17–28.

QUICKE, D. L. J. and USHERWOOD, P. N. R. (1990) Spider toxins as lead structures for novel pesticides. In HODGSON, E. and KUHR, R. J. (eds) *Safer Insecticides: Development and Use,* pp. 385–452. New York: Marcel Dekker.

RAPPE, C., NYGREN, M., LINSTROM, G., BUSER, H. R., *et al.* (1987) Overview on environmental fate of chlorinated dioxins and dibenzofurans; sources levels and isomeric pattern in various matrices. *Chemosphere* 16, 1603–18.

RATCLIFFE, D. A. (1967) Decrease in eggshell weight in certain birds of prey. *Nature* 215, 208–10.

RATCLIFFE, D. A. (1993) *The Peregrine Falcon,* 2nd edn. Calton: T & D Poyser.

RATTNER, B. A., MELANCON, M. J., CUSTER, T. W., HOTHAN, R. L., *et al.* (1993) Biomonitoring environmental contamination with pipping black-crowned night heron embryos: induction of cytochrome P450. *Environmental Toxicology and Chemistry* 12, 1719–32.

REICHERT, W. L., FRENCH, B., L., STEIN, J. E. and VARANASI, U. (1991) ^{32}P postlabelling analysis of the persistence of bulky hydrophobic xenobiotic-DNA adducts in the liver of English sole, a marine fish. In *Proceedings of the 82nd Meeting of the American Association for Cancer Research,* Houston, TX, 87.

RENZONI, A., FOCARDI, S., FOSSI, M. C., LEONZIO, C. AND MAYAL, J. (1986) Comparison between concentrations of mercury and other contaminants in Cory's shearwater. *Environmental Pollution* 40A, 17–35.

RICHARDS, P., JOHNSON, M., RAY, D. and WALKER, C. H. (1999) Novel protein targets for organophosphorous compounds. In REINER, E., SIMEON-RUDOLF, V., DOCTOR, B. P., FURLONG, C. E., *et al.* (eds) *Esterases Reacting with Organophosphorous Compounds. Chemico-Biological Interactions* (Special issue) 119–120, 503–12.

RIVIERE, J.-L. and CABANNE, F. (1987) Animal and plant cytochrome P-450 systems. *Biochimie* 69, 743–52.

ROBINSON, J., RICHARDSON, A., CRABTREE, A. N., COULSON, J. C. and POTTS, G. R. (1967a) Organochlorine residues in marine organisms. *Nature* 214, 1307–11.

ROBINSON, J., RICHARDSON, A. and BROWN, V. K. H. (1967b) Pharmacodynamics of dieldrin in pigeons. *Nature* 213, 734–5.

RONIS, M. J. J. and MASON, Z. (1996) The metabolism of testosterone by the periwinkle *in vitro* and *in vivo.* Effects of TBT. *Marine Environmental Research* 42, 161–6.

RONIS, M. J. J. and WALKER, C. H. (1989) The microsomal monooxygenases of birds. *Reviews in Biochemical Toxicology* 10, 301–84.

RONIS, M. J. J., BORLAKOGLU, J., WALKER, C. H., HANSON, T. and STEGEMAN J. J. (1989) Expression of orthologues to rat P4501A1 and IIB1 in sea birds from the Irish sea 1978–1988. Evidence for environmental induction. *Marine Environmental Research* 28, 123–30.

ROSENBERG, D. W. and DRUMMOND, G. S. (1983) Direct *in vitro* effects of TBTO on hepatic cytochrome P450. *Biochemistry and Pharmacology* 32, 3823–9.

ROSS, P. S., DE SWART, R. L., REIJNDERS, P. J. H., VAN LOVEREN, H., *et al.* (1995) Contaminant-related suppression of delayed-type hypersensitivity and antibody responses in harbour seals fed herring from the Baltic Sea. *Environmental Health Perspectives* 103, 162.

ROY, N. K., STABILE, J., SEEB, J. E., HABICH, C., *et al.* (1999) High frequency of K-ras mutations in pink salmon embryos exposed to Exxon Valdez oil. *Environmental Toxicology and Chemistry* 18, 1521–8.

SAFE, S. (1984) Polychlorinated biphenyls and polybrominated biphenyls; biochemistry toxicology and mode of action. *CRC Critical Reviews in Toxicology* 13, 319–95.

SAFE, S. (1990) PCBs, PCDDs PCDFs and related compounds: environmental and mechanistic considerations which support the development of toxic equivalency figures. *CRC Critical Reviews in Toxicology* 24, 1–63.

SALGADO, V. L. (1999) Resistant target sites and insecticide discovery. In BROOKS, G. T. and ROBERTS, T. R. (eds) *Pesticide Chemistry and Bioscience: The Food-Environment Challenge*, pp. 236–46. Cambridge: Royal Society of Chemistry.

SCHUÜRMANN, G. and MARKERT, B. (1998) *Ecotoxicology*. Chichester: John Wiley.

SCHWARZENBACH, R. P., GSCHWEND, P. M. and IMBODEN, D. M. (1993) *Environmental Organic Chemistry*. New York: J. Wiley.

SECORD, A. L., MCCARTY, J. P., ECHOLS, K. R., MEADOWS, J. C., *et al.* (1999) PCBs and 2,3,7,8-TCDD equivalents in tree swallows from the upper Hudson river, New York State USA. *Environmental Toxicology and Chemistry* 18, 2519–25.

SIBLY, R. M., Newton, I. and WALKER, C. H. (2000) Effects of dieldrin on population growth rates of UK sparrowhawks. *Journal of Applied Ecology* 37, 540–6.

SJUT, V. (ed.) (1997) *Molecular Mechanisms of Resistance to Agrochemicals Chemistry of Plant Protection*. 13. Berlin: Springer-Verlag.

SLEIGHT, S. D. (1979) Polybrominatedbiphenyls: a recent environmental pollutant. In *Animals as Monitors of Environmental Pollutants*, pp. 366–74. Washington DC: National Academy of Sciences.

SODERGREN, A. (ed.) (1991) Environmental Fate and Effects of Bleached Pulp Mill Effluents. Report no. 4031. Lund, Sweden: Swedish Environmental Protection Agency.

SOMERVILLE, L. and GREAVES, M. P. (1987) *Pesticide Effects on Soil Microflora*. London: Taylor & Francis.

SOMERVILLE, L. and WALKER, C. H. (eds) (1990) *Pesticide Effects on Terrestrial Wildlife*. London: Taylor & Francis.

SUETT, D. L. (1986) Accelerated degradation of carbofuran in previously treated soil in the United Kingdom. *Crop Protection* 5, 165–9.

SUMPTER, J. S. and JOPLING, S. (1995) Vitellogenesis as a biomarker for estrogenic contamination of the aquatic environment. *Environmental Health Perspectives* (supplement 7), 173–8.

SUNDSTROM, G., HUTZINGER, O. and SAFE, S. (1976) The metabolism of chlorobiphenyls – a review. *Chemosphere* 5, 267–98.

SUSSMAN, J. L., HAREL, M., FROLOW, F., OEFNER, C., *et al.* (1991) Atomic structure of acetylcholinesterase from Torpedo californica; a prototypic acetylcholine-binding protein. *Science* 253, 872–9.

SUSSMAN, J. L., HAREL, M. and SILMAN, I. (1993) Three-dimensional structure of acetylcholinesterase and of its complexes with anticholinesterase drugs. *Chemical–Biolgical Interactions* 87, 187–97.

SUTER, G. W. (ed.) (1993) *Ecological Risk Assessment*. Boca Raton, FL: Lewis.

TANABE, S. and TATSUKAWA, R. (1992) Chemical modernisation and vulnerability of cetaceans: increasing threat of organochlorine contaminants. In WALKER, C. H. and LIVINGSTONE, D. R. (eds) *Persistent Pollutants in Marine Ecosystems*, pp. 161–80. Oxford: Pergamon Press.

TANFORD, C. (1980) *The Hydrophobic Effect*, 2nd edn. New York: Wiley-Interscience.

THAIN, J. E. and WALDOCK, M. J. (1986) The impact of TBT antifouling paints on molluscan fisheries. *Water Science Technology* 18, 193–202.

THIJSSEN, H. H. W. (1995) Warfarin-based rodenticides: mode of action and mechanism of resistance. *Pesticide Science* 43, 73–8.

THOMPSON, H. M. and WALKER, C. H. (1994) Serum B esterases as indicators of exposure to pesticides. In FOSSI, M. C. and LEONZIO, C. (eds) *Non-Destructive Biomarkers in Vertebrates,* pp. 35–60. Boca Raton, FL: Lewis.

TILLETT, D. E., *et al.* (1992) PCB residues and egg mortality in double crested cormorants from the Great Lakes. *Environmental Toxicology and Chemistry* 11, 1281–8.

TIMBRELL, J. A. (1999) *Principles of Biochemical Toxicology,* 3rd edn. London: Taylor & Francis.

TOMLIN, C. D. S. (1997) *The Pesticide Manual,* 11th edn. Farnham: British Crop Protection Council.

TRAGER, W. F. (1989) Isotope effects as mechanistic probes of cytochrome P450-catalysed reactions. In BAILLIE, T. A. and JONES J. R. (eds) *Synthesis and Application of Isotopically Labelled Compounds,* pp. 333–340. Proceedings of the III International Symposium. Amsterdam: Elsevier.

TURTLE, E. E., TAYLOR, A., WRIGHT, E. N., THEARLE, R. J. P., *et al.* (1963) The effects on birds of certain chlorinated insecticides used as seed dressings. *Journal of Science, Food and Agriculture* 14, 567–77.

USEPA (1980) Ambient Water Quality for Mercury US Environmental Protection Agency, Criteria and Standards Division (EPA-600/479–049).

VAILLANT, C., MONOD, G., VOLATAIRE, Y. and RIVIERE, J.-L. (1989) Measurement and induction of cytochrome P450 and monooxygenases in a primary culture of rainbow trout hepatocytes. *Comptes Rendus de L'Academie des Sciences* 308, 83–8.

VAN GELDER, G. A. and CUNNINGHAM, W. L. (1975) The effects of low level dieldrin exposure on the EEG and learning ability of the squirrel monkey. *Toxicology and Applied Pharmacology* 33 (Abstract no. 50), 142.

VAN ZUTPHEN, L. F. M. and BALLS, M. (eds) (1997) *Animal Alternatives, Welfare and Ethics.* Amsterdam: Elsevier.

VARANASI, U., *et al.* (1992) Chlorinated and aromatic hydrocarbons in bottom sediments, fish and marine mammals in US coastal waters; laboratory and field studies of metabolism and accumulation. In WALKER, C. H. and LIVINGSTONE, D. R. (eds) *Persistent Pollutants in Marine Ecosystems,* pp. 83–118. Oxford: Pergamon Press.

VERHALLEN, E. Y., VAN DEN BERG, M. and BOSVELD, A. T. C. (1997) Interactive effects of the EROD-inducing potency of PHAHs in the chicken embryo hepatocyte assay. *Environmental Toxicology and Chemistry* 16, 277–82.

VIGHI, M., GARLANDA, M. M. AND CALAMARI, D. (1991) QSARs for the toxicity of OP pesticides to bees. *The Science of the Total Environment* 109/110, 605–22.

WAID, J. S. (ed.) (1985–87) *PBCs in the Environment,* vols I–III. Cleveland: CRC Press.

WALKER, C. H. (1969) Reductive dechlorination of *p,p'*-DDT by pigeon liver microsomes. *Life Sciences* 8, 111–15.

WALKER, C. H. (1974) *Comparative Aspects of the Metabolism of Pesticides Environmental Quality and Safety,* vol. 3, pp. 113–53. Stuttgart: Georg Thieme; New York: Academic Press.

WALKER, C. H. (1975) Variations in the intake and elimination of pollutants. In MORIARTY, F. (ed.) *Organochlorine Insecticides: Persistent Organic Pollutants,* pp. 73–131. London: Academic Press.

Walker, C. H. (1978) Species differences in microsomal monooxygenase activities and their relationship to biological half-lives. *Drug Metabolism Reviews* 7 (2), 295–323.

WALKER, C. H. (1980) Species variations in some hepatic microsomal enzymes that metabolise xenobiotics. *Progress in Drug Metabolism* 5, 118–64.

WALKER, C. H. (1981) The correlation between *in vivo* and *in vitro* metabolism of pesticides in vertebrates. *Progress in Pesticide Biochemistry* 1, 247–86.

WALKER, C. H. (1983) Pesticides and birds – mechanisms of selective toxicity. *Agriculture, Environment and Ecosystems* **9**, 211–26.

WALKER, C. H. (1987) Kinetic models for predicting bioaccumulation of pollutants in ecosystems. *Environmental Pollution* **44**, 227–40.

WALKER, C. H. (1989) The development of an improved system of nomenclature and classification of esterases. In REINER, E., ALDRIDGE, W. N., HOSKIN F. C. G. (eds) *Enzymes Hydrolysing Organophosphorous Compounds*, pp. 236–45. Chichester: Ellis Horwood Ltd.

WALKER, C. H. (1990a) Persistent pollutants in fish-eating sea birds – bioaccumulation, metabolism and effects. *Aquatic Toxicology* **17**, 293–324.

WALKER, C. H. (1990b) Kinetic models to predict bioaccumulation of pollutants. *Functional Ecology* **4**, 295–301.

WALKER, C. H. (ed.) (1991) *The Role Of Enzymes in Regulating the Toxicity of Xenobiotics*. Proceedings from Colloquium of the Pharmacological Biochemistry Group. Biochemical Society 638th Meeting, Reading University, 10–12 April 1991. *Biochemical Society Transactions* **19**, 731–67.

WALKER, C. H. (1994a) Comparative toxicology. In HODGSON, E. and LEVI, P. (eds) *Introduction to Biochemical Toxicology*, 2nd edn, pp. 193–218. Norwalk, CT: Appleton and Lange.

WALKER, C. H. (1994b) Interactions between pesticides and esterases in humans. In MACKNESS, M. I. and CLERC, M. (eds) *Esterases, Lipases, and Phosphlipases: from Structure to Clinical*, pp. 91–8. NATO ASI Series. Series A. Life Sciences. New York: Plenum Press.

WALKER, C. H. (1998a) Avian forms of cytochrome P450. In LIVINGSTONE, D. R. and STEGEMAN, J. J. (eds) *Forms and Functions of Cytochrome P450. Comparative Biochemistry and Physiology* **121C** (supplement), 65–72.

WALKER, C. H. (1998b) Alternative approaches and tests in ecotoxicology. *ATLA (Alternatives to Laboratory Animals)* **26**, 649–77.

WALKER, C. H. (1998c) The use of biomarkers to measure the interactive effects of chemicals. *Ecotoxicology and Environmental Safety* **40**, 65–70.

WALKER, C. H. and JEFFERIES, D. J. (1978) The post mortem reductive dechlorination of p,p'-DDT in avian tissues. *Pesticide Biochemistry and Physiology* **9**, 203–10.

WALKER, C. H. and JOHNSTON, G. O. (1989) Interactive effects of pollutants at the kinetic level; implications for the marine environment. *Marine Environmental Research* **28**, 521–5.

WALKER, C. H. and LIVINGSTONE, D. R. (eds) (1992) *Persistent Pollutants in Marine Ecosystems*. Oxford: Pergamon Press.

WALKER, C. H. and NEWTON, I. (1998) Effects of cyclodiene insecticides on the sparrowhawk in Britain – a reappraisal of the evidence. *Ecotoxicology* **7**, 185–9.

WALKER, C. H. and NEWTON, I. (1999) Effects of cyclodiene insecticides on raptors in Britain – correction and updating of an earlier paper by Walker and Newton in *Ecotoxicology* **7**, 185–9. *Ecotoxicology* **8**, 425–30.

WALKER, C. H. and OESCH, F. (1983) Enzymes in selective toxicity. In JAKOBY, W. and CALDWELL, J. (eds) *Biological Basis of Detoxication,* pp. 349–68. Academic Press Inc.

WALKER, C. H., BENTLEY, P. and OESCH, F. (1978) Phylogenetic distribution of epoxide hydratase in different vertebrate species, strains, and tissues using three substrates. *Biochimica et Biophysica Acta* **539**, 427–34.

WALKER, C. H., HOPKIN, S. P., SIBLY, R. M. and PEAKALL, D. R. (1996) and (2000) *Principles of Ecotoxicology*, 1st and 2nd editions. London: Taylor & Francis.

WALKER, C. H., KAISER, K., KLEIN, K., LAGADIC, L., *et al.* (1998) Biomarker strategies to evaluate the environmental effects of chemicals. *Environmental Health Perspectives* **106** (supplement 2), 613–20.

WARNER, R. F., PETERSON, K. K. and BORGMAN, L. (1966) Behavioural pathology in fish; a quantitative study of sublethal pesticide intoxication. In Moore, N. W. (ed.) *Pesticides in the Environment and their Effects on Wildlife*, supplement to vol. 3. *Journal of Applied Ecology* **3** (supplement), 223–48.

WATERS, R., JONES, N. J. and MORSE, H. R. (1994) DNA adducts in aquatic organisms. In Abstracts of British Toxicology Society Meeting Churchill College, Cambridge University, p. 38.

WEBB, R. E., RANDOLPH, W. C. and Horsfall, F. Jr (1972) Hepatic benzo(*a*)pyrene monooxygenase activity in endrin susceptible and resistant pine mice. *Life Sciences* **11** (part 2), 477–84.

WELLS, M. R., LUDKE, J. L. and YARBOROUGH, J. D. (1973) Epoxidation and fate of ^{14}C-labelled aldrin in insecticide resistant and susceptible populations of mosquito fish. *Journal of Agriculture Food and Chemistry* **21**, 428–9.

WHYTE, J. J., VAN DEN HEUVEL, M. R. M., CLEMONS, J. H. M., HUESTIS, S. Y., *et al.* (1998) Mammalian and teleost cell line bioassays and chemically derived 2,3,7,8-TCDD equivalent concentrations in lake trout from Lake Superior and Lake Ontario North America. *Environmental Toxicology and Chemistry* **17**, 2214–26.

WIEMEYER, S. N. and PORTER, R. D. (1970) DDE thins eggshells of captive American kestrels. *Nature* **227**, 737–8.

WIEMEYER, S. N., BUNCK, C. M. and STAFFORD, C. J. (1993) Environmental contaminants in bald eagles – 1980–1984 – and further interpretations of relationships in productivity and shell thickness. *Archives of Environmental Contaminants and Toxicology* **24**, 213–44.

WILLIAMS, R. J., BROOKE, D. N., CLARE, R. W., MATTHIESSEN, P. AND MITCHELL, R. D. J. (1996) *Rosemaund Pesticide Transport Study*. 1987–1993. Report no. 129. Wallingford, UK: Institute of Hydrology (NERC).

WOBESER, G., NIELSEN, N. O. and Schliefer, B. (1976) Mercury and mink. 1. Experimental methylmercury intoxication. *Canadian Journal of Comprehensive Medicine* **40**, 34–45.

WOLFE, M. F. SCHWARZBACH, S. and SULAIMAN, R. A. (1998) The effects of mercury on wildlife: a comprehensive review. *Environmental Toxicology and Chemistry* **17**, 146–60.

WRIGHT, J. F., (1995) Development and use of a system for predicting the macroinvertebrate fauna found in flowing water. *Australian Journal of Ecology* **20**, 181–97.

Index